BIOLOGY AND ECOLOGY OF EDIBLE MARINE GASTROPOD MOLLUSCS

BIOLOGY AND ECOLOGY OF EDIBLE MARINE GASTROPOD MOLLUSCS

Ramasamy Santhanam, PhD

Apple Academic Press Inc.
3333 Mistwell Crescent
Oakville, ON L6L 0A2
Canada

Apple Academic Press Inc.
9 Spinnaker Way
Waretown, NJ 08758
USA

© 2019 by Apple Academic Press, Inc.

First issued in paperback 2021

Exclusive worldwide distribution by CRC Press, a member of Taylor & Francis Group
No claim to original U.S. Government works

ISBN 13: 978-1-77-463521-6 (pbk)
ISBN 13: 978-1-77-188638-3 (hbk)

Library and Archives Canada Cataloguing in Publication

Santhanam, Ramasamy, 1946-, author
Biology and ecology of edible marine gastropod molluscs / Ramasamy Santhanam, PhD.

(Biology and ecology of marine life)
Includes bibliographical references and index.
Issued in print and electronic formats.
ISBN 978-1-77188-638-3 (hardcover).--ISBN 978-1-315-09943-9 (PDF)

1. Sea snails. I. Title. II. Series: Santhanam, Ramasamy, 1946- . Biology and ecology of marine life.

| QL430.4.S36 2018 | 594'.3 | C2018-902025-3 | C2018-902026-1 |

CIP data on file with US Library of Congress

Apple Academic Press also publishes its books in a variety of electronic formats. Some content that appears in print may not be available in electronic format. For information about Apple Academic Press products, visit our website at **www.appleacademicpress.com** and the CRC Press website at **www.crcpress.com**

BIOLOGY AND ECOLOGY OF MARINE LIFE BOOK SERIES

Series Author:

Ramasamy Santhanam, PhD
Former Dean, Fisheries College and Research Institute,
Tamil Nadu Veterinary and Animal Sciences University,
Thoothukudi 628008, India
E-mail: rsanthaanamin@yahoo.co.in; rsanthaanamin@gmail.com

Books in the Series:

- Biology and Culture of Portunid Crabs of the World Seas
- Biology and Ecology of Edible Marine Bivalve Molluscs
- Biology and Ecology of Edible Marine Gastropod Molluscs
- Biology and Ecology of Venomous Marine Snails
- Biology and Ecology of Venomous Stingrays
- Biology and Ecology of Toxic Pufferfish

ABOUT THE AUTHOR

Ramasamy Santhanam, PhD

 Dr. Ramasamy Santhanam is the former Dean of Fisheries College and Research Institute, Tamil Nadu Veterinary and Animal Sciences University, India. His fields of specialization are marine biology and fisheries environment. Presently, he is serving as a resource person for various universities of India. He has also served as an expert for the "Environment Management Capacity Building," a World Bank-aided project of the Department of Ocean Development, India. He has been a member of the American Fisheries Society, United States; World Aquaculture Society, United States; Global Fisheries Ecosystem Management Network (GFEMN), United States; and the International Union for Conservation of Nature's (IUCN) Commission on Ecosystem Management, Switzerland. To his credit, Dr. Santhanam has 25 books on Marine Biology/Fisheries Science and 70 research papers.

CONTENTS

LIST OF ABBREVIATIONS

AbHV	abalone herpes virus
EAA	essential amino acids
FW	fresh weight
GAG	glycosaminoglycan
GSK-3	glycogen synthase kinase 3
HIV-1	human immunodeficiency virus type 1
KLH	keyhole limpet hemocyanin
LC-PUFA	long-chain polyunsaturated fatty acid
MIC	minimum inhibitory concentration
MUFA	monounsaturated fatty acid
NO	nitric oxide
PL	phospholipid
PSP	paralytic shellfish poisoning
PUFA	polyunsaturated fatty acid
SFA	saturated fatty acid
TBT	tributyltin
TT	thrombin time
UFA	unsaturated fatty acid

PREFACE

The gastropods (Phylum: Mollusca; Class: Gastropoda) play important diverse roles in nature and are also sensitive indicators of environmental change. During a long evolution, this class of animals has colonized the majority of marine, freshwater, and terrestrial environments. There are about 50,000 described species of marine gastropods distributed throughout the world from intertidal zones to the ocean depths. Abalone, conchs, and periwinkle snails are popular and highly sought after food items throughout the world. Shells of these gastropods occupy an important role in the commercial shell craft industries, especially in South India. Apart from nutritional values, several species of these edible, marine gastropods have potential industrial and therapeutical applications. Mariculture techniques have also been developed for certain species of these gastropods.

Knowledge about the biology and ecology edible marine gastropod molluscs of world seas and oceans is very much limited, and a comprehensive book on these aspects is still wanting. Keeping this in consideration, an attempt has now been made. In this book, aspects, namely, (1) biology and ecology, (2) profile (habitat, geographical distribution, morphology, food and feeding, reproduction, etc.), (3) nutritional values, (4) pharmaceutical and industrial applications, (5) aquaculture, and (6) diseases and parasites of about 250 species of edible marine gastropod molluscs have been covered in an easy-to-read style with neat illustrations. It is hoped that the present publication would be of great use for the students and researchers of disciplines such as Fisheries Science, Marine Biology, Aquatic Biology and Fisheries, and Zoology, besides serving as a standard reference book for the libraries of all universities and libraries.

I am grateful to Dr. G. Sanjeeviraj, formerly Professor and Head, Department of Fisheries Extension, Fisheries College & Research Institute, Tamil Nadu Veterinary & Animal Sciences University, Thoothukudi, India, for his valuable suggestions on the manuscript. I sincerely thank all my international friends who were very kind enough to collect and send the shells of some species for the present purpose. I also thank Mrs. Albin Panimalar Ramesh for her help in photography and secretarial assistance. The author would like to express his indebtedness to Dr. S. Ramesh, Professor, Ratnam Institute of Pharmacy, Nellore, Andhra Pradesh, India for his suggestions on the chapter dealing with pharmaceutical values of marine gastropods. Suggestions from the users are welcome.

—Ramasamy Santhanam

CHAPTER 1

INTRODUCTION

CONTENTS

ABSTRACT

The characteristics of the class Gastropoda along with the human uses/ commercial importance of edible marine gastropod Molluscs are given in this chapter.

The marine gastropods have been associated with humans since the dawn of civilization. Cowrie shells were once used as currency in many Polynesian cultures. These animals were gathered for food, and their shells were used as tools, ornaments, and later as money. Although the venom from the "sting" of some cone snails is fatal to humans, several bioactive compounds with therapeutical values have been developed from their venom. The most characteristic feature of gastropods is torsion of their soft bodies during early larval stages, producing a crossing of their nerve connectives, bending of the intestine, and twisting of the mantle cavity (together with associated structures, including the ctenidia, anus, kidney openings, etc.) anteriorly over the gastropod head (Fryda, 2013). The maximum lifespan of different species of marine gastropods has been reported to vary from 10 to 40 years as the lifespan of the queen conch, the limpet *Patella vulgata* and the periwinkle, *Littorina littorea* is 40, 16, and 10 years, respectively. Owing to their great demand for meat and as an ornamental shell for shell handicrafts, these gastropods are receiving considerable attention in current years (Flores-Garza et al., 2012).

Gastropods in marine fisheries: The total harvest of marine gastropods in the world is relatively stable, with about 103,000 MT in 1996. Among the edible gastropods, the abalones were important not only as food to coastal First Nations peoples but also played an important part in their spiritual and cultural society. *Haliotis rubra* (blacklip) and *Haliotis laevigata* (greenlip) constitute a major fishery (especially in Tasmania of Australia), and their landings fluctuate between 10,000 and 15,000 MT/year. Another strong abalone fishery is located in California (USA). Conch fisheries are represented by *Strombus* and *Turbo* which are harvested in the tropical belt. The common periwinkle *L. littorea* is mainly harvested in Ireland, but it is also commercially fished in Northeast America. The gastropod fisheries in Chile involve the main species *Concholepas concholepas*. The European whelk (*Buccinum undatum*) fisheries which began in the 1960s have now increased in the recent years, to fulfill the high demand of the Southeast

Asian market (Berthou et al., http://archimer.ifremer.fr/doc/2001/publication-529.pdf, p 24).

Aquaculture of marine gastropods: Aquaculture production of marine Molluscs is largely due to bivalves and compared to this, the contribution of gastropods is not substantial. In 1996, the world aquacultural production of gastropods was less than 2.5% of a total yield of about 137,000 MT. Overall abalone aquaculture, which reached about 2200 MT in 1996, was mainly due to the production of the varicolored abalone (*Haliotis diversicolor*) in Korea, and of the red abalone (*Haliotis rufescens*) in California. Other important abalone species include in decreasing order, the Southern green (*Haliotis fulgens*) and pink (*Haliotis corrugata*) abalones in the Eastern Pacific; giant (*Haliotis gigantea*) and disk (*Haliotis discus*) abalones in the Western Pacific; perlemoen abalone (*Haliotis midae*) in Southern Africa; European ormer (*Haliotis tuberculata*) in Western Europe; and blacklip abalone (*H. rubra*) in Australia. The aquaculture production of the donkey's ear abalone (*Haliotis asinina*) in the Philippines is also worthy of mention. Other gastropods like *Strombus* spp. and *Turbo cornutus* have shown a very limited culture activity with a world production of 27 MT in 1996 (Berthou et al., http://archimer.ifremer.fr/doc/2001/publication-529.pdf, p. 24).

Nutritional value: Many species of marine gastropods are used as food items throughout the world. Conchs in particular are highly sought after due to their mild flavor, especially in the Caribbean and the Florida Keys. Escargot, abalone, and periwinkle snails are also popular food items. A total of about 300 species of marine gastropods are harvested commercially. The meat of edible marine gastropods contains little fat, but it is rich in protein with many nutrients required in a healthy and well-balanced diet. The omega-3 fatty acids (polyunsaturated fatty acids) of edible marine gastropods are essential for normal development and function of the retina; helpful to regulate nerve transmission; regulating blood pressure, blood clotting, body temperature, inflammation or hypersensitivity reactions, immune and allergic responses; and maintaining normal kidney function and fluid balance (Dong, https://wsg.washington.edu/aquaculture/pdfs/Nutritional-Value-of-Shellfish.pdf). These gastropods are harvested for food in many parts of the world, and they are considered a delicacy in France, Korea, Japan, and China. Hence, there is potential for using the underutilized snail meat from marine resources to help fill the gap in current animal protein shortage (https://researcharchive.lincoln.ac.nz/bitstream/handle/10182/2863/ Shi_MApplSci.pdf;jsessionid=8CEB19C444DA715A3BE8BE8640 F37503?sequence=5).

Ornamental value: Due to the ornamentation of the shells, several species of marine gastropods are sold all over the world as curios and souvenirs. Their beautiful shells are also used in the development of products relating to jewelry and shellcraft/handicraft. In the west, the abalone shells are mostly used in art industry to make seashell crafts or jewelry, such as abalone shell engagement ring, beads, necklace, buttons, bracelet, cabochon, hair clips, pendant, and so on. There are about 5000 species of marine Molluscs (bivalves and gastropods) worldwide involved in this market. In Northeastern Brazil, 85 species of marine gastropods are sold individually as decorative pieces or incorporated into utilitarian objects. The gastropod families most represented in the trade of ornamental products include Conidae and Strombidae. Large gastropod shells of king helmet *Cassis tuberosa*, the Hebrew volute *Voluta ebraea*, and the Goliath conch *Eustrombus goliath* are mainly used as part of table lamps. The majority of the ornamental species involved are harvested from the Atlantic Ocean (68%), but many are imported from Indo-Pacific countries, such as cowries *Monetaria moneta*. Among the species harvested in the Atlantic Ocean, 11% are endemic to Brazil as, for example, the Brazilian chank (*Turbinella laevigata*) (Venkatesan, http://eprints.cmfri. org.in/8671/1/Marine_Ornamental_Fish_Culture.pdf).

Non-nacreous natural pearls from marine gastropods: Certain species of edible marine gastropods like the queen conch, *Lobatus gigas* (=*Strombus gigas*), horse conch, *Pleuroploca gigantea* and melo, *Melo volutus* have been reported to produce nonnacreous pearls naturally. These pearls are commonly called conch pearls, Horse conch pearls, and Melo pearls. In the queen conch, the conch pearls are a by-product of commercial conch farming (Scarratt and Hann, http://www.ssef.ch/uploads/ media/2004_Scarratt_Pearls_from_the_lion_s_paw_scallop_01.pdf).

Pharmaceutical value: The abalone shell has several health benefits, and in China, it is a valuable medicine. The abalone nacre of the shell has active ingredients such as calcium carbonate, keratin, and choline, and this shell is commonly used in the treatment of hypertension-induced dizziness, headaches, and so on. The body extract of the colored eablone, *H. diversicolor* has shown antibacterial effect and the hydrolysate of its inner shell possesses hepatoprotective effect. Its acidic extract also showed significant anticoagulant effect. Several species of edible marine gastropods have been reported to contain novel bioactive compounds which are of great use in the development of new drugs (vide Chapter 5).

Dye production: Edible species of marine gastropods are the sources of purple dye which is are used to beautify clothing and other items made from cloth (Verhecken, 1989) (vide Chapter 6).

Lime production: The shell of several species of edible marine gastropods is valued for the commercial lime production (Tabugo et al., 2013).

Aquarium values: Abalones, *Haliotis* spp., gold ring cowrie, *Cypraea annulus*, Nassarius snail, *Nassarius* sp., Queen conch, *Eustrombus gigas*, Sand conch, *Strombidae*, tiger cowrie, *Cypraea tigris*, turbo snail, *Turbo* sp., and so on possess beautiful shell and are of great aquarium value.

Human impacts and conservation of marine gastropods: It is estimated that billions of dollars each year are earned by collecting and selling ornamental gastropod shells around the world. This world-wide seashell trade has devastated populations of gastropods, and this has led some countries to ban the import and export of ornamental gastropod shells. The more endangered gastropod species have been fully protected in some areas, making it illegal to collect or harvest these animals. In Caribbean countries, farming of gastropods like the queen conch is in progress for satisfying the public demand for meat and shells. Since most of the marine gastropods are very intimately connected with coral reef ecosystem either for food, shelter, or reproduction, it is necessary save the coral reef ecosystem which in turn conserves the gastropods. Further, the biology of most of the gastropods is not yet fully studied. Therefore, the species which are already in the list of scarce and in jeopardy species require particular attention. Severe rules and regulation need to be applied so that such species would be protected from extinction. Scientific technologies such as transplantation and sea ranching have also been developed for the conservation and development of natural stocks of marine gastropods.

KEYWORDS

- marine gastropods
- fisheries
- nutritional values
- ornamental values
- pharmaceutical values
- industrial uses
- conservation

CHAPTER 2

BIOLOGY AND ECOLOGY OF EDIBLE MARINE GASTROPOD MOLLUSCS

CONTENTS

ABSTRACT

The different aspects of the biology and ecology of edible marine gastropod Molluscs such as their distribution, habitat, shell morphology, internal anatomy, physiology, and life cycle are given in this chapter.

The edible, marine gastropods are extremely diverse in morphology, feeding behavior, reproductive strategies, habitat range, size, and so on. Further, they have the widest range of ecological niches of all Molluscs. For the protection and sustainable fishery of these animals in world seas and oceans, information on the biology and distribution of this group is indispensable.

2.1   DISTRIBUTION AND HABITAT OF MARINE GASTROPODS

2.1.1   DISTRIBUTION

The marine gastropods are found distributed in all oceans worldwide and are more common in temperate and tropical waters. They occupy all marine habitats ranging from the deepest ocean basins to the supra-littoral.

2.1.2   HABITAT

Most marine gastropods are benthic and mainly epifaunal. Some gastropod species have also lived in the deep sea (e.g., faunas associated with hydrothermal vents), and a few such as the violet snails (Janthinidae) and the sea lizards (*Glaucus*) are planktonic drifting on the surface. Various benthic habitats of gastropods including tide pools, coral reefs, and rocky reefs.

2.2   BIOLOGY OF MARINE GASTROPODS

The name gastropod means "stomach-foot." Marine gastropods have a single valve (shell) and a muscular foot. These gastropods are characterized by the possession of a single (often coiled) shell. Externally, the gastropods appear to be bilaterally symmetrical. However, they are one of the most successful clades of asymmetric organisms. These animals range in size from less than 1 mm to over 80 cm.

2.2.1 SHELLS

The shells of gastropods are essentially exoskeletons. The primary functions of these shells are to protect the animals from predators and provide a place for muscle attachment. Secreted by the mantle, the shell is composed mainly of calcium carbonate precipitated into an organic matrix known as conchiolin. If the opening (aperture) of a shell is on the right side when the shell is held with spire pointing upwards, the shell is said to be "dextral or right-hand" shell. On the other hand, the left-hand coiled shells are referred to as "sinistral or left-hand" shell. Over 90% of gastropods have a dextral shell, and the dextral species may produce sinistral shells and vice-versa occasionally.

2.2.1.1 SHELL MORPHOLOGY

The morphology of the gastropod shells is extremely varied as they possess various shapes and ornament, with size ranging from about 1 mm up to more than 1 m. The shell and its ornament may be broadly linked to the mode of gastropod life. Generally, the most ornamented shells occur in tropical marine environments.

Shell views: Morphologically, a gastropod shell can be viewed in the following standard ways:

i)   Apertural view: In this view, the shell is shown in its full length with its aperture facing the viewer and the apex at the top.

ii) Lateral view: It is the intermediate view between apertural and abapertural views.

iii) Abapertural view: In this view, the shell is shown in its full length with its aperture 180° away from the viewer, and with the apex at the top.

iv) Apical view (or dorsal view): The shell is shown looking down directly onto the apex.

v)   Basal view (or umbilical view): The shell is shown viewed directly from the base.

1   2   3   4   5

Gastropod shell views: 1. Apertural; 2. Lateral; 3. Abapertural; 4. Apical; 5. Basal

2.2.1.2 MORPHOLOGICAL VARIATIONS IN SHELLS

Depending upon the raised or depressed spire, number and shape of the whorls, nature of the spiral angle and size of the last whorl, a variety of forms of gastropod shells have been reported. Some of them are given below.

i)   Cone shaped (obconic): The spire is short and apex is sharp. The body whorl is large. Shell is conical with parallel lips, for example, *Conus*.

ii)   Biconic (two conical shapes touching their bases and tapering at both ends): For example, *Fasciolaria*.

iii)   Fusiform (spindle shaped): The shell is spindle shaped. The body whorl is thick in the middle and tapering near the bottom and the top, for example, *Fusinus*.

iv)   Short, bucciniform: For example, *Buccinum*.

v)   Babylonic (elongated shell with the whorls angulated or shouldered on their upper part): For example, *Babylonia*.

vi)   Turbinated (conical, with rounded base): Spire angle is acute and spire is sharp. Base is flat, for example, *Turbo*.

vii)   Oval-conical (pyramidal, conical with a flat base) or trochiform: Spire is acute and body whorl is globular with convex base, for example, *Tectus*.

viii) Globular (spherical): The spire is small and sharp. The body whorl is large and round with rounded aperture, for example, *Naticarius*.

ix)   Hemispherical: For example, *Nerita*.

x)   Cylindrical: After gradual increase in the diameter of the spire, the diameter remains constant or it may reduce near the base, for example, *Oliva*.

xi) Convolute: Aperture is as long as the shell, nearly or quite concealing the spire, for example, *Cypraea.*
xii) Boat-shaped (slipper-shaped): For example, *Crepidula.*
xiii) Patelliform (conical or limpet-shaped): The apex is sharp and shell is cap-like, for example, *Patella.*
xiv) Auriform (ear-shaped): The shell has very short spire and the aperture is very large, for example, *Haliotis.*

Cone-shaped   Biconic   Fusiform   Bucciniform

Babylonic   Turbinated   Oval-conical

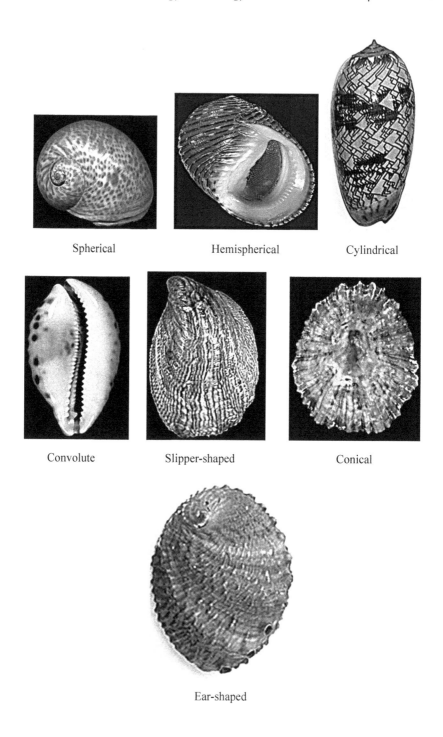

Spherical   Hemispherical   Cylindrical

Convolute   Slipper-shaped   Conical

Ear-shaped

2.2.1.3 *STRUCTURE OF GASTROPOD SHELL*

The gastropod animal lives normally inside a univalve shell which is attached to the soft body by muscles. This attachment may leave distinct scars called muscle scars on the inner shell surface. The soft parts include the broad flattened muscular foot, a distinct head with one or two pairs of tentacles, a pair of eyes, and mouth along with teeth like structures and radula (file-like tongue), digestive, circulatory, respiratory, nervous, urinary, and genital systems. The soft parts are enclosed in a bag-like structure formed by the mantle which secretes the shell. Gastropod shells are attached to the soft body by muscles, which may leave distinct scars (muscle scars) on the inner shell surface. Most gastropod shells are composed of an outer organic layer (periostracum) and an inner, mostly much thicker, calcified layer. The inner layers of gastropod shells consist of minute calcium carbonate crystals (aragonite or calcite) in an organic matrix.

The gastropod shell which in a majority of species is spirally coiled. Each coil is known as a whorl. The number of whorls varies widely in different species. In majority of the cases, the whorls are in contact with each other, while in a few cases, they are separated from each other.

The line of contact of the whorls is termed as suture line. In the embryonic stage, the shell of the animal starts as a small pointed structure known as protoconch (and the part of the shell growing after metamorphosis is termed as teleoconch). As the animal grows older, the whorls increase in size as well as in number by addition of calcareous substances secreted by the mantle. All the whorls together from below the protoconch (except the last whorl), constitute the spire of the shell. The last whorl in which the animal lives is generally larger than rest of the whorls and is known as the body whorl.

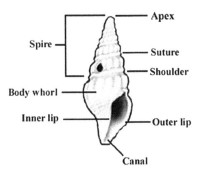

1, spire; 2, body whorl; 3, inner lip; 4, canal; 5, outer lip; 6, shoulder; 7, suture

External morphology of a gastropod shell

The pointed part of the shell is known as apex, and the part of the shell opposite is termed as base. Depending on the number and size of the whorls, the spire may be long, short, horizontal, or depressed.

The opening of the shell is known as aperture. When the aperture is on the right side, then the shell is called "right-handed" or dextral (i.e., the shell opening is to the right when the spire is held upwards). On the other hand, when the aperture is on the left side, the shell is called "left-handed" or sinistral which is rare. Only a limited number of taxa such as *Busycon sinistrum*, the few species of the Fasciolariid subgenus Sinistralia and most of the species of the family Triphoridae, are normally sinistral (left handed—the shell opening is to the left when the spire is held upwards). Like most other gastropods, the shell of *Turbinella pyrum* is almost always right-handed, or dextral, in its shell-coiling, but very rarely a left-handed shell is found (one in about 200,000 individuals).

The inner part of the whorls fuse together to form a pillar like structure called the columella which is extending from the apex to the base. The shell with a columella is known as imperforate. In other forms, the inner part of the whorls instead of fused together are left as a tube-like hollow space, which extends from the apex and opens at the base of the shell is called umbilicus. A shell with an umbilicus is known as perforate.

The shape and size of the aperture vary in different forms and the operculum, a horny plate which closes the aperture varies accordingly. This operculum is present in all living gastropods during their larval stages but is lost in some adults (e.g., limpets). The operculum is mostly horny (corneous) and may be tightly or loosely coiled or concentric. Some gastropod groups have calcareous opercula. The operculum is attached to the dorsal part of the foot and closes the aperture when the animal withdraws into its shell. The margin of the aperture is named peristome. The inner side of the peristomal margin is known as inner lip, while the outer side is called outer lip. The peristomal margin may be provided with two notches, one at the anterior end and the other at the posterior end. These are known as anterior or siphonal canal and posterior or anal canal, respectively.

The shells of the gastropod are ornamented with different ornamental elements like reticulate sculptures, needles, spines, and so on. Many gastropod shells are also decorated with shades of different colors.

2.2.2 INTERNAL ANATOMY

The visceral mass of gastropods is mostly enclosed, together with the mantle cavity, in a calcareous shell. These soft-bodied animals use the pressure of their blood and muscles for movements of different organs.

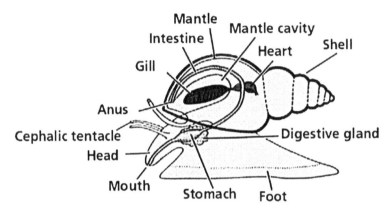

1, foot; 2, digestive gland; 3, shell; 4, heart; 5, mantle cavity; 6, mantle; 7, intestine; 8, gill; 9, anus; 10, cephalic tentacle; 11, head; 12, mouth; 13, stomach

Internal anatomy

Locomotion: A gastropod "crawls" by gliding along on its foot. The foot is a broad, flat muscle. It also attaches to rocks and other substrata. As the gastropod moves, waves of fine muscular contractions sweep along the foot. The contractions lift the animal's foot, then returns it to the surface a little farther ahead, pulling the animal forward.

Food and feeding: The feeding habits of gastropods are extremely varied. Based on their feeding habits, they may be grazers, browsers, suspension feeders, scavengers, detritivores, or carnivores. Herbivorous gastropods graze primarily on algae with their powerful radula. Others swallow detritus to digest decomposing plant matter. They are mainly archaeogastropods such as abalones (Haliotidae), top shells (Trochidae), and nerits (Neritidae); they also include mesogastropods such as periwinkles (Littorinidae) or conchs (Strombidae). Some gastropods feed also on particles by means of elongate gill filaments (e.g., the slipper limpets, *Crepidula*), but others set mucus strings or nets to trap the food, like the worm shells (Vermetidae). Carnivorous gastropods may be partial predators, eating only portions of colonial organisms, or may consume entire solitary prey, which are frequently bivalves. Predatory gastropods drill holes through the shell by means of their radula and use chemical secretions to soften the shell and paralyze the prey (Naticidae, Muricidae). Cone shells (Conidae) inject powerful neurotoxins through hollow, arrow-like teeth before swallowing the prey whole.

Digestive system: The digestive system in gastropods starts with a mouth containing a tooth-bearing ribbon (radula). The organization of gastropod radulae and stomach, as well as additional parts of the digestive system, reflects their different feeding habits. The anus opens into the mantle cavity.

Respiration: In most marine gastropods, one or two gills (ctenidia) located in the mantle cavity are used for respiration.

Circulatory system: It comprises a contractile heart with one or two auricles, and a ventricle, as well as a system of arteries and veins. Gastropod blood transports oxygen using the copper-bearing pigment hemocyanin.

Excretion: One or, rarely, two kidneys help in excretion through the mantle cavity.

Nervous system: The gastropod nerve system includes paired ganglia which are linked with different sensory receptors by connectives and commissures.

Reproductive system: The morphology of the reproductive organs is highly diverse in gastropods. Most gastropods have separate sexes but some groups (mainly heterobranchs) are hermaphroditic.

Imposex: It is a reproductive abnormality of marine gastropods caused by a toxic compound tributyltin (TBT) present in antifouling paints. The term "imposex" or "pseudohermaphroditism" relates to the "superimposition of male features in females" and was first described in dog whelk (*Nucella lapillus*). In certain species of *gastropods* with separate genders, the females presented a penis and/or vas deferens. So far, imposex and *intersex* (a similar phenomenon) have been described in about 150 species of marine gastropods. The effects of imposex can vary, depending on the species. In some species, this phenomenon does not impair reproduction. In some other species, it can lead to population decline as a consequence of reproduction failure. Imposex has been used as a highly sensitive bioindicator of TBT pollution worldwide (Terlizzi et al., 2004). More developed stages of imposex can lead to the sterilization and premature death of the females, affecting the entire population (http://www.marbef.org/wiki/TBT_and_Imposex; http://www.ospar.org/site/assets/files/34390/tbt_example_sheet.pdf).

Life cycle: Most hermaphroditic gastropods do not normally engage in self-fertilization. Basal gastropods release their gametes into the water column where they undergo development. Derived gastropods use a penis to copulate or exchange spermatophores and produce eggs surrounded by protective capsules or jelly. The first larval stage is typically a trochophore that gets transformed into a veliger which settles and undergoes metamorphosis to form a juvenile snail. While brooding of developing embryos is widely distributed throughout the gastropods, there are sporadic occurrences of hermaphrodism in the nonheterobranch species. The basal groups have nonfeeding larvae, while veligers of many neritopsines, caenogastropods, and heterobranchs are planktotrophic. Lifespan of marine gastropods is highly variable.

2.2.3 LIFE CYCLE OF DISC ABALONE, HALIOTIS DISCUS HANNAI

Abalones are solitary animals and are found aggregating only during the spawning season. They reach sexual maturity after 3 years generally. The exact time of spawning, however, varies by location and other environmental cues. They are broadcast spawners, that is, they release their gametes (eggs and sperm) into the open water where fertilization occurs. Mature females can release up to 1 million eggs per spawning although fertilization and survival rates are very low.

Fertilized eggs   4-cell stage   32-cell stag

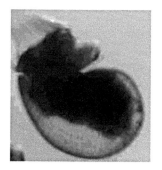
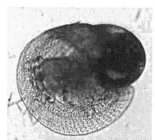

Hatched larva   Ready-to-attach larva

The fertilized eggs are dispersed by water currents until they hatch into motile larvae known as veligers. The veligers undergo several development stages for 4–6 days until they settle. Once settled, the juvenile abalone are called "spat," and they graze on the microalgae present on the substrate. Spat may select a suitable substrate depending on the nature of food availability. Once the spat have reached approximately 5 mm in size, they find their ways into rock crevices for protection and come out at night to feed on drifting seaweed.

KEYWORDS

- **marine gastropods**
- **biology**
- **ecology**
- **physiology**
- **life cycle**

PROFILE OF EDIBLE MARINE GASTROPOD MOLLUSCS

CONTENTS

ABSTRACT

The profile of the different species of edible gastropod molluscs with aspects such as common name, distribution, habitat, description, biology, fisheries/aquaculture, and other uses are given in this chapter.

3.1 ORDER ARCHAEOGASTROPODA

Haliotidae (abalones): Shell of the species of this family is ear shaped, depressed, and loosely coiled. Spire is eccentric. A spiral row of holes is seen on body whorl. Aperture is occupying most of the underside. Interior side is nacreous. There is no operculum. Maximum size is about 6–7 cm long. The animals feed on macroalgae, and colors of their shells may also reflect their diet. Abalones are widely considered to be a delicacy, and there are many farms around the world farming bigger species. Their highly iridescent shells after polishing are used to make jewelries and decorative items. Abalone fishing has a high market value and is considered an activity.

Haliotis asinina (Linnaeus, 1758)

Common name(s): Donkey's ear abalone or ass's abalone.

Global distribution: Indo-West Pacific.

Habitat: Intertidal and upper subtidal zones; inhabits the undersides of boulders and corals; depth range 0–10 m.

Description: Shell of this species has a distinctly elongated contour resembling a donkey's ear. Its outer surface is smooth and almost totally covered by the mantle in life, avoiding encrustations of other animals. Shell has five to seven ovate open holes on the left side of the body whorl. These holes collectively make up the selenizone which form as the shell grows.

Its spire is slightly conspicuous, with a mostly posterior apex. It has a pair of bipectinate ctenidia which are well-developed epipodial tentacles and a rhipidoglossate radula. Color of the shell may range between green olive and brown externally, with triangular patches. Interior surface of the shell is strongly iridescent, with shades of pink and green. It has the fastest growth rate of all the abalones and can grow to a maximum shell length of 12 cm.

Biology

Locomotion/behavior: It is not aggregating in dense populations.

Food and feeding: It is very active at night and feeds on epilithic algae by grazing. The red algae *Gracilariopsis heteroclada* has been reported to promote high growth rates especially in abalone farming in the Philippines.

Reproduction: Sexual maturity occurs in 12 months. This species has a pelagobenthic life cycle including a period of 3 to 4 days in the plankton. Biomineralization begins immediately after hatching, with the formation of a larval shell (protoconch). The protoconch remains inactive until the animal contacts a specific cue which initiates metamorphosis. The transition from protoconch to teleoconch (juvenile/adult shell) is clearly visible at metamorphosis. The early teloconch (postlarval shell) is more robust and opaque than the larval shell but has no pigmentation.

Fisheries/aquaculture: Commercial fisheries exit for this species. It is actively collected in the Southeast Asian countries, for its shell and large flesh. Commercial aquaculture of this species is growing.

Other uses (if any): The adductor muscle of this species is a delicacy in Philippines and China.

Haliotis australis (Gmelin, 1791)

Common name(s): Queen paua, austral abalone, Australian abalone.

Global distribution: Southwest Pacific: New Zealand.

Habitat: Forms large aggregations on reefs in shallow subtidal coastal habitats.

Description: Shell of this species is thin, oval and is quite convex. Sculpture consists of faint spirals and a close strong radiating corrugation. There six to eight perforations which are circular with elevated edges. Right margin is a little straighter. Back of the shell is convex, not carinated at the row of holes. Color pattern is light yellowish-brown, red on the spire, or light green flamed with red. Surface of shell has almost obsolete spiral cords, and regular, close, radiating folds. Between the row of holes and the columellar margin, there are three strong spiral ribs. Spire is a slightly elevated with three whorls. Inside it is corrugated like the exterior, silvery with blue, green, and red reflections. Size of the shell varies between 40 mm and 100 mm.

Biology

Locomotion/behavior: It is a sedentary animal.

Other uses (if any): It is used as a food source in areas of its occurrence.

Haliotis corrugata (Wood, 1828)

Common name(s): Pink abalone.

Global distribution: Native to the Northeast of the Pacific Ocean; North America.

Habitat: Intertidal zone to about 50 m depth; coves and bays as well as along the coast exposed to waves.

Description: Shell of this species is circular, arched and is almost round. It is also thick with rough and undulating surface. Edge of the shell is irregularly trimmed because of the undulations. External shell coloration is green to brown-red. There are two to four open holes which are large, round and high in its surface. Inner region of the shell is iridescent, with predominant shades of pink dark green. Shell surface may also be covered with large amount of algae and other marine animals such as barnacles. Shell grows to a maximum size of 20 cm.

Biology

Food and feeding: It feeds on red algae especially *Sargassum* sp.

Predator(s): Sea otters.

Fisheries/aquaculture: It was commercially fished for the industry of food and jewelry in the early 1940s and after declining in 1995, with its overfishing.

Other uses (if any): It is used as a food source in California.

Haliotis cracherodii (Leach, 1814)

Common name(s): Black abalone.

Global distribution: Eastern Pacific: USA to Mexico.

Habitat: Rocky substrates in intertidal waters; from the high tide line to a depth of 6 m.

Description: Shell of this species is oval and evenly convex, and the two sides are equally curved. Exterior of the shell is smoother and is free from seaweed and other marine growth. Shell is not carinated at the row of holes. Spire is near the margin. There are five to seven small, open respiratory holes, or pores, in the left side of the shell. These holes are mainly used to carry waste and release eggs and sperm. Coloration of the shell is dark brown, dark green, dark blue or almost black. Silvery interior of the shell is pearly with pink and green iridescence. Rear of the shell is spiraled, and the mantle, foot, and tentacles are black. Its shell length can reach a maximum of 29 cm.

Biology

Locomotion/behavior: Black abalone uses its foot to move freely over rock, primarily when immersed in water or at night.

Food and feeding: Black abalones are herbivorous feeding mostly on drifting brown algae and kelp. They also scrape diatoms and other microscopic algae from rocks.

Reproduction: This species attains sexual maturity at 3 years when it reaches a size of 4 cm. Spawning occurs in spring and early summer. It is a broadcast spawner. Larvae are free-swimming for between 5 and 14 days before they settle onto hard substrate, where they then metamorphose into their adult form, develop a shell and settle onto a rock. Its lifespan is 25–75 years.

Predators: Sea otters (such as the southern sea otter, *Enhydra lutris*), fish (such as the California sheephead, *Semicossyphus pulcher*), and invertebrates, including crustaceans such as the striped shore crab, *Pachygrapsus crassipes*, and spiny lobsters.

Conservation status: It is a critically endangered species as per the IUCN Red List 2004.

Fisheries/aquaculture: Commercial fisheries exist for this species. Humans have harvested black abalones along the California Coast for at least 10,000 years.

Other uses (if any): It is used as a food source in California. The meat is eaten fresh or dried; the shell is used as ornament or served as spoons or bowls.

Haliotis discus discus (Reeve, 1846)

Common name(s): Disk abalone.

Global distribution: Southern Hokkaido and southward, Japan; Southern Korea.

Habitat: On rocks; intertidal to 20 m.

Description: In this species, there are four to six holes and one screw groove on the right side of the hole row. When it is viewed from the side, the top of the shell is higher than other abalones. Commo shell size of this species is 20 cm.

Biology

Food and feeding: It feeds mainly on brown algae.

Reproduction: The reproductive cycle of this species can be divided into an inactive stage (January–February), early active stage (March–April), late active stage (May–July), ripe stage (August–October), and spent and degenerative stage (November–January). The main spawning period of this species is August to October especially in Korea.

Other uses (if any): Abalone pearls are made with its inner nacre layer.

Haliotis discus hannai (Reeve, 1846)

Common name(s): Japanese abalone.

Global distribution: Northwest Pacific: Japan and Korea; eastern coast of Mainland China.

Habitat: Shallow subtidal habitats, extending from about 1 to 5 m water depth.

Description: In this species, the dome of the shell is not high, and the shell is generally elliptical. It is also thin and has a curved row of five respiratory pores (rounded shell perforations overlying the respiratory cavity). Interior surface of the shell has an iridescent appearance and is covered by a "nacre" or so-called mother of pearl. Epipodium has a brownish striped appearance.

Biology

Locomotion/behavior: The muscular foot of this animal has strong suction power, permitting the animal to clamp tightly to rocky substrata. It is tolerant of cold water temperatures and can survive in adverse periods when the water temperature is below 5°C in Japan.

Fisheries/aquaculture: Commercial fisheries exist for this species. Farms cultivating the Japanese abalone have been established in many parts of Japan, particularly in the colder water locations. It is also the most commonly cultured species of abalone in Korea. In China, several abalone species are cultured, with *Haliotis discus hannai* being more common in the colder water regions of northern China. This species has been exported to several countries for aquaculture purposes and is now farmed in Chile, Namibia, Ireland, and Hawaii.

Other uses (if any): Its meat is mainly consumed in Japan. Although it is a fairly slow growing species of abalone, it is highly favored in the Japanese market and commands the highest prices of any abalone species world-wide. The calreticulin homolog of this species has been reported to play a potential role in host immunity (Udayantha et al., 2016).

Haliotis diversicolor (Reeve, 1846)

Common name(s): Multicolored abalone or Taiwan abalone.

Global distribution: Northwest Pacific: Taiwan.

Habitat: Intertidal zone; deep rock crevice gaps; depth range 0–23 m.

Description: Shell of this species is long and oval. Two sides of the shell are equally curved, and the back is quite convex. The columellar plate is rather narrow, flattened and is sloping inward. Spire is very near the margin and surface of the shell is spirally lirate. Lines are unequal, rounded, and crossed by low folds. There are 7–9 oval perforations which are used for breathing and excretion. Its color pattern is reddish-brown, scarlet, and green in irregular patches and streaks. Inner surface is silvery with light green and red reflections. Size of the shell varies between 2.5 and 11.0 cm.

Biology

Food and feeding: It feeds on algae-based diet.

Reproduction: This species is a gonochoric and broadcast spawner. Embryos develop into planktonic trochophore larvae and later into juvenile veligers before becoming fully grown adults.

Fisheries/aquaculture: Commercial fisheries exist for this species.

Other uses (if any): The flesh of this species is consumed in Hawaii.

Haliotis fulgens (Philippi, 1845)

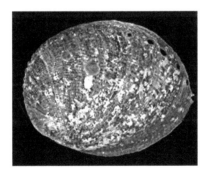

Common name(s): Green abalone.

Global distribution: Southern California and the west coast of the Baja California Peninsula, Mexico.

Habitat: Rock crevices, under rocks and other cryptic cavities; shallow water on open/exposed coast from low intertidal; depth range 0–18 m.

Description: Shell of this species is large, oval, quite convex shell with many low, flat-topped ribs which run parallel to the 5–7 open respiratory pores. It is sculptured with rounded spiral lirae, nearly equal in size. The spire does not project above the general curve of the back. The cavity of the spire is small, almost concealed. Epipodium is present along the side of the foot. The head and epipodial tentacles are olive green, but the epipodial fringes are a mottled cream and brown color, with knobby tubercles scattered on the surface, and a frilly edge. Exterior shell of this species is usually reddish-brown. Inside of the shell is an iridescent blue and green. Maximum size of the adult shell is 25 cm.

Biology

Locomotion/behavior: While juveniles move out at night to forage and return to home site at dawn, adults stay on home site.

Food and feeding: It is a herbivore feeding mostly on macroalgae. Its other food sources include detritus or epiphytic diatoms.

Predators: Sea otters, starfish, large fishes and octopuses.

Reproduction: It has separate sexes. Maturity is reached at 6–13 cm (5–7 years). It breeds year-round. Embryos develop into planktonic trochophore larvae and later into juvenile veligers before becoming fully grown adults. Lifespan is more than 30 years.

Fisheries/aquaculture: Commercial fisheries exist for this species. It has a high cultivation potential, and there is growing interest in farming and marketing this highly prized product, especially in the USA and Chile.

Other uses (if any): It is widely consumed in Hawaii and California.

Haliotis gigantea (Gmelin, 1791)

Common name(s): Giant abalone or Siebold's abalone.

Global distribution: Endemic to the waters off Japan and Korea.

Habitat: Depth range, 10–15 m.

Description: Shell of this species is large, rounded, and heavier. Back of the shell is quite convex and highest in the middle. It is solid but not very thick. Surface of the shell shows coarse, low, unequal spiral cords, and broad wave-like undulations. Dorsal side is like an old folded cliff. There are four open perforations which are situated in high tubercles upon a strong dorsal angle. At the extremity of the row of holes, the margin is triangular shaped. Holes are often elevated, tubular, leading to a short crenelated crown upon the anterior dorsal area. Spire is very small, quite low. Inside there are shallow spiral sulci and indentations. There is no muscle scar. Exterior coloration of the shell is reddish-brown, radiately streaked more or less with chocolate and green. Interior nacre is light colored or silvery, to a high degree iridescent, reflections of emerald green and red predominating. There is no muscle scar. Size of the shell varies between 8 and 20 cm.

Haliotis iris (Gmelin, 1791)

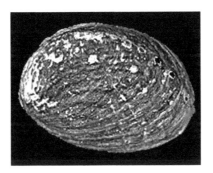

Common name(s): Paua, Blackfoot paua, rainbow abalone or black paua.

Global distribution: Southwest Pacific: New Zealand.

Habitat: Shallow (subtidal) cool water on rocks and reefs at depths less than 6 m; areas of seaweed accumulation and good water movement.

Description: Shell of this species is oval and convex in both sides which are equally curved. Oblique spire is very short and contains two whorls and 5–7 perforations. Surface is pitted. Lip is continuous and is produced beyond the body whorl. The muscle impression is distinct and roughened. Columellar plate is broad, passing into the expanded continuation of the outer lip above. The cavity of the spire is small. Outer surface is pale brown or light olive-green. Inner surface is dark metallic blue and green, with yellow reflections. Size of the shell of this species varies between 80 mm and 180 mm.

Biology

Locomotion/behavior: It is sedentary and can go for months without changing its position.

Food and feeding: It is a herbivore feeding on algae.

Reproduction: Paua are broadcast spawners and spawning is said to be annual.

Predator: The New Zealand eagle ray, *Myliobatis tenuicaudatus*.

Fisheries/aquaculture: Commercial fisheries exist for this species. In New Zealand, paua are currently farmed not only for their meat but also for pearls grown within their shell (Anon. http://www.lbaaf.co.nz/nz-aquaculture-species/modelled-species/paua/). Shells of this type are used for the production of jewelry. At present, this type of fishing is prohibited in view of its small size. There is also a large recreational fishery for this species.

As biocontrol agents: This species has the potential value as biocontrol agents. Caging experiments conducted in the wharf piles of New Zealand showed that this species reduced established biofouling cover by >55% and largely prevented the accumulation of new biofouling over 3 months (Atalah et al., 2014).

Other uses (if any): The paua is renowned throughout the world (especially in New Zealand) not only for its beautiful shell but also for its edible dark, full flavored flesh. Shells of this species are used for the production of jewelry in New Zealand.

Haliotis kamtschatkana (Dall, 1878) (=*Haliotis assimilis*)

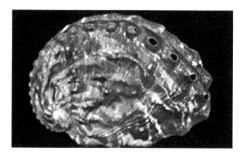

Common name(s): Northern abalone, pinto abalone or threaded abalone.

Global distribution: Northeast Pacific Ocean; North America, Canada, Baja California, Mexico and Korea.

Habitat: Rocky shores; intertidally or subtidally near kelp; depth range 0–100 m; juveniles (10–70 mm shell length) in crevices and other cryptic habitats; adults (>70 mm shell length) on exposed rock surfaces.

Description: Shell of this species is thin, flattened, and ear shaped. Surface is covered with uneven spiral cords, often almost obsolete, and strongly elevated undulations or lumps. Columellar shelf is narrow, flattened, and sloping inward. Shell has 3–6 elevated respiratory holes. These holes collectively make up what is known as the selenizone which form as the shell grows. The Exterior shell is green-brown but can have white or blue coloration. Silvery interior of the shell is iridescent. Epipodium is lacy and green-brown in color. Tentacles surrounding the foot and extending out of the shell are used to sense food and predators. It has an adult shell size of about 8 cm, but it can rarely grow as large as 15 cm.

Biology

Food and feeding: It is a herbivore feeding on macroalgae. Juveniles, however, feed on benthic diatoms and microalgae.

Reproduction: These abalones become sexually mature at a shell length of about 5 cm and all are mature at 7 cm, which requires 2–5 years of growth. They may have ripe gonads throughout the year. Fecundity ranges from 156,985 eggs for 57 mm to 11.56 million eggs for 139 mm. They spawn synchronously, with groups of males and females in close proximity to each other in shallow depths broadcasting their gametes into the water column. Fertilization success depends on the local density of adults. The planktonic phase is short and temperature-dependent (10–14 days at 14–10°C). Larval dispersal is limited. Settlement is thought to occur on encrusting algae. Abalone growth can vary considerably between areas depending on the extent of exposure to wave action and availability and quality of food. As juveniles approach maturity, their diet changes from benthic diatoms and microalgae to drift macroalgae. Lifespan is about 15 years. Population size of this species has declined due to overharvesting, illegal harvesting, predation by recovering sea otters, and disease. Poaching of pinto abalone is a lucrative enterprise and is likely placing continued stress on the remaining abalone populations.

Predator(s): Major natural predators of this species include sea otters and crabs.

Conservation status: It has been listed as "Endangered" by the IUCN Red List of Endangered Species since 2006.

Other uses (if any): This is an important food item in California and Northern Pacific.

Haliotis laevigata (Donovan, 1808)

Common name(s): Smooth Australian abalone, greenlip abalone or whitened ear shell.

Global distribution: Eastern Indian Ocean; endemic to Southeastern and Southwestern Australia.

Habitat: Subtidally attached to rocks, especially in moderately sheltered environments.

Description: Shell of this species is large, thin, and oval shaped. A green ring around the foot at the bottom of the shell is the distinctive feature of this species. Shell is nearly smooth. There are 12 perforations which are very small and their borders are not raised outside. Surface of the shell is sculptured with nearly obsolete spiral threads and cords. Spire is moderately elevated. Whorls number is about 2½. Columellar plate is rather wide, sloping inward, flattened, and obliquely truncated at the base. Cavity of the spire is large and rather shallow. Perforations are unusually small. Exterior coloration of the shell consists of continuous oblique stripes of scarlet and whitish. Inner surface is silvery. Nacre is almost smooth but shows traces of spiral sulci and is very minutely wrinkled. Shell measures up to 18 cm.

Fisheries/aquaculture: Commercial fisheries exist for this species. Mariculture of this species is in progress in Australia.

Other uses (if any): It is commercially harvested for food in the northern coast and Bass Strait Islands of Australia.

Haliotis madaka (Habe, 1977)

Common name(s): Giant abalone.

Global distribution: Japan and Korea.

Habitat: Depth range 20–30 m.

Description: Shell of this species is large, rounded, and thin with a heavy sculpture made of large ribs, folds on growth marks, and sparse bumps. Brown surface is covered with periostracum. Holes are strongly elevated. Surface between the row of holes and the concave columella is large, wide, with a carina. Size of the shell varies between 60 mm and 246 mm.

Biology

Locomotion/behavior: The behavioral character of this species has been examined around the artificial fish reef settled at 30 m depth. In the daytime, the abalones were found to firmly attach at the same position. In the nighttime, they lifted their shell slightly and occasionally changed their body axis to catch the drifted algal food.

Food and feeding: It is a herbivore feeding on macroalgae.

Haliotis midae (Linnaeus, 1758)

Common name(s): South African abalone, perlemoen abalone, or midas ear abalone.

Global distribution: Southeast Atlantic: South Africa.

Habitat: Rocky reefs between the intertidal and subtidal zones up to a depth of 10 m.

Description: This species has a large, flattened, ear-shaped shell with a wide opening at the base. Shell covers the dorsal part of the body leaving some fleshy parts protruding from the side. Body whorl is strongly angled at the position of the perforations, perpendicularly descending from the angle to the columellar margin. Surface of the shell shows strong, elevated, radiating wrinkles, or lamellae. There are 6–11 perforations which are small and subcircular. Through these holes water escapes after aerating the gills. Muscle scar is large and rounded. Columellar plate is rather broad and is sloping inward. Cavity of the spire is large, showing about 1½ whorls from below. Color of the shell is yellowish-gray. Folds are stained with coral-red. Inner surface is pearly and multicolored. Size of the shell varies between 12 and 25 cm.

Biology

Food and feeding: It is a herbivore feeding mainly on macoalgae such as kelp and red- and green algae. When the food comes into contact with its foot, it traps it by raising the front part of the body and extending the far end of the s foot. If a large kelp is caught, the neighboring abalones also join in feeding. In the culture of this species sea weeds are alternated with synthetic food as food sources. Based on algal availability and frequency of occurrence in gut samples, the algalfood items of this species are *Ralfsia verrucosa*, *Ulva* spp., *Plocamium corallorhiza*, *Calliblepharis fimbriata*, *Spyridia cupressina*, *Hypnea rosea*, and *Hypnea spicifer*.

Reproduction: Males and females of this species can be distinguished by looking at their gonads (reproductive parts). The males have cream-colored gonads, while females have green-colored gonads. These abalones reach sexual maturity at around 20–25-mm shell width or around 7 years of age. The breeding season of this species is between March and October, with spawning peaking between April and June. It is a broadcast spawner, and there is no mating behavior. Spawning is initiated by the release of sperm cloud stimulating the females to release their eggs through small breathing holes. As these gametes interact, fertilization takes place. Fertilized eggs hatch into trocophore larva moving actively in the water column and shortly develop into veliger larvae, followed by substrate settlement and metamorphosis. Temperatures have been found to exert a large impact on larval development. High temperatures are largely associated with prevention or even termination of larval development, while low temperatures are associated with longer duration of the larval stage. The favorable temperature for the spawning of this species is about 18°C. The lifespan of this species is about 30 years.

Fisheries/aquaculture: Commercial fisheries exist for this species. It is also a potential species for aquaculture in South Africa. While *Haliotis midae* could reach a maximum size of about 20 cm shell length at an age of about 30 years in the wild, farm production is largely concentrating on an average size of 10 cm after 5 years.

Other uses (if any): This economically important species is highly utilized and has a high market value especially in South Africa.

Haliotis ovina (Gmelin, 1791)

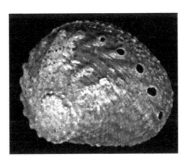

Common name(s): Sheep's ear abalone or oval abalone.

Global distribution: Indo-West Pacific.

Habitat: Attached at corals or under rock ledges; shallow subtidal zone; depth range 1–56 m.

Description: Shell of this species is rounded-oval and quite flat. Upper surface shows strong radiating folds. Columellar plate is flat and very wide. Whorls of the spire contain a corona of tubercles. Body whirl shows radiating folds, sometimes ending in a series of knobs around the middle of the upper surface. There are 4–5 circular perforations which are tubular and elevated. These perforations are situated upon a low keel. Exterior shell coloration is green or reddish, radiately painted with white. Inner surface is silvery. Size of the shell varies between 2 and 10 cm.

Fisheries/aquaculture: Commercial fisheries exist for this species.

Other uses (if any): This species is actively collected for its edible flesh in the Philippines. As its shells are very attractive and beautifully colored, they are often used for home decorations, jewelry making, and knife-handle inlays.

Haliotis pourtalesii (Dall, 1881)

Common name(s): Pourtale's abalone.

Global distribution: Western Central Atlantic: North Carolina to Suriname.

Habitat: Bottom with rocks, stones, sand and shell debris, or reef; depth range 40–370 m.

Description: Shell of this species has 22 to 27 wavy spiral cords on its outer surface which is colored a waxy yellow to light-brown, with a few irregular patches of reddish orange. A light-orange band runs from each hole to the edge of the shell. Inside is pearly white. Spire is rather elevated, consisting of about two and a half whorls. Apex is small and prominent. There are 25 respiratory holes, of which 5 remain open. Outside the row of hole, the usual sulcus is strongly marked. The coil of the spire is rather close and the margin of the columella is flattened. The radula of this species is peculiar in possessing a very narrow lateral tooth. The cephalic tentacles show concentric rings. The shell is relatively small, varying from 18 mm to a maximum length of 30 mm.

Other uses (if any): This species is actively collected for its edible flesh in North America.

Haliotis roei (Gray, 1826)

Common name(s): Roe's abalone.

Global distribution: Endemic to Australia and occurs off Western Australia to Victoria.

Habitat: Between intertidal and subtidal habitats.

Description: Shell of this species has a short-oval shape. Distance of the apex from the nearest margin is somewhat over one-fifth the greatest length of the shell. Sculpture of the shell consists of strong unequal spiral cords crossed by radiating folds. There are seven to nine perforations which are rather small and a little raised. Right side is straighter than the rounded left margin, and the back is depressed. Exterior color of the shell is scarlet-red, more or less marbled with olive-green and is painted with broad white rays. Spiral riblets are numerous, unequal, and are separated by deeply cut grooves. Spire is rather elevated. Inner surface of the shell is silvery, very iridescent, with pink, green, and steel-blue reflections. Columellar plate is narrow and is obliquely truncated at its base. Size of the shell varies between 5 and 12 cm.

Biology

Food and feeding: It is herbivorous, feeding on a variety of driftin macroalgae. Algae consumed by this species varied seasonally and between platforms. Volumes of gut contents were found to be greatest in winter, corresponding with the time of maximum algal food availability.

Reproduction: Males are more common than females especially in Western Australia. These abalones mature at a size of about 4 cm. Females of 12 cm size produce about 9 million eggs. This species has a short period of intense spawning in July and August, followed by low levels of spawning until December. This species has been reported to grow rapidly to more than 4 cm in its first year. In the second year, it reaches 6 and 7 cm in the third year. While the legal size for its capture is 6 cm, 7-cm-sized animals are normally collected in commercial fishery. Abalones of the subtidal populations reach faster growth rates and a larger maximum size, and there is no significant difference in growth between sexes.

Fisheries/aquaculture: This species is farmed in Australia.

Other uses (if any): It is a delicious food item in Australia.

Haliotis rubra (Péron, 1816)

Common name(s): Blacklip abalone.

Global distribution: Eastern Indian Ocean: Australia.

Habitat: Crevices, caves, fissures and on vertical rock surfaces, from low tide to 50 m; small animals live under boulders, moving to more exposed situations as they grow.

Description: This species has a distinctive dark band along the edge of its shell, but the shell is often fouled with algae and plants, making the abalone difficult to find. It has a large, much depressed shell which is rounded and oval shaped. Distance of the apex from the margin is one-fifth the length of the shell. There are 6–8 perforations which are elevated and circular. Holes at the top of the shell are normally used for breathing. Water is drawn in the front of the shell, passed over the gills where oxygen is extracted, and then pumped out of the holes at the top. Exterior surface of the shell is either dark red with few radiating angular white patches or dull red and green, streaked, and mottled. Inside is corrugated like the outer surface, silvery, and very brilliantly iridescent. Columellar plate is broad, flat, and obliquely truncated at its base. Cavity of the spire is wide, open, but shallow. Size of the shell varies between 4 and 20 cm.

Biology

Locomotion/behavior: This species is usually seen aggregating. It prefers to feed at night.

Food and feeding: This species grazes on drift algae and algae growing on rock surfaces. It captures drift algae or overhanging kelp fronds by raising the front of the foot and trapping it. In areas where water movement is poor, or algae is scarce, then the abalone grows only slowly.

Reproduction: In this species, broadcast spawning takes place when fishes release their eggs and sperm into the water, where fertilization occurs externally. Embryos develop into lecithotrophic larvae and later into juvenile veligers before becoming fully grown adults. Spawning takes place between February and April and again between October and December. The growth of the shell has been reported as 2.5 cm at an age of 1 year, 13 cm at 5 years, and 14.5 cm at 6 years. Lifespan of this species is about 15 years with a maximum length of 15–22 cm.

Predators: The predators of this species vary according to the size of the abalone. Young abalone during their crevice-living phase is preyed upon by whelks, crabs, octopus, and wrasses. On the other hand, adult abalone is eaten by fish, octopus, rock lobster, stingrays, and starfish.

Association: This species has been found associated with the members of the family Eusiridae (Crustacea, Amphipoda) (Lowry and Stoddart, 1998).

Fisheries/aquaculture: Commercial fisheries exist for this species. It is also cultivated in Australia.

Other uses (if any): It is an important food item in Australia. Its large shell is very beautiful and pearly inside and is often used in jewelry.

Haliotis rubra conicopora (Péron, 1816) (=*Haliotis conicopora*)

Common name(s): Brownlip abalone.

Global distribution: Endemic to Australia; South Australia and Tasmania.

Habitat: Amongst boulders and in cracks, caves and crevices in reef and rocky intertidal environments; depth range 2–30 m; prefers substrata of granite (occasionally limestone).

Description: This subspecies is to be genetically identical to *Haliotis rubra*, and therefore, it is conspecific. It has rigid, oval-shaped protective shell which is brown in color, often covered in algae and/or pieces of seaweed. These abalones get their name from the color of the "lip" or edge of their muscular foot which is distinctly brown in this subspecies. On the inside, the shell is smooth and shiny with mother-of-pearl coloration. Common size of the shell varies from 10 to 16 cm. Lifespan of this subspecies is about 13 years.

Biology

Food and feeding: It is a herbivore feeding on algae. It uses its rasp-like radula (tongue) to graze on drifting red algae.

Reproduction: These abalones reach maturity at about 3 years old and 8–9 cm in shell length. Spawning season is April–June especially for Western Australia. Fecundity (number of eggs in a single spawning) for this species has been reported to be 5 million eggs. Brownlip abalones are broadcast spawners, whereby males and females release their sperm or eggs into the water at the same time. Its maximum grow out size is 25 cm (wild).

Fisheries/aquaculture: A commercial fishery exists for this species in Western Australia. Brownlip abalone has been identified as one of Australia's largest and possibly fastest growing abalone species and is attracting increasing fishing effort and harvest pressure. This species now represents more than 20% of the South Coast's greenlip/brownlip catch.

Haliotis rufescens (Swainson, 1822)

Common name(s): Red abalone.

Global distribution: Eastern Pacific: Mexico and USA; introduced in the Northeast Atlantic, Southeast and western Pacific and Hawaii.

Habitat: Sand and gravel bottoms, from low intertidal to 24 m in depth; mainly found in or in near rocky macroalgal substrata.

Description: Its shell length can reach a maximum of 31 cm making it the largest species of abalone in the world. Shell is large, thick, and dome shaped. There are 3–4 oval holes or respiratory pores. These holes collectively make up the selenizone which form as the shell grows. Usually, the epipodium is black, and it is smooth and broadly scalloped along the edge. Tentacles are black. Shell surface is generally brick-red. Inside of the shell is strongly iridescent. Underside of the foot is yellowish-white in color. This species was used in the study of the microscopic development of nacre. This species also has one of the best meat-to-shell ratios of any of the meat abalone.

Biology

Food and feeding: They feed on kelp (red algae) species such as giant kelp (*Macrocystis pyrifera*), feather boa kelp (*Egregia menziesii*), and bull kelp (*Nereocystis luetkeana*). Juveniles, however, eat coralline algae, bacteria, and diatoms.

Reproduction: These abalones are broadcast spawners. Spawning occurs in spring and early summer. Embryos develop into planktonic trochophore larvae and later into juvenile veligers before becoming fully grown adults.

Fisheries/aquaculture: Commercial fishery exists for this species. Fishing for these abalone populations peaked in the 1950s and 1960s in California and was followed by a decline which was largely due to disease and the recovery of sea otter populations. The California Fish and Game Commission ended fishing for these abalones in 1997. In Northern California, commercial fishing was only legal for 3 years during the World War II. As a result, a recreational fishery still exists in northern California. Since scuba diving to harvest these abalones is banned, the fishery of this species consists of shore pickers searching the rocks at low tide. At present, the minimum legal size for this species is 18 cm, and three specimens may be taken per day. There is also an annual legal limit of 18 abalone per person (Wikipedia). This species is well suited for farming in both land- and ocean-based operations.

Other uses (if any): The flesh of this species is widely consumed in Australia, China, Japan, Mexico, and the United States. This species is most desirable commercially for its size and light meat color.

Haliotis scalaris (Leach, 1814)

Common name(s): Staircase abalone, ridged abalone or ringed ear abalone.

Global distribution: Endemic to Australia; occurs from Southwest (Western Australia) to Victoria including Tasmania.

Habitat: Under-boulder or crevice.

Description: Shell of this species is moderately large but thin, depressed and is irregularly oval shaped. It is showing a strong spiral rib on each side of the row of five to six open perforations, and prominent elevated radiating lamellae around the spire. Its exterior color pattern is reddish or variegated olive and green. Its inner surface is silvery and very iridescent, with excavations corresponding to the elevations of the outer surface. Columellar plate is narrow, and obliquely truncated below. This species has been reported to deposit two fine rings a year in the spire, one in about December and the other from May to July, correspondingly to extreme summer and winter sea temperatures. Additional fine rings, and in particular brown rings, are also deposited in response to boring annelids and drilling muricid gastropods (Shepherd and Huchette, 1997). Size of the shell varies between 6 and 10 cm.

Haliotis sorenseni (Bartsch, 1940)

Common name(s): White abalone.

Global distribution: California.

Habitat: Open low and high-relief rock or boulder habitat that is interspersed with sand channels; depth range 25–30 m.

Description: Shell of this species is oval shaped and very thin. Bottom of its foot is orange, and epipodium (a sensory extension of its foot that has tentacles) is a mottled orange-tan. Shell reaches a maximum size of 25 cm. Its lifespan is about 40 years.

Biology

Food and feeding: It is a herbivore, grazing mainly on macroalgae, such as *Laminaria farlowii* and *Agarum fimbriatum*, and also other species of red algae.

Reproduction: Like many other gastropods, white abalone has a complex life cycle involving larval stages. Fertilized eggs hatch into larvae which finally metamorphose into the adult form and settle from the plankton to a hard substrate. As a broadcast spawner, white abalone reproduces by releasing their eggs and sperm into the surrounding water. If fertilized, the eggs hatch after only 1 day, but high concentrations of sperm are required in order to get an egg fertilized. Therefore, aggregations of adult male and female abalone are necessary for the successful fertilization in these abalones.

Conservation status: The white abalone is an endangered species in the United States.

Other uses (if any): It is widely consumed in California. The white abalone has the most tender and flavorful meat among all the abalone species. Considered a delicacy in California, these white abalones have almost declined to zero level due to increasing overfishing, and its fishery was closed in 1996. Therefore, it is being maricultured in order to produce young that can be sea ranched in the ocean, in order to restore the population levels before it becomes extinct.

Haliotis spadicea (Donovan, 1808)

Common name(s): Blood-spotted abalone.

Global distribution: South Africa.

Habitat: Rocky shores.

Description: Prominent features of this species are the peculiar form (depressed shell with an oblong-ovate shape), narrowed at the anterior end, reddish or chocolate surface with some white blotches, brilliantly pearly inner surface, smooth except for radiating folds, and coppery red stain within the cavity of the spire. Perforations (5–8) are close together and are almost perfectly circular. Columellar shelf or plate, which slopes outward, is rather narrow and convex on its face. Apex is nearer to the margin than in *H. midae*, another Cape species. Shell grows to approximately 7 cm in length.

Other uses (if any): The meat of this abalone has been consumed by people living along the coast in the past, although abalone fishing is now banned in South Africa.

Haliotis tuberculata (Linnaeus, 1758)

Common name(s): European abalone (ormer).

Global distribution: Britain and Ireland; Channel Isles.

Habitat: Prefers shallow sublittoral rocky habitats that contain fissures and crevices in bedrock; rock ledges and the underside of boulders lying on bedrock or sand.

Description: Shell of this species is ear shaped, with few rapidly increasing whorls, which are open below, so that the margin below is only somewhat bent in. Spire is variable in size, generally small in comparison with the extensive last whorl. At the outer side of the whorls, with the exception of the first 2, there is a series of holes. In most cases, these holes are with raised margins; some of which close to the shell aperture are open and the other holes are closed. Shell is nacreous internally and is colored externally. Ormers grow up to 12.3 cm in length.

Biology

Ecology: Temperatures between 8.5°C and 9.0°C are found to be the lower lethal limit for this species. These abalones can survive short-term immersions in salinities down to 14% and could sustain some growth in salinities down to 24%. Further, these abalone populations prefer low water movement for their spawning. This, adaptive behavior would minimize dispersion of gametes and larvae (Mgaya, 1995).

Food and feeding: These abalones are herbivorous and are browsing on a variety of seaweeds (red algae) such as *Palmaria palmata, Delesseria, Griffithsia, Laminaria* spp., *Uliva lactuca, Chondrus crispus, Enteromorpha intestinalis*, etc.

Reproduction: These abalones are generally gonochoristic, though mosaic hermaphroditic animals have also been reported very rarely. Male abalones mature earlier and at smaller size than female. Males reach sexual maturity when they are 2 years old at about 4 cm in shell length and females at 3 years of age with a size of about 5 cm. Fertilization is external, with both eggs and sperm are released simultaneously, though males tend to release their gametes first, followed by females. Larval and early juvenile phases last for 90 days, and at about 160 days after fertilization, the spat reach a size of 6 mm in shell length. These abalones grow at a slow rate, and it takes a minimum period of 3 years to attain a moderate size of 4.5 cm in shell length. The longevity of this species is more than 15 years (Mgaya, 1995).

Predators: The main predators of ormers are Octopus spp. and starfishes (*Asterias rubens* and *Marthasterias glacialis*). Other fish predators include *Labrus bergylta, Symphodus melops*, and *Blennius* sp.

Other uses (if any): This species is widely consumed in France.

Haliotis virginea (Gmelin, 1791)

Common name(s): White foot paua.

Global distribution: Southwest Pacific: New Zealand.

Habitat: On rocks, at 3–5 m deep.

Description: Shell of this species has a sculpture which is made of narrow radial cords, crossed by thin growth striae. Apex is more eccentric, and shell is lighter and thinner. Coloration of shell is green, mottled with dark brown or red. Size of the shell varies between 3 and 7.5 cm.

Biology

Reproduction: Members of this family are mostly gonochoric and broadcast spawners. Embryos develop into planktonic trochophore larvae and later into juvenile veligers before metamorphosing into fully grown adults.

Other uses (if any): In New Zealand, the flesh of this species is used as food source.

Haliotis walallensis (Stearns, 1899)

Common name(s): Flat abalone.

Global distribution: Eastern Pacific: Canada to USA.

Habitat: Rocks or in crevices in low intertidal to 21 m in depth.

Description: Shell of this species is oval, long, and narrow and is considerably flattened. There are four to eight holes, of which five or six are open. Shell surface sculpture is regular with numerous low, evenly rounded ribs crossed by closely spaced, lamellae-like striae. There is no muscle scar although some may have small clumps of scattered green and brownish nacre in the muscle attachment area. Body is a mottled yellow and brown with tinges of green. Epipodium is colored in yellowish-green with large brown and yellow splotches. Tentacles are dark green and slender. Exterior shell color is dark brick red with occasional mottlings of greenish blue and white. Inside of shell is pale pink, with green reflections. Maximum length of the shell is 17.5 cm.

Biology

Food and feeding: It feeds by grazing on small attached algae.

Reproduction: Members of the family are mostly gonochoric and broadcast spawners. Embryos develop into planktonic trocophore larvae and later into juvenile veligers before becoming fully grown adults.

Other uses (if any): The flesh of this species is used as a food source in BC (Canada) and South California (USA).

Fissurellidae (keyhole limpets): Fissurellids have conical shells and are seen under rocks or in shady areas during low tide, sometimes with their mantle exposed. Keyhole limpets of this family have a hole at the top of the shell, the portal through which waste products are released. This makes them different from the true limpets, which release waste from the mantle beneath the shell. Their flattened shells allow them to easily creep under rocks and crevices to seek food and also to avoid the hot sun. Interior is with a horseshoe-shaped muscle scar. There is no operculum.

Diodora graeca (Linnaeus, 1758)

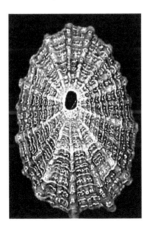

Common name(s): Greek keyhole limpet.

Global distribution: Mediterranean and West Africa to the North Sea (coasts of Ireland and western Britain) and the subarctic North Atlantic.

Habitat: Beneath stones and clinging to rocks in the sublittoral zone up to a depth of 250 m.

Description: Shell of this species is small, flat, oval, and conical. Apex is about one-third of shell length behind anterior margin. It has no whorls and is shield-shaped. Base of the shell is somewhat bent. Its oval aperture (with slight flared posterior lip) is located at the dorsum and is keyhole-shaped. This hole serves as an outlet for water and waste products. The shell has a slight reticulated sculpture with concentric cords crossed by radial ridges. Animal has a broad creeping foot and a developed mantle. Foot of the animal is oval, straight-sided, with 30–35 epipodial tentacles at base. Body is cream colored, deep orange or red. Color of the shell varies from creamy white to yellow white and is often slightly orange tinted. Maximum size of the shell is 2.5 cm.

Biology

Food and feeding: This species is mainly herbivorous and it also feeds on detritus.

Reproduction: Members of the family are mostly gonochoric and broadcast spawners. Embryos develop into planktonic trocophore larvae and later into juvenile veligers before becoming fully grown adults.

Carotenoid content: Carotenoids such as α-, β-, γ-carotene (lutein-free and epoxide form), zeaxanthin, diatoxanthin, mutatoxanthin, and astaxanthin have been identified from this species. β-Carotene and astaxanthin were found to be dominant and the total contents varied from 2.4 to 10.5 μg/g wet weight (Czeczuga, 1980).

Other uses (if any): The flesh of this species is consumed in Montinegro Coast; Mediterranean, and south of the United Kingdom. It is a Crotian cuisine.

Fissurella crassa (Lamarck, 1822)

Common name(s): Thick keyhole limpet.

Global distribution: South Pacific Ocean: Chile; Antarctic.

Habitat: Habitats with differential intertidal wave exposure.

Description: Shell of this species is simple, conical, depressed, and recurved. Apex is perforated a little anterior in an oblong or oval manner. Water for respiration and excretion is drawn in under the edge of the shell and exits through this key-hole-shaped aperture at the apex. Exterior surface is ribbed longitudinally and is slightly striated transversely. Body does not exceed or marginally exceeds that of the shell. Outer radular plate has four cusps. Propodium (anterior end of the foot) has no tentacles. Its maximum reported age is 10 years. Size of the shell varies between 3 and 9 cm.

Biology

Functional hermaphrodite gonad: *Fissurella crassa* with portions of ovary and testicle with ability to generate ova and sperms has been identified. This demonstrates the potential that these animals possess to develop functional hermaphroditism.

Locomotion/behavior: *F. crassa* showed a spatial activity pattern fluctuating between a central place foraging and a ranging strategy. This spatial activity pattern is suggested as the main mechanism of orientation used by the animal to relocate the refuge (Serra et al., 2001).

Food and feeding: Animals are herbivores, using the radula to scrape up the algae from the surface of rocks.

Reproduction: Members of the family are mostly gonochoric and broadcast spawners. Embryos develop into planktonic trocophore larvae and later into juvenile veligers before becoming fully grown adults.

Other uses (if any): Flesh of this species is consumed in west, central, and South America.

Fissurella costata (Lesson, 1831)

Common name(s): Costate keyhole limpet.

Global distribution: Southeast Pacific: Peru and Chile.

Habitat: Rocky intertidal areas (exposed low water).

Description: Shell is simple, conical, depressed, and recurved. Apex is perforated a little anterior in an oblong or oval manner. Exterior surface is ribbed longitudinally and is slightly striated transversely. Body does not exceed or marginally exceeds that of the shell. Outer radular plate has four cusps. Propodium (=the anterior end of the foot) has no tentacles. Size of an adult shell varies between 2.5 and 9.0 cm.

Biology

Food and feeding: It is herbivorous, mainly feeding on algae.

Reproduction: Members of the family are mostly gonochoric and broad-cast spawners. Embryos develop into planktonic trocophore larvae and later into juvenile veligers before becoming fully grown adults.

Other uses (if any): Flesh of this species is consumed in Peur, Chile.

Fissurella cumingi (Reeve, 1849)

Common name(s): Keyhole limpet.

Global distribution: Southeast Pacific: Peru and Chile.

Habitat: Lowermost intertidal zone and immediate subtidal zone; depth range 0–15 m; deep tide pools at low tide; sublittoral rocky zone; on matrices of the tunicate *Pyura praeputialis*.

Description: Shell of this species is simple, conical, depressed, and recurved. Apex is perforated a little anterior in an oblong or oval manner. Exterior surface of the shell is ribbed longitudinally and is slightly striated transversely. Size of an adult shell varies between 8 and 10 cm.

Biology

Food and feeding: Members of the Family Fissurellidae are herbivorous, grazing on algae.

Reproduction: Members of this family are mostly gonochoric and broadcast spawners. Embryos develop into planktonic trochophore larvae and later into juvenile veligers before becoming fully grown adults.

Fisheries/aquaculture: Mariculture of this species is in progress in Chile.

Other uses (if any): Flesh of this species is consumed in Chile.

Fissurella gemmata (Menke, 1847)

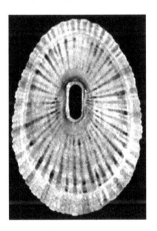

Common name(s): Not assigned.

Global distribution: Mexico.

Habitat: Flat exposed rocks.

Description: Shell of this species is simple, conical, depressed, and recurved. Apex is perforated a little anterior in an oblong or oval manner. Exterior surface of the white sturdy shell is ribbed longitudinally and is slightly striated transversely. Interior shell is cream colored and callus is bordered with gray. Size of specimen: length 3.5 cm; height 1.1 cm.

Fissurella latimarginata (Sowerby, 1835)

Common name(s): Peerski's companion.

Global distribution: Pacific Ocean: off Peru and Chile.

Habitat: On rocks; depth range 2–5 m.

Description: Shell of this species is simple, conical, depressed, and recurved. Apex is perforated a little anterior in an oblong or oval manner. Exterior surface of the white sturdy shell is ribbed longitudinally and is slightly striated transversely. Size of an adult shell varies between 5 and 6 cm.

Fisheries/aquaculture: It is cultivated in Chile.

Other uses (if any): Flesh of this species is consumed in Chile.

Fissurella maxima (Sowerby, 1834)

Common name(s): Giant keyhole limpet or maximum keyhole limpet.

Global distribution: Southeast Pacific: from Huarmey, Peru to Cape Horn, Chile.

Habitat: Low intertidal and high subtidal levels, under *Lessonia* sp. leaves on exposed rocky shores; encrusts on algae, and on the undersides of large flat rocks at low tide; depth range 0–4 m.

Description: Shell of this species is simple, conical, depressed, and recurved. Apex is perforated a little anterior in an oblong or oval manner. Exterior surface is ribbed longitudinally and is slightly striated transversely. Size of an adult shell varies between 6 and 14 cm. In younger shells, internal margin is proportionately broader than in those which are more fully grown. In some individuals, this margin shows well-developed crystalline structure.

Biology

Food and feeding: It is a herbivore consuming diatoms.

Reproduction: It is a dioecious species with a sex ratio of 1:1 in different size classes. Maturity is attained when the animal's shell length reaches a size of 5–6.5 cm. Ovaries are green and testis are median brown to yellowish white. Main spawning period of this species is late November–December (late spring–early summer) and a secondary period in July–August (winter). Members of the family are broadcast spawners. Embryos develop into planktonic trocophore larvae and later into juvenile veligers before becoming fully grown adults.

Fisheries/aquaculture: Commercial fishery exists for this species.

Other uses (if any): Flesh of this species is consumed in Chile.

Fissurella nigra (Lesson, 1831)

Common name(s): Black keyhole limpet.

Global distribution: Western Atlantic and Eastern Pacific.

Habitat: Intertidal mean and lower intertidal areas; up to 70–80 cm above the lower boundary of the tides.

Description: Shell of this species is large, oval, and wider at the back. Surface of the shell presents marked circular crossing grooves. Exterior surface of the shell is of purplish-black color. Interior is thick, white, and smooth. Size of the shell varies between 4.5 and 11 cm.

Biology

Reproduction: It is a gonochoristic (sexes are separate) species. There are no external sexually dimorphic features. The testis of the male is beige-colored and the ovary of females is bright green. This species has a unimodal reproductive cycle with one breeding season between October and December, coinciding with the period when gametes are spawned. Fertilization is external. Embryos develop into the planktonic trochophore larvae approximately 48-h postfertilization, and from this larvae, the veliger stage appeared 72-h postfertilization—nocturnal. Spontaneous spawning occurred between August and November (most abundant in October and November) in adults kept under laboratory conditions. Spawning was also induced in October by combining thermal shock with desiccation. This method proved successful, provoking massive spawning (Perez et al., 2007).

Fisheries/aquaculture: Various species of keyhole limpet of the *Fissurella* genus constitute a multispecific fisheries activity in Chile. *Fissurella nigra* is currently extracted by local fishermen in southern Chile.

Other uses (if any): Flesh of this species is consumed in Chile.

Fissurella picta (Gmelin, 1791)

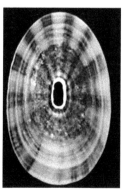

Common name(s): Painted keyhole limpet.

Global distribution: Pacific Ocean: along Ecuador; Atlantic Ocean: along the Falkland Islands.

Habitat: Depth range: 0 to 105 m.

Description: Shell of this species is simple, conical, depressed, and recurved. Apex is perforated a little anterior in an oblong or oval manner. Exterior surface is ribbed longitudinally and is slightly striated transversely. Size of an adult shell varies between 3.5 and 10.0 cm.

Biology

Food and feeding: It is not strictly a herbivore, although it prefers algae such as *Ulva* sp., *Polysophonia* sp., and *Gelidium* sp. However, under laboratory conditions *Gracilaria chilensis* proved to be the best source of energy available for growth in juveniles and limpets fed on *Ulva* sp. showed a negative energy balance. Specimens maintained in suspended farming systems and fed with the artificial diet (*G. chilensis*) reached an average commercial size of 53 mm in period of about 3 years (López et al., 2003).

Other uses (if any): This species is of significant commercial and edible value and considerable ecological importance in southern Chile.

Megathura crenulata (Sowerby, 1825)

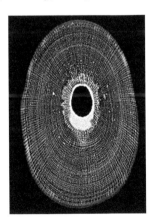

Common name(s): Great keyhole limpet or giant keyhole limpet.

Global distribution: Western North America; from Southern California to the Baja California peninsula in Mexico.

Habitat: Rocky coast; intertidal zone up to a depth of 33 m.

Description: This species is one of the largest keyhole limpets. Mantle covers the shell except for the keyhole on top. Exposed mantle of this species is black, gray, or whitish, with a purple or black edge.

Biology

Food and feeding: This species consumes a variety of food items such as filamentous cyanobacteria, diatoms, brown and red algae such as seaweeds, seagrass, forams, hydrozoans, bryozoans, nematodes, crustaceans, bivalves, gastropods, and tunicates. The larger part of its diet, however, is composed of brown and red algae, tunicates, hydrozoans of the genus *Eudendrium*, and bryozoans of the genus Crisia.

Other uses (if any): This species has been used for experimental studies on gamete agglutination. Its blood contains a hemocyanin that appears blue due to its copper content.

Neritidae (nerites): Shells of the species of this family are thick and somewhat hemispherical with a relatively low spire and a very large, rounded body whorl. Aperture is semicircular, without a siphonal canal. Inner lip is protruding as a septum that narrows the aperture. Inner walls of the spire are resorbed. Operculum is calcified, with a projecting peg. These animals feed on algae.

Clithon retropictum (Martens, 1879)

Common name(s): Not assigned.

Global distribution: Taiwan; South Korea.

Habitat: Brackish water as well as freshwater.

Description: The spirals are low but the shells are mostly corroded, and the shells have many fins, mostly brown, of the largest species in the genus. The outer edge of the shell mouth without teeth, inner lip smooth and well developed, the inner lip shell on the edge of a wide and several smaller granular protrusions, granular protrusions below the number of small dentate processes. Cover slightly raised, near the outer lip with a brown band.

Nerita albicilla (Linnaeus, 1758)

Common name(s): Blotched nerite or oxpalate nerite.

Global distribution: Indian Ocean and Western Pacific.

Habitat: Shelters under rocks on open coasts between mid and low tide; depth range 0–7 m.

Description: Shell of this species is with spire which is depressed, and its width is greater than height. It is sculptured with broad, low, spiral ribs. Columellar deck is pustulose, particularly at outer edge. Columella is concave centrally, smooth or with two or three weak teeth centrally. Outer lip is with one or two strong teeth posteriorly, sometimes one anteriorly, and 12–15 weak lirae. Exterior shell surface is dull, black with white flecks or patches, often formed into two irregular spiral bands. Columella deck is white or yellow, and aperture is white. Operculum is calcareous and is finely pustulose with gray or yellowish color. Maximum size of the shell is 3.5 cm.

Biology

Reproduction: Members of this family are mostly gonochoric and broadcast spawners. Embryos develop into planktonic trocophore larvae and later into juvenile veligers before becoming fully grown adults.

Isopreropodine: A nitrogenous oxindole alkaloid, isopteropodine has been collected from this species (Martin et al., 1986). This compound may have therapeutic use with its antiproliferative and cytotoxic effects to induce apoptosis in human breast cancer, sarcoma as well as lympho-blastic leukemia cell lines.

Fisheries/aquaculture: Commercial fishery exists for this species.

Other uses (if any): It is collected for food by coastal dwellers. Its shell is used for the shell trade.

Nerita articulata (Reeve, 1855)

Common name(s): Lined nerite snail.

Global distribution: Indo-West Pacific; Southeast Asia.

Habitat: Rocky shores; mangrove tree trunks and roots, monsoon canal walls, muddy banks, and rocky areas in or near mangroves.

Description: Shell of this species is sturdy oval and spire does not stick out. Color of the shell is beige, grayish, or pinkish with fine, spiraling black ribs. Flat underside is smooth, white, sometimes with yellow patches. There are small notched "teeth," usually three, on the straight edge at the shell opening. Operculum is thick and is evenly covered in tiny bumps, pinkish with black portions. Body is pale with fine black bands on the foot and long thin black tentacles. Common size of the shell is 2–3 cm.

Biology

Food and feeding: It grazes on algae. It seems to return to the same spot after a feeding bout. It is possibly due to its, mucus trail that it follows back to its resting spot.

Reproduction: Members of this family are mostly gonochoric and broadcast spawners Embryos develop into planktonic trocophore larvae and later into juvenile veligers before becoming fully grown adults.

Fisheries/aquaculture: The species is known to be eaten by Australian aboriginal people and is likely consumed locally elsewhere.

Nerita balteata (Reeve, 1855) (=*Nerita lineata*)

Common name(s): Not assigned.

Global distribution: Indo-West Pacific: Tropical Australia; Peninsular Malaysia.

Habitat: Mangrove forests; on wave breaker rocks and crevices at low tide; reef in the intertidal zone.

Description: This species has a fairly large, rounded, finely grooved shell with a sharp outer lip. It is sculptured with thin spiral ribs, which are brown to dark purple in color, set on a white background. It has a calcareous operculum attached to the foot. Such a structure seals off the retracted animal in the shell, thereby protecting it against predation. Length of the shell is 1.2–3.4 cm.

Biology

Food and feeding: It is herbivorous, feeding on algae, especially that associated with mangrove roots and trunk bases.

Other uses (if any): This species can be used as bioindicator of heavy metals in its environment, because it accumulates cadmium, nickel, and lead in its shell and copper, zinc, and iron in its soft tissues.

Nerita chamaeleon (Linnaeus, 1758)

Common name(s): Chameleon nerite snail.

Global distribution: Western Pacific.

Habitat: Intertidal rocky shores, mud flats, and sandy substrates; breakwaters and seawalls.

Description: Shell of this species is sturdy and rounded. Its spire sticks out a little. There are thick, spiraling ribs, which are smooth and are regularly spaced. Unlike Chameleon, this snail can't change the color of its shell. But different individual snails on the shore may have different shell colors and patterns. Flat underside is white. This side may have ridges and a few small rounded bumps. Small ridged teeth (2–4) are seen on the straight edge at the shell opening. Larger shells are often with at least one large rounded tooth on one side of the shell opening. Operculum is thick and is evenly covered in tiny bumps. Body is pale with fine black bands on the foot and tentacles are long, thin, and black. Maximum length of the shell is 3.5 cm.

Biology

Locomotion/behavior: It is often seen in large groups and is more active at night or cool days at low tide.

Reproduction: Members of this family are mostly gonochoric and broadcast spawners. Embryos develop into planktonic trochophore larvae and later into juvenile veligers before becoming fully grown adults.

Other uses (if any): It is collected for food by coastal dwellers and its shell is used in the shell trade.

Nerita polita (Linnaeus, 1758)

Common name(s): Polished nerite snail.

Global distribution: Indo-Pacific.

Habitat: Wet fine sandy shore; littoral fringe of rocky shores; depth range 0–2 m.

Description: Shell of this species is very thick and spherical. Shell has a depressed spire. Its width is greater than height. Exterior shell is glossy; usually smooth apart from fine axial growth lines, but occasionally with fine spiral ribbing. Flat underside is white, smooth, and glossy, sometimes slightly wrinkled. Tiny notched "teeth" are seen on the straight edge at the shell opening, the uppermost one is squarish. Operculum is thick; and greenish gray smooth except for a border of fine bars at the edge.

Body is pale with fine black bands. Shell size is up to 3 cm high and 4 cm wide.

Biology

Locomotion/behavior: It is mostly active during nocturnal low tide and digs into the sand during high tide and diurnal low tide. Further, these snails inhabiting different shores are behaviorally adapted to the actual morphology of the coast and its exposure to wave action.

Food and feeding: It feeds on microalgae on exposed rock surfaces around midtide level, but burrows a few centimeters into the sand when not feeding.

Reproduction: Members of this family are mostly gonochoric and broadcast spawners. Life cycle: Embryos develop into planktonic trocophore larvae and later into juvenile veligers before becoming fully grown adults.

Fisheries/aquaculture: Commercial fishery exists for this species.

Other uses (if any): In some places, it is prized as food and for its attractive shell.

Nerita insculpta (Récluz, 1841)

Common name(s): Sculptured nerite.

Global distribution: Western Pacific: Taiwan and Indonesia.

Habitat: Rocky reef and mangrove areas in intertidal zone; depth range 200–230 m.

Description: Shell of this species is small and stout. Spire is low and sculpture consists of 10 broadly rounded spiral ribs separated by narrow interspaces and crossed by fine growth lines. Outer lip is crenulated by spiral ribs. Inner lip and columellar deck are smooth. Traces of dark spots are seen along the spiral ribs. Surface and inner surface of shell is brown. Shell height is 1–1.5 cm.

Biology

Reproduction: Members of the family are mostly gonochoric and broadcast spawners. Embryos develop into planktonic trocophore larvae and later into juvenile veligers before becoming fully grown adults.

Patellidae (limpets): Shell of the species of this family is conical. Its sculpture is essentially radial. Interior shell is with a horseshoe-shaped muscle scar. There is no operculum. True gills are replaced by a fringe of respiratory tentacles in these limpets.

Cymbula safiana (Lamarck, 1819) (=*Patella nigra*)

Common name(s): Saffian limpet.

Global distribution: Northeast Atlantic and Mediterranean Sea.

Habitat: Middle-rocky coastline.

Description: This species is with a conical conch shell which is not twisted on itself with striations resembling a limpet. A hole is seen in the vertex of the shell where it excretes. Shell size varies between 2.5 and 12 cm.

Biology

Food and feeding: It feeds on microalgae.

Reproduction: Members of the family are mostly gonochoric and broad-cast spawners. Embryos develop into planktonic trocophore larvae and later into juvenile veligers before becoming fully grown adults.

Other uses (if any): It is a food item in Northern Morocco, Senegal, and Málaga.

Patella caerulea (Linnaeus, 1758)

Common name(s): Mediterranean limpet or rayed Mediterranean limpet.

Global distribution: Mediterranean Sea; Atlantic Ocean: off the Canary Islands, Madeira and the Azores.

Habitat: Natural and artificial habitats.

Description: This species has a thin shell which is depressed, spreading, usually more or less distinctly six or seven angled. Riblets rather fine and notably unequal.

Biology

Reproduction: Members of the family are mostly gonochoric and broadcast spawners. Embryos develop into planktonic trocophore larvae and later into juvenile veligers before becoming fully grown adults.

Other uses (if any): This is an important food item in Mediterranean and Adritic Sea. These limpets are important structuring agents of intertidal assemblages, controlling the distribution of algae, bulldozing small sessile animals, or consuming sessile and mobile prey.

Patella candei (d'Orbigny, 1840)

Common name(s): No common name.

Global distribution: Atlantic Ocean: off the Canary Islands and Madeira.

Habitat: Middle and upper littoral; remaining fixed to the rocky substratum during the day.

Description: Shell of this species is conical orbicular with adult specimens varying in size, usually in the range of 2–8 cm in diameter. Shells are generally compressed and rounded, with a subcentral apex tending forward. Ribs are poorly marked, with no vertical striations, or with them only slightly pronounced. Shell has the external surface with gray to brown color with greenish tones and dark radial bands. Inner surface is brownish, with an iridescent bluish hue or a yellowish (or brownish) edge and a whitish central area.

Biology

Locomotion/behavior: Feeding at night by scraping the areas with algae, returning to the place before dawn.

Food and feeding: It is a herbivore feeding on algae.

Reproduction: *Patella candei gomesii* is comprising two ecomorphs, namely, "smooth limpet," which is inhabiting the eulittoral zone, and the "fly limpet," which is mainly seen higher in the shore, on the splash zone of exposed areas. The annual reproductive cycles of these ecomorphs were found to be similar. Both sexes showed synchronous patterns and matured most of the year. In April, significant increases were observed in the relative volume occupied by previtellogenic and vitellogenic cells in females, and by spermatogonia, spermatids, and spermatocytes in males. The maximum mature oocytes and spermatozoa were observed in July. Based on these observations, it is concluded that the breeding season of this species is during the whole year peaking in the summer when reproductive condition is highest and the main spawning would occur (João et al., 2005).

Fisheries/aquaculture: Intense capture of this species for human consumption in several regions, including in the Azores has been reported.

Other uses (if any): It forms an important food source in Azores, Madeira, and Canaries.

Patella ferruginea (Gmelin, 1791)

Common name(s): Ribbed Mediterranean limpet or ferreous limpet.

Global distribution: Mediterranean Sea: Spain; Northern coast of Africa between the Straits of Gibraltar (Ceuta) and Cape Bon and the island of Zembra (Tunisia).

Habitat: Upper mediolittoral level, where algae cover is very light and an imperceptible microbial biofilm predominates; areas exposed to the waves; artificial harbor stones.

Description: This species has a large and sturdy shell with pronounced broad ribs (between 30 and 50), making the edge very sinuous. Ribs are nodular and somewhat irregular. Shell is often found eroded and colonized by epibiont organisms such as barnacles and algae. External color of the shell is rusty to cream colored, and marble white on the inside, with a darker central zone (muscle scar) and dark brown inner edge. Foot is yellowish orange at the base and dark gray on the sides. Cephalic region is also dark in color, with the blackish tentacles standing out. *Patella ferruginea* is considered a long-lived and slow growing species. It does not reach sexual maturity until the end of the second year of life and there is good evidence that it can live for more than 10 years. Some authors have suggested that it can even reach 35 years. Maximum size is 6 cm.

Biology

Locomotion/behavior: Adults are sedentary and move only to feed, traveling short distances, preferably at high tide, and at night. They exhibit homing behavior as they finish feeding they return to the same resting place (i.e., they exhibit homing behavior).

Food and feeding: It feeds on diatoms, cyanobacteria, and other algae.

Reproduction: A marked gender segregation by size has been observed in this species. That is, all are males in specimens in the range of 2.5–4 cm, but at larger sizes the proportion of females gradually increases. This species is also said to be protrandrous hermaphrodite, and these individuals can change sex in both directions, and not just from male to female. The annual reproductive cycle is concentrated during August to November, and they are totally sexually inactive during the rest of the year. Spawning takes place in late November, which is timed to coincide with the maturity of the animals. After a short swimming larval stage, the juveniles settle in the same habitat as the adults.

Predators: The robustness of its shell and the adhesive force of its foot protect these molluscs from many predators. Its main predators are *Eriphia* *verrucosa* and *Pachygrapsus marmoratus* crabs and the gastropod *Stramonita haemastoma.*

Conservation status: It is the most endangered marine species on the list of the European Council Directive 92/43/EEC on the Conservation of Natural Habitat of Wild Fauna and Flora (1992). Human pressure is probably the main contributing factor to the currently endangered status of this species. Further, it is possibly the most threatened species in the Mediterranean. Its Mediterranean range has progressively contracted to a few restricted areas, and the species is now threatened with extinction. Programs relating to environmental education to avoid its collection for fishing, food, or for fun as decorative objects need to be conducted, and further research on the reproductive biology of this species is also to be carried out to assess the future programs of its conservation (Guerra-García et al., 2004).

Other uses (if any): It is an important food source in Northern Morocco.

Patella rustica (Linnaeus, 1758) (=*Patella lusitanica*)

Common name(s): Rustic limpet or Lusitanian limpet.

Global distribution: Northeast Atlantic and Mediterranean Sea: Spain.

Habitat: Supralitoral, especially in areas hit, splash and good lighting.

Description: Shell of this species is thick, oval, and high conical with acute apex and numerous radial ribs and concentric growth bands. Its exterior color is grayish with black dotted on the ribs. Interior side of the shell is with dark radios that travel a more or less long stretch. Maximum diameter of the shell is 4–5 cm and height is 1.5 cm. Soft body of the animal has a head with two sensory tentacles, one sucker-shaped foot and a visceral mass constricted in the shell. Edge of the mantle, which is made up of the gills of the pallae, is embedded in a continuous groove of the inner shell.

Biology

Locomotion/behavior: The limpets exposed to the air have a shell which is larger and sharper than those that live a little lower along the coast.

Food and feeding: It is a herbivorous species feeding on tiny, shallow algae and cyanobacteria that cover the rock. It practices the "return to shelter," that is, it leaves its cubicle to scrape small plants, particularly when the sea is agitated. It leaves only 10–30 cm and returns to the same location. It is reported to make two trips per day if water movements are favorable.

Reproduction: It is hermaphrodite protandre, that is, it is male at first and then it is female. Male limpets attain sexual maturity at 2 years old, and it becomes female when it is about 4 years old. Fertilization is external. Breeding takes place in the fall and the sea scatters the eggs. Each fertilized egg gives a ciliate larva, called a trochophore, which is planktonic for a few days and then settles, often in a puddle. By becoming adult, it moves on an exposed rock. Prusina et al. (2014) in their studies reported that the males and females differed in size distribution, with females becoming more prevalent from about 2.8 cm onwards. The shell length at which 50% of males were sexually mature was 1.3 cm. According to them, it has only one reproductive cycle per year with a spawning peak between November and December for both sexes, and gonad redevelopment from January. Prusina et al. (2015) in their studies reported that 90.8% of collected shells were less than 4 years of age and only two shells (1.6%) were older than 6 years. They also reported that the shells of this species were degraded to different degrees by microbial bioerosion. The damage was most often restricted to the oldest parts of the shell, that is, apex of the shell. The microboring organisms identified were pseudofilamentous and filamentous cyanobacteria *Hormathonema paulocellulare*, *Hyella caespitosa*, *Mastigocoleus testarum*, and *Leptolyngbya* sp. The overall intensity of infestation was generally low, but it increased in severity with shell age and length.

Associated species: Different species of algae and invertebrates may be fixed on the shell of the limpets. The species of the biocenosis of the upper intertidal rock include the blue periwinkle (*Melarhaphe neritoides*), starry barnacle (*Chthamalus stellatus*), the monodonte (*Osilinus turbinatus*) and algae (*Porphyra leucosticta, Rissoella verruculosa, Bangia atropurpurea,* and *Lithophyllum papillosum*).

Fisheries/aquaculture:

Other uses (if any): It is an important food source in Portugal-Mediterranean Mediterráneo; Andalucía (Málaga, Almería); and Galicia.

Patella ulyssiponensis (Gmelin, 1791)

Common name(s): Rough limpet or China limpet.

Global distribution: From the Mediterranean to the British Isles.

Habitat: Rocky shores; prefer gullies and rock pools; most common on exposed shores avoiding extreme shelter and low salinities.

Description: Shell of this species is conical. Its apex is noticeably anterior to midline. Radiating ridges are finer than in *Patella vulgata* and are in a characteristic pattern of alternating single and triple ridges, particularly at somewhat straightened posterior margin. Animal has cream-orange foot with white pallial tentacles which are arranged in two series of different sizes. Shell is porcellaneous interior and is bluish-white with yellow-orange color centrally. Maximum size of the shell is 5 × 4 × 2 cm.

Other uses (if any): In Madeira Island in Portugal, this species is eaten after cooking in a pan with garlic and lemon juice. It is also an important food item in Britain and Spain.

Patella vulgata (Linnaeus, 1758)

Common name(s): Common limpet or common European limpet.

Global distribution: Northeast Atlantic Ocean, from Nordland, Norway to the Mediterranean.

Habitat: Exposed, rocky locations from the high tidal zone and shallow subtidal zone.

Description: Shell of this species is conical with irregular radiating costae. Apex is central or slightly anterior. Apical angle is smaller (70–80°) in high shore specimens and is larger (105–115°) in low shore and young animals. Its pallial tentacles that encircle the edge of the shell are transparent. Color of the shell exterior is gray, with a tint of yellow. Inside is greenish gray. Maximum shell size is 7 × 5 × 3 cm.

Biology

Locomotion/behavior: Common limpets exhibit strong homing behavior, returning to home scars after foraging trips. Foraging behavior in this species is based on microhabitat. That is, individuals forage diurnally when submerged on horizontal rock, while individuals on vertical rock forage nocturnally when emerged.

Food and feeding: It grazes on diatoms, blue-green algae, spore lings of macroalgae, and detritus.

Other uses (if any): It is an important food item in Belgium, France, British Isles, and Norway to Spain.

Trochidae (trochids): The trochids are called as top shells as their shells which are conical to globose (with a flattened base) largely resemble toy tops. They have a thin, corneous, and round operculum. Most species feed on algae and detritus, but some smaller species may filter plankton from the water column as their food. The bigger species are often collected for food. As their shells are iridescent when polished, they are collected for making decorative items or cut into buttons.

Monodonta edulis (Lowe, 1842) (=*Phorcus sauciatus*)

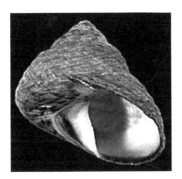 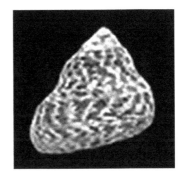

Common name(s): Lined top shell.

Global distribution: European waters.

Habitat: Reef shore of the intertidal zone.

Description: Shell of this species is solid, opaque, and broad conical with five or more teleoconch whorls. First teleoconch whorl is usually eroded, the second with 6–7 spiral cords; the third with 7–8 spiral cords; and the following whorls are with less pronounced spiral cords, separated by a distinct suture. There are numerous fine growth lines which are discernible from the third teleoconch whorl onward. Base of the shell is wide, convex, rounded, and depressed at the umbilical area, with 10–13 flat concentric ridges. Umbilicus closed even in juveniles and is covered with a white thickened callus. Aperture is rounded in profile; inner, basal, and outer lip are sharp; outer lip is with a black spotted, green border; and columella is white, with a single, wide, rounded basal columella tooth. Interior of the aperture is nacreous. Color of the shell is green-grayish or light-brown, with dark-brown or reddish elongated bands following the spiral cords. Color of base is similar to that of the whorls, with some of the bands projecting toward it. Operculum is circular, multispiral, transparent, and orange colored. Size of the shell varies between 1.1 and 2.7 cm.

Other uses (if any): It is an important food source in Canaries, Madeira and North Africa.

Monodonta labio (Linnaeus, 1758)

Common name(s): Toothed top shell or lipped periwinkle.

Global distribution: Central and East Indian Ocean, Indo-China, Indo-Malaysian Oceania, the Philippines; the Persian Gulf, West Indian Ocean to Micronesia, Western Pacific, Micronesia, and Australia.

Habitat: On or under rocks and coral in the lower intertidal zone; often seen in groups on boulders, stones and seawalls; tolerates a wide variety of substrata; depth range 0–4 m.

Description: Pattern of the shell is largely resembling a stone pavement. Shell of this species is heavy and coarse with rough, grained surface with moderate sutures between rounded whorls. Body whorl is swollen and the penultimate whorl is somewhat less. Aperture has a nacreous interior. There is a single large tooth-shaped structure at the shell opening which is white and smooth. Operculum which is made of a horn-like material is thin and yellow with concentric rings. The flexible operculum allows the animal to withdraw deep into the coils of the shell, thereby protecting the animal from prying claws of hungry crabs. Body is pale and edge of the mantle is fringed with long tentacles. There is a large foot which is pale on the underside and is mottled greenish on the upper side. A pair of long tentacles is seen at the head. Color of the shell varies from a dark reddish brown to pale brown, with spiraled dashes of cream or pink. Length of the shell size varies between 1.5 and 4.5 cm.

Biology

Locomotion/behavior: It is more active at night.

Food and feeding: This species is a herbivorous snail grazing on microalgae.

Reproduction: Members of the family are mostly gonochoric and broadcast spawners. Embryos develop into planktonic trocophore larvae and later into juvenile veligers before becoming fully grown adults.

Fisheries/aquaculture: Commercial fishery exists for this species.

Other uses (if any): It is locally collected by coastal people for food. Shell is used in shellcraft.

Phorcus turbinatus (Born, 1778) (=*Monodonta turbinata*)

 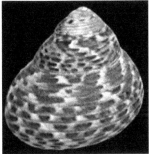

Common name(s): Turbinate monodont.

Global distribution: Endemic to the Mediterranean.

Habitat: In coastal areas, on rocks, in spaces beaten by the tide, along with other molluscs and seaweed.

Description: Shell of this species is very solid, thick, and imperforate with a conical shape. Its conoid spire is more or less elevated. Apex is eroded. There are 6 whorls which are slightly convex, with impressed spiral lines between the series of blotches, the last generally descending anteriorly. Base of the shell is eroded in front of the aperture. Aperture is very oblique. Outer lip is thick and smooth. It is pearly and iridescent within. Columella is flattened on the face and pearly. Exterior shell is whitish, tinged with gray, yellowish, or greenish and is tessellated with numerous spiral series of reddish, purple, or chocolate subquadrangular blotches. Size of the shell varies between 1.5 and 4.3 cm.

Biology

Locomotion/behavior: Often it remains out of the water near the shore, by keeping some water in the shell to avoid dehydration.

Food and feeding: It feeds on algae.

Reproduction: Their eggs and larvae are part of the plankton for a while. After some time, they go to the bottom and lead to adult.

Symbiosis or association: Their empty shells are used by hermit crabs, such as sand and rock witches.

Bile pigment in the shell: A blue-green pigment has been isolated from shells of this species. The pigment behaved as a highly polar compound and gave the Gmelin reaction. The zinc complex of the pigment exhibited yellow-green fluorescence in U.V. light. It is suggested that this pigment is a biladiene probably related to the bilatriene present in the foot of this species (Bannister et al., 1968).

Other uses (if any): It is an important food item in Portugal and Mediterranean and is eaten raw in some places.

Phorcus articulatus (Lamarck, 1822) (=*Monodonta articulata*)

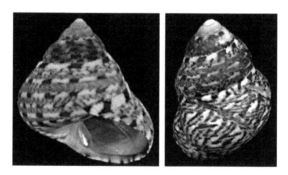

Common name(s): Articulate monodont.

Global distribution: Mediterranean Sea.

Habitat: Infralittoral rocky shore.

Description: Shell of this species is imperforate in adult, and perforate when immature. It is also heavy and thick with an elongate-conical shape. Its exterior color is greenish or whitish, spirally traversed by bands composed of alternating white and black, purplish or red squarish spots. Intervals between the bands are longitudinally closely lineolate with blackish coloration. Spire is elevated. Shell has about 6 whorls. Upper ones are slightly convex and last one is constricted and concave below the suture becoming convex. Spiral has grooves or lines but become obsolete in adults. Aperture is smaller and less oblique. Height of the shell varies between 1.5 and 2.8 cm and its diameter between 2.1 and 2.4 cm.

Biology

Reproduction: Members of the family are mostly gonochoric and broadcast spawners. Embryos develop into planktonic trocophore larvae and later into juvenile veligers before becoming fully grown adults.

Other uses (if any): It is used as food in Mediterranean.

Phorcus lineatus (da Costa, 1778) (*=Osilinus lineatus; Monodonta lineata*)

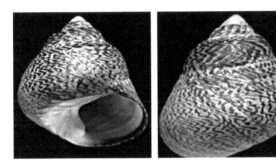

Common name(s): Toothed top shell, thick top shell or lined top shell.

Global distribution: Northeast Atlantic, West Channel.

Habitat: Pools and crevices of the upper mediolittoral zone; under the rocks; on rocky or boulder shores, avoiding excessive exposure and deposits of shingle or sand.

Description: Shell of this species is solid and conical, with five to six tumid, rapidly expanding whorls. Columella is thick, with prominent tooth near base. Inner lip is reflected over umbilicus which remains as slight depression or chink in adults shells. Exterior shell is cream or horn colored and is heavily overlain with reddish-brown, purple, or green zigzag streaks. Base of shell is less heavily pigmented. Iridescent nacre is seen on inner surface. Cephalic tentacles are with lobed sides. Foot is with finely papillate margin and tuberculate sides. There are three pairs of epipodial tentacles, each with two sensory papillae at base. Operculum is flat and polygyrous (around 15 turns). This species has lost its gill, which was replaced by a kind of highly vascularized organ used to capture oxygen. This organ allows it to spend long periods dry. Lifespan of this species is 10 years. Maximum size is 3.0 × 2.5 cm.

Biology

Locomotion/behavior: This species is rather nocturnal. During the day, it takes refuge in the depths of the crevasses.

Food and feeding: It is herbivorous grazing on the microscopic algae and the algae debris on the rocks.

Reproduction: The monodonts are hermaphrodite. Sexual maturity is reached at 2 years. There is a clear and strong period of growth before the breeding season which takes place in late spring (May) and until full summer. Gametes are broadcast in water during rallies that take place in late spring. Fertilization is external. From eggs larvae appear and they lead a planktonic life for a few days before landing on the bottom and then the young gastropods migrate up the foreshore. When the larvae attach (a few weeks after the eggs hatch), they measure 1 to 3 mm. They measure 6–9 mm after 1 year, 8–13 mm at the end of 2 years, 13–15 mm at the end of 3 years, and about 24 mm at 10 years.

Association: This species is often seen with common periwinkle with whom it shares the bottom crevices and sometimes with *Littorina saxatilis*. It can also be seen with *Gibbula umbilicalis* in pools of mediolittoral. Empty shells *Phorcus lineatus* are often occupied by the hermit crab, *Clibanarius erythropus*. They may also be colonized by barnacles.

Other uses (if any): It is an important food item from England to Portugal.

Umbonium vestiarium (Linnaeus, 1758)

Common name(s): Button tops.

Global distribution: Indo-West Pacific: East Africa to Indonesia; Philippines and Queensland.

Habitat: Shallow subtidal water; low tide mark; sand bars or sandy shores; fine sandy bottoms; brackish; depth range 0–5 m.

Description: Shell of this species is small, thin, circular, and glossy and is much wider than long. Spire is low, with faintly convex, somewhat embracing whorls and shallowly incised suture. Periphery of body whorl is regularly rounded. Base of shell is flattened, with a very large, smooth callus plug filling completely the umbilicus. Entire surface of shell is smooth and polished, devoid of concentric grooves on the spire whorls. Operculum is thin and is made of a horn-like material with yellow concentric rings. The flexible operculum allows the animal to withdraw deep into the coils of the shell. Outer lip of the aperture sharp, smooth inside. Columella is smooth and is strongly curved anteriorly. Color: outside of shell is extremely variable in pattern and coloration, in shades of gray, brown, olive green, pink, red, yellow, or even white. Umbilical callus usually with a different color. Body is pale speckled. Edge of the mantle is fringed with long tentacles. Foot is long and leaf like. Tiny eyes are seen on long stalks. Long tentacles are finely banded, with two tubular siphons, one with fringes. Maximum shell width is 1.5 cm and length is 2 cm.

Biology

Locomotion/behavior: The long mobile foot of this species is used to burrow rapidly into wet loose fine sand. To escape predators, button snails make a short, spiraling leap then quickly bury themselves into the sand again. When disturbed, submerged button snails may pop up and float on the water surface.

Food and feeding: More like bivalves, button snails filter feed for detritus and plankton. It has an inhalant siphon fringed with short tentacles which is used to suck in water, and an exhalant siphon which expels the water. When there is no enough food in the water, it may use its right tentacle and long foot to gather edible bits on the sand surface.

Reproduction: Members of this family are mostly gonochoric and broadcast spawners. Life cycle: Embryos develop into planktonic trochophore larvae and later into juvenile veligers before becoming fully grown adults.

Association: Empty buttons shells are favorite homes of tiny hermit crabs.

Conservation status: Button snails are now listed as "Vulnerable" on the Red List of threatened animals of Singapore.

Other uses (if any): This species is commonly used for food in the Philippines. In local markets, vendors traditionally give the buyer aromatic thorns (*Acacia pennata*) to pry the meat out of the small shell. Shells of this species are sold as cheap curios. They are also used in the shellcraft industry, to make dolls and decorative items.

3.2   ORDER MESOGASTROPODA

Calyptraeidae (slipper shells and cups and saucers): The calyptraeids, or slipper snails, resemble limpets with their flattened or conical shells. The "slipper" in the name "slipper limpet" is based on the appearance of the inside of the shell, which with its half-shelf resembles a western bedroom slipper. They are hermaphrodites, starting out as males but finally become females later. They are mostly filter feeders, and living with the hermit crabs.

Crepidula fornicata (Linnaeus, 1758)

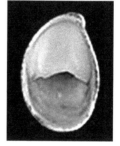

Common name(s): Common slipper shell, boat shell, quarterdeck shell, fornicating slipper snail or Atlantic slipper limpet.

Global distribution: Native to the western Atlantic Ocean (Eastern coast of North America)

Habitat: Intertidal, infralittoral, and circalittoral and in estuaries; sometimes living stacked on top of one another, on rocks, on horseshoe crabs, shells, and on dock pilings; depth range 0–70 m.

Description: Shell of this species is oval, up to 5 cm in length, with a much reduced spire. Its large aperture has a shelf, or septum, extending half its length. Shell is smooth with irregular growth lines and white, cream, yellow, or pinkish in color with streaks or blotches of red or brown. These slipper limpets are commonly seen in curved chains of up to 12 individuals. Large shells are found at the bottom of the chain, with the shells becoming progressively smaller toward the top. Although they have a foot for locomotion, by the time they reach maturity, they attach themselves to hard substrate and remain stationary. Size of the shell is 2–5 cm.

Biology

Locomotion/behavior: Groups of individuals are often seen heaped up and fastened together.

Food and feeding: It feeds on plankton and minute detrital food items through either suspension or deposit feeding.

Reproduction: *Crepidula fornicata* is a protandrous hermaphrodite (animals start their lives as males and then subsequently may change sex and develop into females). Although breeding may occur between February and October, peak activity occurs in May and June. Most females spawn twice in a year. Females can lay around 11,000 eggs at a time in 50 egg capsules. Incubation of the eggs takes 2–4 weeks followed by a planktotrophic larval phase lasting 4–5 weeks. Due to the length of its planktonic phase, its dispersal rate is high. The spat settle in isolation or on top of an established chain of *C. fornicata*. Males reach sexual maturity 2 months after settlement. If a male develops directly into a female, sexual maturity may be reached in 10 months.

Other uses (if any): Slipper limpets are a versatile food. They have the flavor to serve alone as a main course, an appetizer or be incorporated into many different dishes. Before, during, and after cooking slipper limpets produce a large amount of liquid which may be used as a clam juice substitute. Therefore, this shellfish and its recipes could become commercially important in the years to come.

Crepidula onyx (Sowerby, 1824)

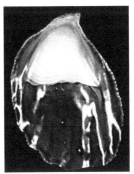

Common name(s): Onyx slipper snail.

Global distribution: Introduced from America; from Essex along south coast to Bristol Channel, Belfast Lough, Co. Kerry, Helgoland and Southwest Netherlands.

Habitat: On shells and stones on soft substrata; often cast ashore by storms.

Description: Shell of this species is depressed, elongate, and cap shaped. Spire is set posteriorly and to right. Shell is smooth, with irregular growth lines. Aperture is elongate-oval, often slightly concave on spire side. A large shelf-like internal partition is seen posteriorly. In the body, the mantle cavity is much deeper and the gill is longer. Penis is not bilobed. Color of body is yellowish, with dark pigment on the snout, tentacles, mantle edge, and penis. Coloration of the shell is cream or pinkish, with streaks and blotches of reddish brown. Shell size is up to 2.5 × 5 cm.

Biology

Locomotion/behavior: The animals form curled stacks of up to 12 individuals. Large shells are at the bottom, becoming progressively smaller toward the apex of the chain.

Crepidula plana (Say, 1822)

Common name(s): Eastern white slipper snail.

Global distribution: Native to the east coast of North America; ranges from Canada to North Carolina.

Habitat: Epibenthic habitats in oyster reefs; intertidal to subtidal with low wave exposure, and slow flowing water conditions; found inside empty shells of whelks and moon snails; also inside large gastropod shells that are inhabited by hermit crabs; depth range 0–110 m.

Description: Shell of this species is flat and white. Its internal septum is flat, and there are no muscle scars inside the shell. Body is also white. Length of the shell is from 1.5 to 4.3 cm.

Biology

Ecological adaptations: It is a eurythermic species. It tolerates a temperature range between 5°C and 35°C. It is oligohaline to euryhaline. This species tolerates salinities between 15 and 38 ppt. It prefers well-oxygenated conditions.

Food and feeding: It is a suspension feeder.

Reproduction: Males are smaller than females. Longevity for male is 6 months and that of female is 2 years. However, the size and sex change of this species often depends on the environmental conditions. It is a protandric hermaphrodite which is reproductively active year round. It has as many as 6 reproductive periods per year, producing from 100 to 200 young. Reproductive seasonality peaks during the summer for this species. Eggs in Crepidula cf. plana are brooded. Typical larval development time is 21 days, but can range from 14 to 28 days.

Fisheries/aquaculture: It is a potential species for mariculture.

Crucibulum scutellatum (Wood, 1828)

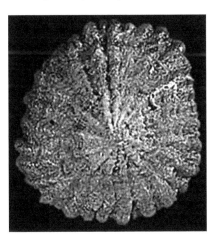

Common name(s): Cup-and-saucer snail.

Global distribution: Panama, Cedros Island, Baja California.

Habitat: On stones or other shells on interidal mud flats to offshore depths of 27 m.

Description: Shell of this species has coarse, scaly, radial ribs which may be latticed by concentric sculpture. Internal cap is attached along one side as well as at apex. Coloration of the shell is brownish. Size of the shell ranges from 1.5 to 6.5 cm.

Crucibulum umbrella (Deshayes, 1830)

Common name(s): Cup-and-saucer snail.

Global distribution: Gulf of California to Panama.

Habitat: Intertidal.

Description: This species largely resembles *Crucibulum scutellatum*. It however differs from the latter in the internal cup which is attached only at the apex and stands free in the middle of the shell. Common size: diameter 6 cm; height 2 cm.

Cerithiidae (ceriths): Shell of tis family members is sharply conical, with a high, many-whorled spire and rather small aperture. Sculpture is variable. Aperture is with a siphonal canal. Outer lip is somewhat expanded. Operculum is ovate and corneous, with a few spiral coils. These species are often gregarious, feeding on algae and detritus.

Cerithium vulgatum (Bruguière, 1792)

Common name(s): Common cerithe.

Global distribution: Mediterranean and Atlantic.

Habitat: On stones on sandy or muddy bottoms.

Description: This species has an elongated, pointed and hard shell, with numerous turns and sharp protrusions sculptures. Exterior shell has grayish color. Often their empty shells are occupied by small hermit crabs. Maximum size of the shell is up to 8 cm in length.

Biology

Food and feeding: It is a surface deposit-feeder.

Reproduction: Members of the family are mostly gonochoric and broadcast spawners. Embryos develop into planktonic trocophore larvae and later into juvenile veligers before becoming fully grown adults.

Other uses (if any): This species has also been reported to play an important role in metal cycling in shallow marine ecosystems. It detoxifies metals by incorporating them into insoluble granules formed in the digestive gland (Nicolaidou and Nott, 1999).

Cypraeidae (cowries): The cypraeids, commonly known as cowries, have smooth and glossy egg-shaped shells. Shells of some species may have interesting patterns, while others may be very plain. Spire of shell is concealed under body whorl. Surface of the shell is highly polished and smooth. Aperture is long and narrow, and channeled at both ends. Both lips are with teeth. There is no operculum. Due to their smooth and pretty shells, many cowries are collected for making decorative items.

Cypraea tigris (Cate, 1961)

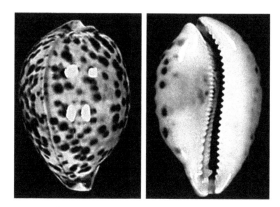

Common name(s): Tiger cowrie.

Global distribution: Indo-Pacific region: from Africa to Micronesia and Polynesia; Coral Sea and Philippines; Australia.

Habitat: Associated with live coral colonies, such as Acropora; on the reefs or the sandy sea bottom; depth range 10–40 m.

Description: Shell of this species is large, heavy, glossy, roughly egg-shaped, and dextral. A blurred red line may be present along the length of the shell at the midline on the dorsal surface. Lower margins are rounded. Shell opening is lined with tooth-like serrations. Upper or dorsal side is white, pale bluish-white, or buff, densely covered with dark brown or blackish barely circular spots. Ventral side is white or whitish. Exterior surface of the mantle has numerous pin-like projections that are white tipped. Maximum length of the shell is 15 cm.

Biology

Food and feeding: It is carnivorous eating coral and various invertebrates. Juveniles, however, eat algae.

Other uses (if any): It is an important food item in Hawaii. The shells of this species are very popular among shell collectors, and are used as a decorative item. The shell of this species is also believed to help to facilitate childbirth and some women in Japan hold a shell of this species during childbirth.

Luria lurida (Linnaeus, 1758)

Common name(s): Brown cowry, cowrie porcelain.

Global distribution: Mediterranean, Atlantic, Channel and North Sea.

Habitat: Small semidark caves, crevices or under rocks up to about 60 m. of depth; rocky seabed and corals; among the rhizomes of Posidonia.

Description: Shell of this species is elongated (cylindrical to ovoid), smooth and shiny. It has a spiral winding about an axis, but which remains hidden by the last turn Dorsal side is pale brown or reddish, with three transversal darker bands alternating with narrower clearer bands. At the extremities of the shell, there are two separate dark brown spots. Aperture is wide, with several teeth. Mantle is dark brown and may cover the entire shell. Male is usually smaller and elongated than the female. Shell size ranges from 1.4 to 6.6 cm.

Biology

Locomotion/behavior: As they are afraid of the light, they hide themselves in their habitats during the day.

Food and feeding: It is a carnivorous and predatory species feeding during the night mainly on sponges of *Verongia aerophoba*, *Aiplysina aerophoba*, *Chondrilla nucula*, and *Tethya auranti*.

Reproduction: This species is with separate sexes and internal fertilization. After mating, each fertilized female may lay up to 1000 protective capsules which may contain up to 600 eggs. Eggs are laid on hard substrates. Each light yellow ootheca will darken and become brown during embryonic development. Embryos develop into planktonic trochophore larvae and later into juvenile veligers before becoming fully grown adults.

Other uses (if any): It is an important food item for the people of the Montinegro coast.

Mauritia maculifera (Schilder, 1932) (=*Cypraea. maculifera*)

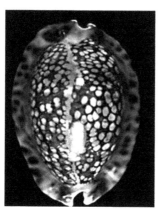 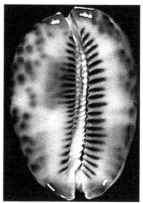

Common name(s): Blotched cowry or reticulated cowry.

Global distribution: Indian Ocean: Chagos and Seychelles; Pacific Ocean (from Southeast Asia, Philippines, Fiji, and Micronesia through western Polynesia and Hawaii).

Habitat: Shallow water, subtidal and low intertidal; under slabs and rocks, coral reefs or in gullies and holes of the algal crests; depth range 1–15 m.

Description: Shell of this species is smooth and shiny. Aperture is long and narrow, with several dark brown teeth. This species can be distinguished by a characteristic brown columellar spot. In the living cowries, mantle is transparent, with blue sensorial papillae and may cover the entire shell. Juveniles are gray with wavy brown lines. Dorsal surface of the shells is dark brown, with distinct large bluish dots. On the edges, there are large brown spots. Shell base is white or pale brown or pale pinkish. Maximum shell length is 10.0 cm.

Biology

Food and feeding: As they are afraid of the light, they start feeding at dusk mainly on sponges or coral polyps.

Reproduction: Members of the family are mostly gonochoric and broad-cast spawners. Embryos develop into planktonic trocophore larvae and later into juvenile veligers before becoming fully grown adults.

Other uses (if any): This species is an important food item in Hawaii and throughout Indo-Pacific. It is collected mainly for its shell which is used in shellcraft.

Mauritia mauritiana (Linnaeus, 1758) (=*Cypraea mauritiana*)

Common name(s): Humpback cowry, chocolate cowry, mourning cowry or Mauritius cowry.

Global distribution: Indian Ocean: along Southeast Africa; western Pacific Ocean (western and northern Australia, Malaysia, Philippines, and Hawaii).

Habitat: Low intertidal water, usually under rocks or in rocky crevices; prefer areas with strong water movement; depth range 2–50 m.

Description: Shell of this species is smooth and shiny. Dorsal surface is dark brown, with distinct large yellowish or amber dots. Edges of the dorsal side and the base are completely dark brown. Aperture is long and narrow, with several dark brown teeth and clear spacing. Mantle is completely black, without sensorial papillae. Because of this rough habitat, the foot of the animal is very strong to hold onto the rock even under rough habitat including the crashing surf. Size of the shell ranges from 4 to 13 cm.

Other uses (if any): It is an important food item in Hawaii, Japan, and Indo-Pacific. Shell of this species has ornamental values.

Monetaria annulus (Linnaeus, 1758) (=*Cypraea annulus*)

 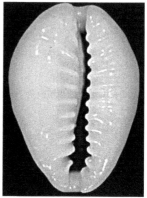

Common name(s): Ring cowrie, gold ringer, Yellow-ringed cowrie or gold-ringed cowrie.

Global distribution: Indo-Pacific.

Distribution: Throughout the Indo-West Pacific region: from East Africa to the Central-Pacific Ocean.

Habitat: Intertidal reef dweller; rock pools and shallow reef; under chunks of dead coral, rocks and in crevices; ocean side reef flat; depth range 1–8 m.

Description: Shell of this species is smoothly rounded over the dorsal surface. The ivory or yellowish colored shell with its bright orange ring on the dorsal surface gives the common name, "Yellow ringed cowrie" to this cowrie. Mantle is gray/black with a fine delicate pattern. Papillae are short and yellow/gray in color. Egg masses are yellowish-white. Mean shell length of females was significantly larger than that of males. Size of the shell varies between 1 and 5 cm.

Biology

Food and feeding: It is a primarily carnivore.

Reproduction: Sexual dimorphism has been reported in this species. Reproductive stages of this species are as follows: Stage 1 animals (initial maturity), Stage 2 animals (peak maturity), and Stage 3 animals (spawned). There are two reproductive seasons (spring and from late summer to winter) for this species. Members of the family are mostly gonochoric and broadcast spawners. Life cycle: Embryos develop into planktonic trochophore larvae and later into juvenile veligers before becoming fully grown adults.

Fisheries/aquaculture: Commercial fishery exists for this species.

Other uses (if any): Shells of this species have ornamental values. They are mainly collected for shellcraft. Shell (along with *Cypraea moneta*) was formerly used as a currency in many areas of the world.

Monetaria caputserpentis (Linnaeus, 1758)

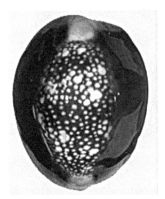 

Common name(s): Serpent's-head cowry or porcelain head of snake.

Global distribution: Indo-Pacific; Red Sea to Pacific Ocean; Hawaii to East Africa.

Habitat: Corals, rock reefs and rocky shores from the intertidal zone down to depths of 200 m; shallow subtidal zone beyond the breakers.

Description: Shell of this species is oval and slightly triangular and flattened on the edges. Exterior shell is dark brown (chocolate). On the back and in the center of the shell, there are small white dots. Two light markings characteristics lie ahead of and behind the shell. Juveniles have a light gray shell with a broad brown band in the middle. Maximum size of the shell is 4 cm.

Biology

Locomotion/behavior: This species is diurnal. It can be seen out of the water at low tide.

Food and feeding: It is a herbivore.

Reproduction: This species is of separate sexes. Fertilization is internal. Animals deposit white egg capsules on the substrate. Hybridization with *Erosaria moneta* has been reported in this species.

Other uses (if any): These shells are sold with a purple top, which is done by dipping the dorsum in acid. Shell of this species is used in popular decorations in Marshallese handicrafts.

Monetaria moneta (Linnaeus, 1758)

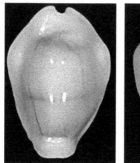 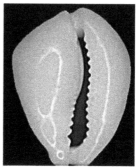

Common name(s): Money cowry.

Global distribution: Tropical Indo-Pacific.

Habitat: Intertidal rocky areas and shallow tide pools among sea weed, coral remains, and empty bivalve shells; exposed reefs at low tide.

Description: This species is called "money cowry" because the shells were historically used in many Pacific and Indian Ocean countries as a form of exchange before coinage was in usage. Shell of this species is irregular and flattened, with very calloused edges and roughly subhexagonal. Color of the shell is pale (from white to dirty beige), but the dorsum is transparent (greenish gray with yellowish margins) with sometimes darker transverse strips and a fine yellow ring. Opening of the shell is wide and white, with pronounced denticules. Some specimens of this species are found with a thin gold ring, but this is rally much less prominent than in *Monetaria annulus*, with which this species may share the same habitat. Mantle of the live animal is mottled with black and dirty white (zebra striped). Shell is small measuring up to 4.5 cm.

Biology

Food and feeding: It is a herbivore feeding on algae and marine vegetation growing on loose rocks and pieces of dead coral.

Other uses (if any): The shells of this species were used as currency in many countries until the 1800. Like the other common owries, this species is often used in Marshallese handicrafts. The shell of this species is used in jewelry and in other decorative items such as baskets and wall hangings. The shell is also still used in divination rituals in some religions.

Littorinidae (periwinkles): The periwinkles are a diverse group of snails found on the upper shores, either on rocks, trees or sea weed. Shell of the members of this family is ovate-conical, without an umbilicus. Aperture is rounded, without a siphonal canal. Columella (the inner side of the aperture) is smooth. Operculum is corneous and thin with relatively few spiral coils. These animals are usually inactive and are exposed to more heat on the upper shore. Hence, they are very heat tolerant. They normally feed on algae, but the individuals inhabiting the trees usually feed on the film of algae growing on the leaves and bark. They are able to breathe air with their simple lungs, which is modified from the mantle cavity.

Austrolittorina antipodum (Philippi, 1847)

Common name(s): Blue-banded periwinkle.

Global distribution: New Zealand.

Habitat: Rocky shores at or above the high intertidal zone; harbors through to exposed coastlines where it is living in the splash zone.

Description: Shell of this species is conical with a white background and blue stripe. However, the color may be eroded off due to changes in environmental conditions. Common size: height 2 cm and width 1.2 cm.

Biology

Locomotion/behavior: Adults live higher up the shore and juveniles normally shelter inside empty barnacle shells.

Food and feeding: It grazes on thin films of seaweed and lichen.

Littoraria angulifera (Lamarck, 1822) (=*Littorina angulifera*)

Common name(s): Mangrove periwinkle.

Global distribution: Caribbean Sea and the Western Atlantic Ocean from Florida South to Brazil; eastern Atlantic between Senegal and Angola.

Habitat: On the branches and prop roots of the red mangrove (Rhizophora mangle) above the sea level.

Description: This species has a small shell of six to seven whorls with a pointed top. Shell is sturdy and is engraved with tiny helical lines with the seams. There is a middle channel at the lower side of the external lip. Its operculum is dull brown. Shell color is reddish, gray, or with dark slanted markings. Rarely, it may be orange or yellow, and the inside is white. It has fairly degenerated ctendia with a vascularized mantle epithelium which support oxygen exchange in air. Maximum shell length is 4 cm.

Biology

Physical tolerances: It favors warmer waters and air temperatures. It is most commonly found in brackish estuaries, with larger individuals in less saline waters.

Locomotion/behavior: It is solitary animal. Its daily movement is related to tides. That is, these snails gather near to the tideline at high tide. During the tide retreat snails spread. The smaller snails remain nearer to the tide-line, whereas others relocate to cool places underneath leaves or behind stems in upper heights of the tree. Although adults travel over substantial vertical distances, juveniles stay near to the tideline as only large animals make a self-lubricating seal, which assists in avoiding drying out.

Food and feeding: It is a herbivore browsing on fungi and algae growing on the mangroves.

Reproduction: *Littoraria angulifera* is ovo-viviparous. Fertilized eggs are brooded inside the periwinkle, and the planktonic veliger larvae are then released. After about 2 months, these larvae develop into pediveliger larvae which undergo metamorphosis and settle. The maximum lifespan of this species may vary with the food availability and environmental factors.

Association: These mangrove periwinkles are associated with several organisms common to mangroves and other intertidal areas.

Predators: These animals may be preyed upon by a variety of birds, fishes, large crabs, and mammals.

Other uses (if any): It is a common food item in Atlantic USA and Canada Bermuda; South Florida; and West Indies. This animal used as a zootherapeutical product for the treatment of chesty cough and shortness of breath in traditional Brazilian medicine in the Northeast of Brazil (Anon. https://sta.uwi.edu/fst/lifesciences/documents/Littoraria_angulifera.pdf).

Littorina irrorata (Say, 1822)

Common name(s): Marsh periwinkle.

Global distribution: Atlantic coast and Gulf Coast of North America, from Massachusetts to Texas.

Habitat: Salt marshes; depth range 0–22 m.

Description: Shell of this species is elongate conic in shape and is longer than wide. There are 8–10 gradually increasing flat whorls which comprise the shell. Among them, the body whorl measures about half of the total height. Aperture is oval with a sharp outer tip and regular grooves on the inside edge. Coloration of the shell is dull grayish white with tiny dashes of reddish brown on the ridges of the spiral. This species uses its gill to get oxygen from the water. Maximum shell length is about 3 cm.

Biology

Physical tolerances: The large range and distribution of this species throughout temperate to tropical latitudes suggests that the species has a wide thermal tolerance (5–45°C). The ability of these animals to attach to a surface via a mucous holdfast and withdraw completely into its shell is said to be a means of escaping extreme temperatures and desiccation. Most populations of this species have been reported to tolerate low salinity waters.

Food and feeding: This periwinkle is herbivorous, grazing on algae, detritus, fungi, and the marsh cordgrass, *Spartina alterniflora*. Its diet, such as fungi, has been reported to encourage the growth of this animal. The animal creates wounds on the grass, *Spartina alterniflora*, which are then infected by fungi, *Phaeosphaeria* and *Mycosphaerella* genera. Such fungi are the preferred diet of this species. These animals may also deposit feces on the wounds that they create, which encourage the growth of the fungi because they are rich in nitrogen and fungal hyphae. Juvenile snails raised on uninfected leaves do not grow and are more likely to die, indicating the importance of the fungi in the diet of this species.

Reproduction: Marsh periwinkles lay individual eggs into the water. These eggs hatch into free-swimming larvae, which develop into small snails in midsummer. The lifespan of this species may vary with food availability and environmental factors.

Predators: The predators of this species include the squareback marsh crab, *Armases cinereum*; Atlantic mud crab, *Panopeus herbstii*; blue crab, *Callinectes sapidus*; and the crown conch, *Melongena corona*. Additional predators include large fishes, birds, and mammals.

Associated Species: Marsh periwinkles may be associated with several organisms common to salt marshes and other intertidal areas.

Other uses (if any): It is an important food item in Atlantic USA and Canada.

Littorina littorea (Linnaeus, 1758)

Common name(s): Common periwinkle.

Global distribution: Native to the Northeastern, and introduced to the Northwestern, Atlantic Ocean.

Habitat: Rocky shores in the higher and middle intertidal zone; small tide pools; muddy habitats such as estuaries; fairly tolerant of brackish water; depth range 0–60 m.

Description: Shell of this species is broadly ovate, thick and is sharply pointed. It contains six to seven whorls with some fine threads and wrinkles. Spire is prominent and pointed. Last whorl may occupy up to 85% of shell height. Shell is smooth, especially in adult specimens. Aperture is ear shaped. Outer lip is arising tangential to last whorl. Inner lip is thick. There is no umbilicus or umbilical groove. Shell is generally black or dark gray-brown and is lighter toward apex, with heavier pigmentation between spiral ridges. Rarely red, orange, or white individuals may occur. Inside of the shell has a chocolate-brown color. Columella region is white. Cephalic tentacles are slightly flat and broad, with many transverse black stripes and, ventrally, there is a single longitudinal line. Maximum size of the shell: width 1.2 cm; length 4 cm; and height 5.2 cm.

Biology

Locomotion/behavior: When exposed to either extreme cold or heat while climbing, this periwinkle withdraws into its shell and start rolling, hoping to hit the water.

Food and feeding: It is an omnivorous species, grazing primarily an algae such as *Ulva lactuca* and *Ulva intestinalis*; *Enteromorpha* sp. and *Porphyra* sp.; and diatoms. But it will also feed on small invertebrates such as barnacle larvae.

Reproduction: It is dioecious and oviparous and is reproducing annually with internal fertilization of egg capsules which are shed directly into the sea. Females lay 10,000 to 100,000 eggs contained in a corneous capsule from which pelagic larvae appear in a development time of 4–7 weeks. These larvae then escape and eventually settle to the bottom. This species can breed year round. It reaches maturity in 18 months at 10 mm. Female specimens ripe from February until end of May, when most are spawning. Male specimens are mainly ripe from January until end of May. Settlement of the young happens during end of May to end of June. Owing to its robust nature, *Littorina littorea* can be highly variable in phenotype with several different morphs present. Its lifespan is 5–20 years.

Fisheries/aquaculture: The fishery of this species has become increasingly important as harvesting approached 1500 t in the late 1990s in Canadian waters alone. It is a considerable by-catch in the rockweed fishery. There are several potential benefits from aquaculturing this species including a more controlled environment, easier harvesting, less damages from predators, as well as saving the natural population from commercial harvesting.

Other uses (if any): The common periwinkle is sold as a delicacy at fish markets in Europe, France, Atlantic coasts of North America, Scotland, and USA. It is also considered a delicacy in African and Asian cuisines. It was introduced to Canada through ballast water or intentionally for food. Its meat is high in protein and omega-3 fatty acids and low in fat. This species has also been found highly suitable as a bio-indicator for contamination of the marine environment. This is largely due to its ability to accumulate trace elements and compounds and consequential behavioral changes.

Littoraria scabra (Linnaeus, 1758) (=*Littorina scabra*)

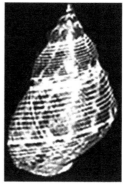

Common name(s): Mangrove periwinkle.

Global distribution: Red Sea; Indian Ocean: Aldabra, Chagos, South Africa, Kenya, Madagascar, the Mascarene Basin, Mauritius, Mozambique, the Seychelles and Tanzania; Pacific Ocean: Hawaii and New Zealand.

Habitat: Mangrove-associated species; tidal creek, mangrove, and mud flat.

Description: This species has a thin strong shell with pointed spire. It has rounded, spirally ribbed whorls and well-defined suture. Sharp angle columella is with the base of outer lip. Shell color varies with dark brown blotches and spots. Adult shell varies between 1.5 and 4.0 cm.

Biology

Food and feeding: This species is mainly a generalist herbivore, which shifts its diets according to food availability. It also has the ability to ingest and assimilate greater quantities of diverse foods such as microalgae, foliose/corticated macrophytes, filamentous algae, mangrove tissues, and zooplankton. Feeding of the animal takes place in the bottom areas of mangrove trees (roots and trunks) during low tides, while top areas (branches and leaves) provide limited food resources for these snails during high tides. However, these animals have been reported to prefer microalgae and bacteria.

Reproduction: It is an ovoviviparous species.

Littorina sitkana (Philippi, 1846)

Common name(s): Sitka periwinkle.

Global distribution: Northeast Pacific: Alaska to western USA; coasts of Siberia, Japan and the Pacific coast of North America extending from Alaska to Oregon.

Habitat: Sheltered intertidal areas; on rocks or among algae such as rock-weed; on pilings and eelgrass; in and around crevices.

Description: Shell of this species is rounded and almost globose shaped. Shell diameter is almost equal to its height. Shell has some strong spiral sculpturing in the form of continuous ridges and furrows. Spire is consisting of three whorls with a white band inside the aperture. Exterior color of the shell is highly variable: commonly very dark, almost black, or dark purple, uniform brown, red-brown, or gray. Paler gray shells with lighter bands or with yellow or orange on lighter parts are also present in this species. Shell interior is brown or orange. There is close link between the shell coloration of this periwinkles and their background substrate, that is, there is a very strong relation between the shades of the periwinkle's shell and the colors of the rock—light colored shells stayed on light shaded rocks and vice versa. Snail body is dark with black tentacles on head. The empty shells are used by small hairy hermit crabs, *Pagurus hirsutiusculus*. Maximum length of the shell is 2.5 cm.

Biology

Locomotion/behavior: Adults are much more active on foggy days. The animal crawls by lifting one half of the foot at a time. Individuals of more exposed coasts are smaller with thinner shells and a larger foot.

Food and feeding: It feeds on diatoms and fine algae. It also eats the black lichen tiny settled barnacles.

Reproduction: Adults of this species are dioecious and may spawn several times in a year. It attains sexual maturity at 5.5–7.0 mm (females) and 4.2–6.0 mm (males). It has been reported to spawn near dawn and dusk in laboratory experiments. After mating, the female lays its eggs in gelatinous masses on rocks and algae. The capsules look like tiny reddish flattened lemons. Each egg mass may contain the eggs of several females. Veliger develops shell within a week inside egg case, then hatches in about 30 days and begins to consume jelly of egg mass and diatoms which may have colonized the jelly surface. In this species, the larvae hatch as juveniles and do not have a pelagic (planktonic) stage.

Predators: The predators of this species include snails such as *Nucella lamellosa*, seastars such as *Evasterias troschelii* and *Leptasterias hexactis*, red rock crabs *Cancer productus*, nemertean worms such as *Amphiporus formidabilis*, fish such cockscomb pricklebacks (*Anoplarchus purpurescens*), pile perch (*Rhacochilus vacca*), clingfish (*Gobiesox meandricus*), and a variety of ducks and surfbirds.

Naticidae (moon shells): Moon shells are snail-like globular forms with a half-moon-shaped aperture. Some species of this family are flattened and disk-like. A thick rib-like callus obscures the umbilicus, and the aperture lip is fringed by a thin sharp edge. In life, the mantle flaps from each side cover the shell, protecting its lustrous finish. These naticids with round shell are nocturnal, and many of them are whitish in color. They hunt for small clams and snails in the sand to feed on. Moon snails lay their egg capsules in sand collars.

Euspira fusca (Blainville, 1825) (=*Lunatia fusca*)

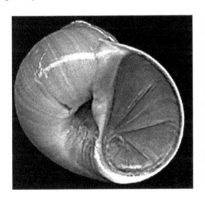 

Common name(s): Dark necklace shell.

Global distribution: Northeast Atlantic Ocean: European waters and the Mediterranean Sea.

Habitat: Sublittoral; muddy sand to a depth of 165 m.

Description: Shell of this species is very much similar to *Euspira pulchella* but with distinct, slightly channeled sutures. Outer lip arises tangential to the last whorl. Umbilicus is elongate, partially occluded by inner lip, which is narrowed by groove passing into umbilicus, but not interrupted. Body as in *E. pulchella*; anterior margin of propodium rounded. Exterior shell is in chestnut-brown color. The moon snail shells are often reused by small hermit crabs as *Pagurus cuanensis*, hairy pagurus. Maximum length is 3.6 cm and diameter 3 cm.

Biology

Food and feeding: It is a carnivore feeding on other gasteropods or bivalves by drilling a hole in their shell.

Reproduction: Members of the family are mostly gonochoric and broadcast spawners. Embryos develop into planktonic trocophore larvae and later into juvenile veligers before becoming fully grown adult.

Other uses (if any): It is an important food item in Spain and Italy.

Euspira heros (Say, 1822)

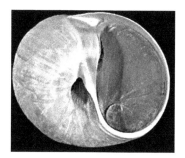 

Common name(s): Northern moonsnail.

Global distribution: Western Atlantic: North America.

Habitat: Sand substrates in bathyal, infralittoral and circalittoral parts and estuary; depth range 0–435 m.

Description: Shell of this species is large and globular. Operculum is large, ear shaped in outline, corneus, and somewhat transparent. Maximum length of the shell is 11.5 cm.

Biology

Food and feeding: It is a predatory species on clams and snails, including other moon snails. Its powerful foot helps the animal to plow under the sand in search of other food molluscs. On finding one, it "drills" a hole into the shell with its radula, releases digestive enzymes, and sucks out the somewhat predigested contents.

Reproduction: Members of the family are mostly gonochoric and broadcast spawners. Embryos develop into planktonic trocophore larvae and later into juvenile veligers before becoming fully grown adults.

Other uses (if any): It is an important food item in Gulf of St Lawrence, Eastern Canada to North Carolina, USA.

Mammilla melanostoma (Gmelin, 1791) (=*Polinices melanostoma*)

Common name(s): Blackmouth moon snail.

Global distribution: Indo-West Pacific: from East Africa (including Madagascar, the Red Sea, and the Gulf of Oman) to eastern Polynesia; north to Japan and south to southern Queensland.

Habitat: Reef quarries; salt water ponds; sandy bottoms; sublittoral, from shallow subtidal levels to a depth of about 20 m.

Description: Shell of this species is mostly white with a red-brown operculum. There is a black staining on the columella. Maximum shell length is 5.5 cm.

Other uses (if any): It is collected for food and for the shell.

Naticarius hebraeus (Martyn, 1786) (=*Naticarius cruentatus*)

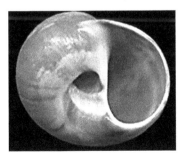

Common name(s): Hebrew moon snail.

Global distribution: Lusitanian; European waters; and Mediterranean Sea.

Habitat: Sandy, sandy-muddy, or fairly coarse detrital bottom in infralittoral zones.

Description: Shell of this species is medium, large, not very thick, robust, and spherical with a flattish spire. Its breadth is greater than height. Suture is barely incised. Surface is smooth, without any sculpture, apart from some marked growth lines. Base has a rather wide and deep umbilicus. Aperture is semicircular with a thin outer lip. Base color of the shell is whitish. Whole surface is decorated with small chestnut patches of different sizes. Umbilicus is hazel in color. Operculum is calcareous white and paucispiral. Common size of the shell is 6 cm.

Other uses (if any): It is an important food source in Spain and Italy.

Naticarius stercusmuscarum (Gmelin, 1791) (=*Naticarius punctatus*)

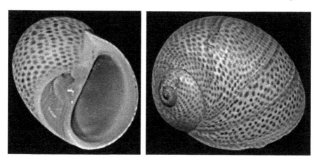

Common name(s): Fly-specked moon snail or common necklace shell.

Global distribution: Mediterranean Sea and in Northwestern Africa.

Habitat: Coastal areas; deeper layers; on hard floors; on sandy and muddy seabeds; rocky bottoms; depths from 5 m.

Description: This species has a medium or large white shell. It is gray toward center of the turns, and yellow-orange toward the upper side and the lower side of the last around. Numerous rust-colored oval spot are arranged in curved radial rows. Mantle and the foot are brownish and dotted with bright spots. It has a well-developed foot with the two cephalic visible appendages. Fossils of this species can be seen in the sediments of Italy and Greece. Shell may reach a size of 25–5.6 cm.

Biology

Ecology: The hermit crab *Pagurus prideaux* adopts housing dead snails.

Food and feeding: It is a successful hunter and its favorite food is mussels. It works with chemical substances (hydrochloric acid), which it uses together with its radula as a drilling machine. It reaches the interior of food molluscs and slurps the soft body of the victim.

Reproduction: In this species, the female lays her eggs in the form of a broad, corrugated band on stony substrate. These bands may reach a diameter of several centimeters. This is achieved by swellable substances, which surround the eggs as protection. The young trochophora larvae emerged from these eggs live in plankton. Subsequently, the veligers appear to metamorphose into adults which lead a sessile life.

Other uses (if any): It is an important food source in Spain. Natica is edible. The tasty meat is also raw.

Neverita duplicata (Say, 1822) (=*Polinices duplicatus*)

Common name(s): Shark eye, Atlantic moon snail.

Global distribution: Western Atlantic.

Habitat: On sandy shores just below the low tide line; offshore banks and soft substrate; depth range 0–58 m.

Description: Shell of this species is spiral in shape, smooth, short, and wide. Central apex of the shell is often a dark blue which can make the shell resemble an eye. Body whorl makes up most of shell. Exterior color of the shell is greenish gray to light brown, sometimes with orangish-brown markings. Base of shell is white. Spire is low but pointed. Siphonal canal is absent. Bottom of shell is with thick brown to purple callus which partly blocks the umbilicus of the shell. Aperture is large and oval. Foot is large and it can envelope the entire snail shell. Maximum size of the shell is 9 cm.

Biology

Food and feeding: It is a carnivorous gastropod feeding on smaller snails and clams by drilling through the shells of those creatures. It secretes an acid onto the victim's shell, then pierces the shell with its radula and then feeds on the soft tissue within. Its favorite food item is the Coquina clam.

Reproduction: Members of the family are mostly gonochoric and broadcast spawners. Embryos develop into planktonic trochophore larvae and later into juvenile veligers before becoming fully grown adults. During breeding season, the females lay egg masses that are surrounded by a sand and mucus mixture for protection.

Neverita lewisii (Gould, 1847) (=*Euspira lewisii*)

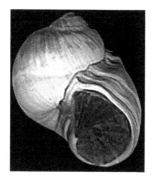 

Common name(s): Lewis's moon snail.

Global distribution: Northeast Pacific: Alaska, Canada, and USA.

Habitat: On sand along the intertidal zone; depth range 0–180 m.

Description: This species has an extremely large foot, which can extend up over the shell and mantle cavity. Part of the propodium contains a black-tipped siphon which leads water into the mantle cavity. Cephalic tentacles, located on its head, are usually visible above the propodium. When the animal retracts its soft parts into the shell, a lot of water is expelled; thus, it is possible to close the shell with its operculum. Maximum length of the shell is 14 cm.

Biology

Food and feeding: It is a strict predator and feeds mainly on bivalve molluscs by drilling a hole in the shell with its radula and feeding on the victim's soft internal tissues.

Reproduction: Like other moon snails, this species lays its eggs in a "sand collar." The eggs which are in thousands hatch into microscopic larvae which feed on plankton until they undergo torsion and metamorphose into the adult stage.

Other uses (if any): It is an important food source in Southeast Alaska to Baja California.

Polinices mammilla (Linnaeus, 1758)

 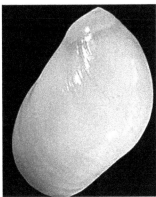

Common name(s): Oval moon snail, white natica, pear-shaped moon snail or white moon snail.

Global distribution: Indo-Pacific.

Habitat: On sandy bottoms; at the low tide mark; depth range 0–20 m.

Description: Shell of this species is thick and oval. As its spiral tip is smoothly sticking out, the overall shape of the shell resembles a teardrop or breast. Exterior shell is white, glossy, and unmarked, but sometimes with large irregular patches of brown, black, orange, or yellow. Underside is completely white, often with a large bump and shallow depression next to the bump. Umbilicus is completely covered by the sliding layer. Operculum is horny, smooth, amber yellow, and semicircular. Body is pure white. Maximum shell length is 6 cm.

Biology

Food and feeding: It is a carnivorous prosobranch which drills holes on its prey such as limpets, barnacles, and other molluscs, especially bivalves.

Reproduction: Members of the family are mostly gonochoric and broadcast spawners. Embryos develop into planktonic trochophore larvae and later into juvenile veligers before becoming fully grown adults.

Other uses (if any): It is collected in large quantities, for food and for the shell. In Thailand, the shells are used in shell trade and for shell craft.

Potamididae (Horn shells or creeper snails): Shell of the members of this family is thick, solid, and high-conical, with many flattened or slightly convex spire whorls. Sculpture is coarse, with spiral grooves or cords and axial ribs, giving a reticulated to nodular aspect. Aperture is small, with a short siphonal canal. Periostracum usually well developed, brownish to corneous. Outer lip often thickened and more or less flaring. Operculum is rounded and corneous, with many spiral coils and a subcentral nucleus. Head is with a pair of tentacles, abruptly narrowing distally and bearing eyes at or above their thickened bases. Foot is rounded in front and obtuse behind. These snails feed on algae and detritus. Some of the bigger species are collected for human food.

Cerithidea obtusa (Lamarck, 1822)

Common name(s): Obtuse horn shell or mud creeper.

Global distribution: Indo-West Pacific: from Madagascar and India to eastern Indonesia; north to the Philippines and south to Australia.

Habitat: Mud tidal banks; Animals often concentrate in the wettest spots, when the mud bottom is partly dry at low tide. Occurs at the low tide mark; freshwater and brackish and mangroves.

Description: Shell of this species is medium sized, with a moderately high conical spire and broad rounded base. Spire whorls are convex, with moderately deep suture, six or seven rounded spiral cords crossed by stronger, broad axial ridges, and forming a pattern of rounded nodules. Apical part of spire is always broken off. Body whorl is wide and rounded at periphery with 12 to 15 fine spiral cords on base. Aperture is wide and subcircular in outline. Outer lip is thickened and flaring, with a tongue-shaped anterior end produced over the siphonal canal. Columella is narrow, without internal spiral ridges. Anterior siphonal canal is short, open and oblique. Exterior shell is brown or dull purplish brown, with a brighter zone just below the suture. Base is plain brown or yellowish with a darker brown zone. Aperture is brownish. Animal's head and base of tentacles are gray with pale yellow spots. Tentacles are pinkish gray and darker at base. Sides of foot are dark gray with red spots and red margin. Maximum length of the shell is 6 cm.

Biology

Ecology: The favorable temperature range of this species is 22–26°C.

Locomotion/behavior: It can crawl up to 7 m of where once it reach the highest tree, it will then jump off the bark and fall onto the mud.

Food and feeding: It is a scavenger.

Reproduction: Members of family are mostly gonochoric and broadcast spawners. Embryos develop into planktonic trocophore larvae and later into juvenile veligers before becoming fully grown adults.

Allergens: This species has been reported to contain numerous major and minor allergenic spots. These allergens may cause severe allergic reactions including anaphylactic shock (Kamarazaman et al., 2016).

Other uses (if any): It is used as food in Southeast Asia. Many Southeast Asians have been seen eating this as a dish particularly in Malaysia, Indonesia and Vietnam.

Cerithidea quoyii (Hombron and Jacquinot, 1848) (=*Cerithidea quadrara*)

Common name(s): Quadrate horn shell.

Global distribution: Vietnam.

Habitat: Brackish waters; mangrove-associated.

Description: Shell of this species is thin, medium, elongate, and conical with spire angle 30–40°. Both spiral and axial ribs which are equal in strength are present on spire. Apex is decollate in adult and is often eroded. Body whorl is rounded. Suture is moderate. Aperture is quadrangular. Apertural lip is flared and slightly thickened. Collumela is thin and twisted with brown color. Head and base of tentacles of the body are pinkish-gray with cream spots. Anterior half of snout is blackish, sometimes with a few yellow spots. Tentacles are pale gray with black rings. Sides of foot are gray, and blackish anteriorly, with small yellowish cream spots. Sole of foot is gray and pinkish toward margin. Mantle is pale pinkish-gray.

Pirenella cingulata (Gmelin, 1791) (=*Cerithidea cingulata*; *Cerithideopsilla cingulata*)

Common name(s): Girdled horn shell.

Global distribution: Indo-West Pacific: from India and Sri Lanka to Papua New Guinea; Japan and Queensland.

Habitat: Brackish or supersalted fishponds; mangrove mudflats; upper bottom layer of mud which is almost liquid.

Description: Shell of this species is black brown with white or yellow spiral cords which formed grains with axial cords. Spire is triangular gimlet with three varices. Spiral canal under the suture is deep. Aperture is elongate and anterior edge of outer lip is long. Outer lip is posteriorly expanded in a flaring, wing-like process. Columella is without spiral ridge. In this species, most shells are sober and lustrous one is few. Maximum length of the shell is 4.5 cm.

Biology

Food and feeding: It grazes on diatoms, bacteria, and detritus.

Reproduction: Members of the family are mostly gonochoric and broadcast spawners. Embryos develop into planktonic trocophore larvae and later into juvenile veligers before becoming fully grown adults.

Other uses (if any): It is extensively collected for food and its shell is used to make lime in the Philippines.

Telescopium telescopium (Linnaeus, 1758)

Common name(s): Horn snail or Telescope snail.

Global distribution: Indo-West Pacific; coasts of the Indian Ocean and the Gulf of Thailand.

Habitat: Freshwater, brackish water, and marine habitats; quiet waters where the substratum is muddy and rich in detritus.

Description: Shell of this species is large, thick, and elongate conical. Apex is often eroded. Body whorl is flattened, about 0.3 of total shell length. Suture is shallow. Aperture is obliquely quadrangular. Apertural lip is sinuate but not flared and thickened. Two thickened spiral lines are present on the base of apertural lip. Collumela is thick and twisted, with brown color. Outer lip is thin and not flared. Operculum is small and circular. Animal is black-gray with a highly extendible proboscis and a dirty-white sole. Snout is large, long, with a pair of cephalic tentacles. Tentacles are sharply constricted at tips. There is a third eye on its mantle margin, in addition to a pair of eyes at the tentacles. Foot is large with whitish sole. Inhalant siphon is thick and interior edge is with one or two orange-pigmented spots. Shell color is dark brown on base and lighter in apex. It can stay out of water for long periods of time. Maximum length of the shell is 15 cm.

Biology

Locomotion/behavior: This species is found to have an amphibious habit. It is often partly buried in mud, with its only top of spire projecting. In the laboratory, the animal was observed crawling to the water edge in the aquarium and keeping part out of water and was not permanently submerged. Further, its tolerance of high temperature and extreme dryness (when it is exposed to low tide) may contribute to the survival of this species in the shallow waters of the mangrove swamps.

Food and feeding: It feeds on algae from the mud surface at low tide by using its proboscis. Further it is a deposit feeder, taking in mud and digesting the detritus and other organic matter in it.

Reproduction: Members of the family are mostly gonochoric and broad-cast spawners. Embryos develop into planktonic trocophore larvae and later into juvenile veligers before becoming fully grown adults.

As a biomonitor/biomarker of pollution: This species has been reported to serve as good biomonitoring material of Pb, Cu, Ni, and Zn pollution in the tropical intertidal mudflats. The digestive caecum of this species was found to accumulate higher concentration of Zn (214.35 µg/g dry weight). On the other hand, the shell and gill of this species accumulated Pb with concentrations of 41.23 µg/g dry weight and 95.76 µg/g dry weight respectively (Yap, 2014; Yap et al., 2009, 2011). The blue secretion of *Telescopium telescopium* has been reported to possess the xenobiotic metabolizing enzyme, namely, arylamine *N*-acetyl transferase. This enzyme could be a biomarker for detection of environmental pollution (Gorain et al., 2014).

Other uses (if any): This species serves as food for humans in Southeast Asia and Indonesia and Philippines.

Tympanotonos fuscatus (Linnaeus, 1758) (=*Tympanotonus fuscatus*)

Common name(s): West African mud creeper or mud-flat periwinkle.

Global distribution: Coasts of Angola, Cape Verde, Gabon; West Africa.

Habitat: Brackish water and freshwater; intertidal regions and mangrove swamps; muddy quiet waters rich in detritus.

Description: It is the only extant species in the genus *Tympanotonos*. Shell of this species is characterized by turrented, granular, and spiny shells with tapering ends. It may reach a size of about 3.5–10 cm.

Biology

Ecology: This species is euryhaline and has the ability to tolerate a wide range of salinities between 0.1 and 25 mg/L.

Locomotion/behavior: This species can survive for a long period in the absence of water. It tends to concentrate under the roots and decaying red mangrove trees and small collection of water during low tide.

Food and feeding: The species is a deposit feeder, feeding on mud and digesting the detritus and other organic matter in the mud.

Reproduction: It is a bisexual species.

Other uses (if any): This species provides a relatively cheap source of animal protein for humans as it is rich in protein (about 21%), vitamins, and minerals. These molluscs are important food delicacy among the riverine communities of the Niger Delta; they are collected from the wild; and their marketing form an important industry. The flesh of this species is also used as bait by fisher folks. The organism is also found to be very medicinal for endemic goiter due to its iodine content. Due to its high calcium, phosphate, and iron content, it is also recommended for pregnant women. The shell of this species is grounded for several purposes such as powder for pimples, cleansing, (e.g., vim for washing) as fertilizers, as calcium source in animal feed. Other uses include building construction, ornamentals and cosmetics.

Strombidae (conchs): Shell of the members of this family is thick and solid, with a relatively large body whorl. Aperture is with a siphonal canal. A distinct notch is seen along the anterior margin of the outer lip. Operculum is thin, corneous, and claw-like with serrated edges. This is used like a "walking stick" to anchor the animal. They have a pair of eyes on long stalks. Living in tropical and subtropical sea areas, they feed on algae and organic debris. Many species are collected by human for food, and their pretty shells are used for making decorative items and ornaments.

Conomurex decorus (Röding, 1798) (=*Strombus decorus*)

Common name(s): Mauritian conch.

Global distribution: Mediterranean.

Habitat: Shallow bays, on mixed (rock, sand, and mud) bottoms.

Description: Shell of this species is biconical, with a moderately high spire and a large body whorl. Spire whorls possess a definite keel, which is continued on the body whorl at a short distance from the suture. Early whorls are also with axial folds. Aperture is elongated. Exterior color of the shell is whitish with brown markings which are organized to form unequal spiral bands. Shell size varies between 3.5 and 8.0 cm.

Biology

Food and feeding: It is herbivorous, feeding on algae.

Reproduction: Larvae of this species are planktotrophic.

Other uses (if any): Although it is not currently a widely consumed species, it is a potential seafood.

Conomurex luhuanus (Linnaeus, 1758) (=*Strombus luhuanus*)

Common name(s): Strawberry conch or conical cone worm.

Global distribution: Indonesia, Falyma, Fiji, Papua New Guinea, Australia, Japan, and Korea.

Habitat: Shallow sand and rubble lagoon interisland reefs; on pinnacles; seagrass areas in intertidal and shallow subtidal areas; depth range 1–16 m.

Description: Shell of this species is solid and conical. There are six spiral whorls with well-developed sutures. Body whorl surface is smooth with poorly developed spiral and axial ribs. Aperture is narrow and white-orange in color. Columella is smooth and black-brown colored. Its stromboid notch is deep and easily distinguished. Spiral canal is short and widely opened. Interior may be slightly lirate. External color of the shell may be either white with an orange, brown/tan pattern of blotches, or completely brown/white. Interior is strong orange, red, or pink colored, and the inner lip border is black or chocolate brown. Size of the shell ranges from 5 to 8 cm.

Biology

Locomotion/behavior: Animals of this species may aggregate in fairly large numbers.

Food and feeding: It has a long proboscis for feeding on algae.

Other uses (if any): It is an important food source in West Pacific, East Australia, and Japan.

Euprotomus aurora (Kronenberg, 2002) (=*Euprotomus aurisdianae; Strombus aurisdianae*)

Common name(s): Diana conch, Diana's ear or ear conch.

Global distribution: Indo-West Pacific.

Habitat: Intertidal and shallow subtidal zones; in shallow water coral reef areas, such as coral sand, grassy sand flats, and dead coral; depth range 0–10 m.

Description: This species has a thick and solid shell which has a nearly elliptical contour. Shell has a high pointed spire and an irregular body whorl, ornamented with large knobs and divergent ridges. Flaring outer lip has a posterior expansion, like a spine, that extends itself posteriorly as far as half the length of the apex. Liration is present near the anterior and posterior ends of the outer lip. Inner lip is smooth with a thin callus. Siphonal canal is strongly bent, and the stromboid notch is deep. Shell color may vary from dull cream to pale gray, with irregular darker spots and lines. Ventral callus and inner lip are commonly glossy white. Aperture is rich orange or pink interiorly, and becomes paler toward the outer lip. Maximum shell length is up to 9 cm.

Biology

Food and feeding: It is a herbivore feeding on algae.

Conservation status: This species is listed as "Critically Endangered" in the Red List of threatened animals of Singapore.

Other uses (if any): It is an important food source in Indo-West Pacific. Shell of this species is commonly used in shellcraft and is sold in local markets in the central and northern Philippines.

Gibberulus gibberulus (Linnaeus, 1758) (=*Strombus gibberulus gibberulus*)

Common name(s): Humpbacked conch.

Global distribution: Indian Ocean, including the coast of Kenya.

Habitat: Clean sand; sandy patches on reef flats.

Description: Shell of this species is smooth and gibbous. Spire is occasionally varicose. Body whorl is grooved at the base. Columella is smooth. Interior of the aperture is radiately striate. Shell is mottled and hieroglyphically marked with yellowish brown and white. Markings are often arranged in a few or numerous interrupted revolving bands. Aperture is tinged violaceous, scarlet, or dark purplish brown. Adult shell size varies between 3 and 7 cm.

Biology

Food and feeding: It feeds on detritus and algae.

Other uses (if any): Besides serving as an important food source, this species is also used in aquaria.

Harpago chiragra (Linnaeus, 1758) (=*Lambis chiragra*)

Common name(s): Chiragra spider conch.

Global distribution: Indo-Pacific.

Habitat: Coral reef areas; seagrass beds; littoral and sublittoral zones; tidal pools and low tide levels; depth range 0–32 m.

Description: This species has a very thick, robust, and heavy shell, with a distinct anterior notch. Its most prominent characteristic features are the six long and curved marginal digitations, expanded from the flaring, thick outer lip and canals. Columella and aperture are lirate. Eyeballs peek out over the operculum in aperture view. Juveniles have thin outer lips on the shell and lack the fingers. Green proboscis bearing the mouth can be seen between the two eye stalks in juvenile. Sexual dimorphism is strongly present in this species. Female individuals are much larger than the males. Shell length of this species varies between 8.5 and 32 cm.

Biology

Locomotion/behavior: Adults of this species are always seen paired, with a large female accompanied by one, or rarely, a couple of smaller males.

Food and feeding: It is a herbivore, feeding on plants and algae.

Reproduction: Members of the family are mostly gonochoric and broadcast spawners. Embryos develop into planktonic trocophore larvae and later into juvenile veligers before becoming fully grown adults.

Other uses (if any): It is an important food source in East Indian Ocean and East Polynesia. Its shell is used in shellcraft.

Laevistrombus canarium (Linnaeus, 1758) (=*Strombus canarium*)

Common name(s): Dog conch, yellow conch, or yellow dog conch.

Global distribution: Indo-Pacific: from India and Sri Lanka to Melanesia, Australia and southern Japan.

Habitat: Muddy and sandy bottoms.

Description: This species has a heavy shell with a rounded outline. Outer surface of the shell is almost completely smooth, except for barely visible spiral lines and occasional varices on the spire. Stromboid notch on the outer lip is inconspicuous. This notch can be observed to the right of the siphonal canal as a shallow, secondary anterior indentation in the lip. Siphonal canal is straight and short. Columella is smooth, without any folds. Adult specimens have a moderately flared, posteriorly protruding outer lip. Body whorl is roundly swollen at the shoulder, with a few anterior spiral grooves. Shell has a medium-to-high cone-shaped spire, with at least five delicately furrowed whorls. Shell color is variable, from golden yellow to light yellowish-brown to gray. Underside of the shell is rarely dark; paler than the top, or totally white. Shell aperture is white. A zigzag network of darker lines may be present on the outside of the shell. Periostracum is yellowish-brown. It is usually thick, reticulated, and fringed over the suture. Operculum which is corneous is dark brown, slightly bent sickle-shaped, with seven to eight weak lateral serrations. Shell length of adult specimens varies from 3 to 7 cm.

Biology

Locomotion/behavior: The dog conch exhibits behaviors including burrowing and a characteristic leaping form of locomotion.

Food and feeding: It is a herbivore grazing on algae and detritus.

Reproduction: It is a gonochoristic and sexually dimorphic species. Fertilization is internal. Larvae spend several days as plankton and are undergoing a series of transformations until they reach complete metamorphosis. The maximum lifespan is 2–2.5 years.

Predators: Predators of this species include carnivorous gastropods such as cone snails and volutes. Its vertebrate predators include macaques and humans.

Fisheries/aquaculture: It is an economically important species in the Indo-West Pacific, and it is suffering population declines due to overfishing and overexploitation.

Other uses (if any): It is an important food source in South India and Sri Lanka where its soft parts are used in a wide variety of dishes. Shells are used for making ornamental products. As its shells are heavy and compact, they are used as sinkers for fishing nets. It is also useful as a bioindicator for organotin pollution monitoring near Malaysian ports.

Lambis lambis (Linnaeus, 1758)

Common name(s): Spider conch or common spider conch.

Global distribution: Indo-Pacific: from East Africa, India, Sri Lanka, Andaman and Nicobar Islands and Australia, to French Polynesia and Japan.

Habitat: Intertidal and found on subtidal reef and seagrass; on reef flats and on coral-rubble bottoms or in mangrove, muddy areas, usually associated with fine red algae; depth range 0–24 m.

Description: It has a very large, robust, and heavy shell. Shell is double cone; turns form a triangle rather pronounced; middle part is wider and the base is narrowing. Last lap is important and deviates markedly from the central axis. Shell of the males usually smaller and with shorter spines on the outer lip. Its most striking characteristic feature is its flared outer lip, ornamented by six hollow marginal digitations. Three anterior most digitations are short and posteriorly bent in male individuals, and longer and dorsally recurved in females. Shell opening is pearly and pinkish with orange or yellow tints. Part of the body is olive-brown with white spots. It has large eyes on long stalks and a thick siphon. Like other conch snails, it has a curved, knife-shaped operculum attached to a long strong foot. Exterior color of the shell is white or cream; and brown, purplish or bluish black patches. Interior is glazed and may be pink, orange, or purple. Maximum length of the shell is 29 cm.

Biology

Locomotion/behavior: It shallowly burrows in sand or gravel and is often occurring in colonies.

Food and feeding: It gazes on animal matter in the mud. It predominantly feeds on sand grains, polychaetes, bivalves, and small crustaceans. It also grazes on fine red algae.

Reproduction: Sexual dimorphism is pronounced in this species. Reproduction is internal. The female lays bright orange egg strings on the substrate. The larvae are pelagic. These larvae are transformed into juvenile veligers before becoming fully grown adults.

Conservation status: The spider conch is listed as "Vulnerable" on the Red List of threatened animals of Singapore where it is affected by human activities such as reclamation and pollution. Trampling by careless visitors and overcollection for their shells can also have an impact on local populations.

Association: The shell of this species is covered with epibionts (algae).

Other uses (if any): This species has an important, commercial fishery in the Southeast Coast of India. These conches are landed as by-catches from crab nets. It is often collected for food by coastal populations. The meat of gastropods is considered as a delicacy, rich in protein, and the fat content is very low. Chutney powder (a side dish for Idli or Dhosai in South India) is prepared from its meat. This chutney powder was found to be microbiologically and organoleptically good until the end of the storage period, and it was safe for human consumption. Shell of this species is used in shellcraft.

Lambis millepeda (Linnaeus, 1758)

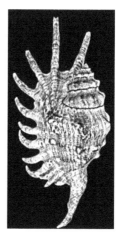

Common name(s): Millipede spider conch.

Global distribution: Indian Ocean: off Madagascar; Southwest Pacific Ocean: from the Philippines to southern Indonesia and Papua New Guinea.

Habitat: Marine: coral reef area; shallow-water bottoms, from low in the intertidal zone to a depth of a few meters.

Description: Lips rugose. Intercalary digitations developed. Posterior digitation simple; length: 90–150 mm. Angle armed with four digitations, the uppermost being an intercalated one. Posterior digitation simple. Lateral digitations, especially the primary, much recurved. Lips with wrinkles moderately developed. Maximum shell length is 16 cm.

Biology

Reproduction: Members of the family are mostly gonochoric and broadcast spawners. Embryos develop into planktonic trocophore larvae and later into juvenile veligers before becoming fully grown adults.

Other uses (if any): It is an important food source in Philippines, Southwest Pacific. It is mainly collected for its shell which is industrially used in lime production and is sold as decorative item.

Lambis scorpius (Linnaeus, 1758)

Common name(s): Scorpion conch or scorpion spider conch.

Global distribution: Indo-Pacific: Indonesia to Polynesia; Japan, and Queensland and New Caledonia.

Habitat: Live coral reef areas and under or among dead coral slabs and boulders; shallow subtidal zones; depth range 0–5 m.

Description: The species name is derived from the anterior finger (on the right side of the shell) the curve of which resembles a scorpion's tail. Shell of this species is thick-walled and strong. Outer lip is thickened and is turned away to the outside. Spiral sculpture is represented by broad, relatively elevated edges and flat, narrow ridges between them. On the surface, there are more or less pronounced bumps. Exterior shell color is white, yellow, or brown. Inner lip is brown with transverse white stripes. Outer lip is purple with a wide brown band and transverse white stripes. Fingers are made last as the animal grows, starting out hollow and gradually filling in. Length of adult shell varies between 9.5 and 22 cm.

Biology

Reproduction: Members of the family are mostly gonochoric and broadcast spawners. Embryos develop into planktonic trochophore larvae and later into juvenile veligers before becoming fully grown adults. Hybrids between this species and the much rarer *Lambis crocata* have been reported.

Fisheries/aquaculture: Commercial fishery exists for this species.

Other uses (if any): It is an important food species in West Pacific. Shells of this species are industrially used in lime production and are sold as decorative items.

Lambis truncata (Lightfoot, 1786)

Common name(s): Giant spider conch,

Global distribution: Indo-Pacific: from Africa to Sudan, Egypt, Yemen, and India, to Southeast Asia and Australia; Hawaii and Japan.

Habitat: On rubble and coarse sand in shallow coral reef areas; on shell of old specimens often worn and encrusted with calcareous algae, vermetid snails, and tubes of polychaete worms; depth range 0–30 m.

Description: It is a large spider conch. This species is similar to *Lambis lambis* but with a more squarish outline with six moderate digitations. Lips are smooth. Body whorl is unarmed at angle. Spines along the mouth are slightly open. Outer surfaces of the shell have a thin layer of periostracum which is light brown colored and is fragile. Inner surface of the shell is shiny and white. Younger shells are creamy white. Columella and lip are mauve brown. Maximum length of the shell is 43 cm.

Biology

Food and feeding: It is a herbivore, feeding on macroscopic or unicellular algae and algal detritus.

Reproduction: Members of the family are mostly gonochoric and broadcast spawners. Embryos develop into planktonic trochophore larvae and later into juvenile veligers before becoming fully grown adults.

Fisheries/aquaculture: Commercial fishery exists for this species.

Other uses (if any): It is an important food species in East Africa and Bay of Bengal and is actively collected by native populations. Shells of this species are industrially used in lime production and are sold as decorative items. In spite of its weight and considerable size, the shell is favored, especially by tourists, due to the beauty of its heavily glazed aperture.

Lentigo lentiginosus (Linnaeus, 1758) (=*Strombus lentiginosus*)

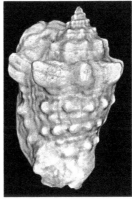

Common name(s): Silver conch or freckled stromb.

Global distribution: Widespread in the Indo-Pacific.

Habitat: Intertidal and subtidal zones; sandy and rubbly areas on lagoon reefs, pinnacles, and shallow portions of the seaward reef; depths of about 3–13 m.

Description: Adult shell of this species is very heavy and thick, with a characteristic deep stromboid notch, and a flared, very thick and posteriorly expanded outer lip. Columella is anteriorly projected, and the siphonal canal is convex. Body whorl has a notably irregular surface, which is ornamented by spiral cords and rows of blunt tubercles which form elevated knobs on the shoulder. Inner lip is smooth, with a large callus which spreads over the spire and over the body whorl. Shell has a tall spire, and each whorl of the spire has a row of heavy knobs and groves. Shell color is white with large irregular brown blotches and dots. Columellar callus has a faint silvery gloss. Lateral margin of the outer lips has a series of tan blotches. Aperture is pink to orange on the interior, becoming paler toward the margins. Maximum shell length of this species is 10 cm.

Biology

Locomotion/behavior: These animals are found in pairs. They bury in the sand during the day. This species is known for its jerky and discontinuous leaping motions.

Food and feeding: It is a herbivore grazing on algae.

Other uses (if any): It is an important food source in Indo-West Pacific. The shells of this species are used in shellcraft, and are commonly sold in local markets around the central Philippines.

Lobatus costatus (Gmelin, 1791) (=*Strombus costatus*)

Common name(s): Milk conch.

Global distribution: Western Atlantic.

Habitat: Shallow sandy areas and meadows of phanerogams; depth range 2–55 m.

Description: Shell of this species is thick and medium sized, and its shape is extremely variable. Turns of the shell are short and the protuberances of the shell are rounded. Sculpture is of projecting nodules on shoulder. Aperture is flared in mature specimens, sometimes with very thick lip. Exterior shell is cream to brown, frequently flecked with other hues. Interior of the shell is smooth and shiny, sometimes with golden highlights. Head of the animal is greenish. An indentation at the base of the shell allows the conch to observe the surroundings with its eyes. It has a powerful muscular foot. Operculum is used by the animal as a lever in order to straighten out as well as to progress by leaps forward. Operculum also serves as a defense against predators. Maximum length of the shell is 23 cm.

Biology

Food and feeding: It is a herbivore feedings on algae and phanerogams as well as organic detritus.

Reproduction: In this species, the sexes are separate and the fertilization is internal. At the time of laying, the females gather at a precise point. The female is laying its eggs, in the form of long cords of jelly that contains microscopic eggs. Wrapped on itself, the crescent-shaped laying of 8–10 cm is wrapped on itself and it resembles a ball of gray wool, dirty sand color. Development includes a long-lasting, planktotrophic stage. The veliger larva has a tiny transparent shell. After 3 weeks in the plankton, the larva is placed on the bottom and metamorphoses into adult.

Association: The shell of this species is often covered with algae, sponge, and coral.

Fisheries/aquaculture: Commercial fishery exists for this species.

Other uses (if any): It is an important food source for the people of South Florida, West Indies, Brazil, and Bermuda.

Lobatus galeatus (Swainson, 1823) (=*Strombus galeatus*)

Common name(s): Eastern Pacific giant conch.

Global distribution: Eastern Pacific: Gulf of California to Peru; South Atlantic.

Habitat: Rocky, sandy bottoms near mangrove areas; depth range 0–30 m.

Description: Shell of this species is very thick and heavy with an oblong outline and a short pointed spire that lacks spines and nodules. Spine is often eroded in adult specimens. Body whorl is very inflated, with numerous spiral ridges and low, slightly noticeable nodules on the shoulder. Periostracum is thick. Outer lip is very flared and posteriorly expanded, not higher than the apex of the spire. Edge of the outer lip edge bears a shallow stromboid notch that is often associated with the undulations originating from the superficial spiral sculpture. Columella is smooth with a well-developed callus. Exterior shell is colored ivory white to light brown, with a darker spire and a brown periostracum. Aperture is bright white, and the outer lip and columellar callus are extensively orange or dull brown in adult specimens. Maximum size of the shell is 23 cm.

Biology

Locomotion/behavior: The animal initially fixes the posterior end of the foot by thrusting the point of its sickle-shaped operculum into the substrate. Then, it extends its foot forward, lifting the shell and throwing it forward in a motion that is called "leaping." It is known to move long distances. It also has a burrowing behavior, in which an individual sinks itself entirely or partially into the substrate. This species spends part of the life time partially buried in the sand.

Predators: The giant conch is preyed upon by invertebrates, such as octopi. Vertebrate predators include rays, triggerfish, and snappers.

Fisheries/aquaculture: It is an economically important species in many areas where it occurs. It is primarily used for subsistence and commercial fishery.

Other uses (if any): It is an important food source for the people of Gulf of California and Ecuador. The shells of this species are used as wind instruments.

Lobatus gigas (Linnaeus, 1758) (=*Strombus gigas*)

Common name(s): Queen conch or pink conch.

Global distribution: Western Atlantic: South Carolina to Brazil.

Habitat: Coral reef and rocky shore; seagrass beds and alternating sand flats; depth range 2–73 m.

Description: Shell of this species is large, heavy, and solid with a short conical spire. Spire is high and strongly angled, with pointed knobs where obscure axial ribs cross angles. Knobs on the last 3 whorls are large and pointed. Interior of the outer lip and the aperture are pinkish, suffused with white or yellow. Outer lip has a broad upper expansion that is generally as high or higher than the spire. Its lower half is somewhat wavy. It is mostly body whorl, with an aperture that is moderately narrow and channeled at both ends. Outer lip is thickened and greatly flared in fully grown specimens. There are 8 to 10 whorls, with blunt nodes on the shoulders. Body of the animal has a long snout, two eyestalks with well-developed eyes, additional sensory tentacles, a strong foot and a corneous, sickle-shaped operculum. Exterior shell is yellowish white with irregular brownish markings. Interior is bright rosy pink. Young specimens have zigzag axial stripes of brown. Maximum reported size of the shell is 35 cm and weight 3 kg, which is achieved at about 3–5 years of age. It is a long-lived species, with a lifespan up to 40 years.

Biology

Locomotion/behavior: Adults potentially avoid predators by extending their foot forward, grabbing onto substrate and hop forward. Nocturnal activities may also be a strategy to avoid attention from visual predators.

Food and feeding: It is a herbivore, feeding on detritus, macroalgae, and epiphytes. The green macroalga *Batophora oerstedi* appears to be a preferred food of this species.

Reproduction: This species has separate sexes and internal fertilization. Adult animals migrate to shallow, warmer inshore waters to mate and lay their eggs. During mating, the muscular foot of the male attaches to the posterior portion of the female's flared lip. The male extends its end into the females' vaginal groove. Copulation occurs both diurnally and nocturnally. After fertilization, female conches lay eggs in the form of gelatinous strings. The sticky external substance of the egg strings enables them to form a compact egg mass with the surrounding sand. On an average 8–9 egg masses with 180,000–460,000 eggs may be produced per season by each adult female. Reproductive season lasts from March to October with a peak from July to September. Females spawn multiple times during this season. Embryos hatch 3–4 days after spawning producing a two-lobed larvae known as veliger. These pelagic larvae live in the water column for 18–28 days before they settle and metamorphose on the sediment. Embryonic shell develops after 16–40 days post hatching. One-year-old juveniles (80–100 mm shell length) are believed to emerge from the sediment and feed in seagrass beds. Sexual maturity is reached at approximately 3.5 years of age (180–270-mm shell length), which is preceded by the development of a flared shell lip.

Association: The shell and soft parts of living queen conch serve as a home to several commensal animals, including slipper snails, porcelain crabs and cardinalfish. Its parasites include coccidians.

Predators: The queen conch is hunted and eaten by several species of large predatory sea snails, and also by starfish, crustaceans, and vertebrates (fish, sea turtles, and humans).

Conservation status: The Queen Conch is now considered an endangered species and listed in Appendix II of the CITES (Convention for the International Trade on Endangered Species).

Fisheries/aquaculture: This species has a high commercial fishery. The fisheries of this species have grown exponentially in the last 30 years, with resulting declines in population and area closures. The queen conch is cultured by extensive systems only in Turks and Caicos. It is cultured more intensively in the Bahamas, Netherlands Antilles (Bonaire), the US Virgin Islands, and in Martinique, where hatcheries have been constructed to repopulate overfished beds. Hatchery reared juveniles are released in either protective cages or enclosures, or directly (Anon. http://www.fao.org/docrep/t8365e/t8365e05.htm).

Other uses (if any): Meat of this species has been consumed for centuries and has traditionally been an important part of the diet in many islands in the West Indies, Bermuda, Bahamas, and Southern Florida. It is consumed raw, marinated, minced, or chopped in a wide variety of dishes, such as salads, chowder, fritters, soups, stew, pâtés, and other local recipes. The meat is also used as a source of bait for fish traps in American Virgin Islands. These queen conchs are also prized for their attractive shells. Its shell is sold as a souvenir and used as a decorative object. Historically, Native Americans and indigenous Caribbean peoples used parts of the shell to create various tools. This species has been reported to produce pearls naturally. Dwindling populations from overfishing and environmental changes have made natural pearls rarer, leading to a multifold increase in value. Cultivation of pearls from this species is also becoming very common.

Lobatus goliath (Schröter, 1805) (=*Eustrombus goliath*)

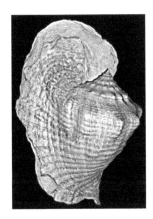

Common name(s): Goliath conch.

Global distribution: Northeastern and Southeastern coast of Brazil.

Habitat: Seagrass beds; depth range 0–50 m.

Description: It is a very large edible sea snail. It has a very large, heavy, and solid shell, with a very conspicuous, widely flaring and thickened outer lip. Stromboid notch is inconspicuous in adult individuals, but it can be identified as a secondary anterior indentation to the right of the siphonal canal. Aperture of the shell is colored tan. It has also a shorter spire and duller spines. Outer lip expands far beyond the length of the spire in the shells of adult individuals. Maximum reported length of an adult shell is 38 cm.

Biology

Fisheries/aquaculture: Commercial fishery exists for this species mainly for its popular and ornamental shells.

Other uses (if any): The flesh of the goliath conch is edible in Brazil. Shell of this species is widely utilized in handicrafts and sold as a souvenir in markets and craft stores in several regions of Brazil.

Lobatus peruvianus (Swainson, 1823) (=*Strombus peruvianus*)

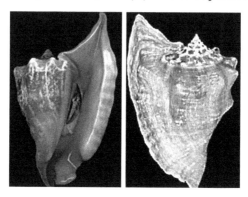

Common name(s): Peruvian conch or cock's comb conch.

Global distribution: Eastern Pacific: Mexico to Peru.

Habitat: Tide pools at about low-tide mark.

Description: Shell is large and strikingly distinct. It is nodulous, heavy, and coiled. Spire is depressed and top is prominent. Outer lip is produced above and attenuated. Margin is reflected and aperture is striated. Common size of the shell is 14 cm.

Lobatus raninus (Gmelin, 1791) (=*Strombus raninus*)

Common name(s): Hawk-wing conch.

Global distribution: West Tropical Atlantic: Caribbean Sea, Gulf of Mexico and Lesser Antilles.

Habitat: Gulf; seagrass beds of shallow waters; offshore reefs; sandy areas; depth range 0–55 m.

Description: Like other species of the same genus, it has a robust, conical, somewhat heavy, and solid shell, with a distinct stromboid notch. Shell is spiral in shape. Spire is medium and pointed. There are about eight whorls. Body whorl is dorsally ornamented by coarse spiral ridges. Posterior expansion of the flaring outer lip is lower than the spire. It has a powerful muscular foot. Operculum is long, horny, and crescent shaped. When disturbed, the animal will retreat into the shell, closing it with the operculum. An indentation at the base of the shell allows the animal to observe the surroundings with its eyes. Head of the living animal has a large proboscis and two eyestalks. Exterior shell color is brownish, with several disperse white spots. This may be overgrown by algae or debris. Interior of the shell is smooth and shiny. Both inner and outer lips are cream or white. Maximum recorded shell length is 13 cm.

Biology

Locomotion/behavior: The operculum is used by the animal as a lever to straighten, as well as to advance by leaps forward.

Food and feeding: It is a herbivore feeding on algae as well as organic detritus.

Reproduction: In this species, the sexes are separate and the fertilization is internal. The veliger larva has a tiny transparent shell. After 3 weeks of planktonic life, the larva reaches bottom and metamorphoses into adult.

Other uses (if any): It is an important food source for the people of Southeast Florida and Brazil.

Margistrombus marginatus (Linnaeus, 1758) (=*Strombus marginatus*)

Common name(s): Marginate conch.

Global distribution: Indo-Pacific: Andaman Sea and Strait of Malacca.

Habitat: Shallow water, on sandy, muddy or rubble bottoms; depth range 0–30 m.

Description: Shell of this species is very beautiful, and it is smooth and moderately elongate. Adult shells have 10 complete whorls. Aperture is narrow and smooth. Columella is polished. Body whorl is traversed by spiral grooves and is stronger toward base. Stromboid notch is weak. Shell is dorsally brown with transverse bands and cream ventrally. In live specimens, the beauty of the shell is hidden as it is normally covered with debris. Shell length is up to 6 cm.

Strombus alatus (Gmelin, 1791)

Common name(s): Florida fighting conch.

Global distribution: Western Central Atlantic: from North Carolina to Texas, Mexico, Belize, and Puerto Rico.

Habitat: Shallow waters; seagrass beds, shallow reefs, and sand and rubble; depth range 0–180 m.

Description: Florida fighting conch contains a small, jagged spire at the top of the shell and about seven whorls. Shell has less prominent subsutural spines and a slightly more projected outer lip. Soft-body, eyestalks, and snout emerge from its shell. Front of the shell has two curved edges. These edges allow the eyestalks to become aware of its surroundings, keeping the rest of its body safe. The snout acts as a trunk to collect food. Maximum length of the shell is 11 cm.

Biology

Locomotion/behavior: This species has been reported to live in deep waters of 10 m or greater during winter and migrates to shallow waters of 2 m during summer to spawn.

Food and feeding: It feeds on diatoms and other epiphytic algae.

Reproduction: It is a broadcast spawner. Free-swimming larvae hatch from the eggs and develop into veligers which later settle down the water column and undergo metamorphosis into juvenile conch.

Fisheries/aquaculture: Commercial fishery exists for this species. It is also commercially cultivated for food market and aquarium market.

Other uses (if any): The flesh of this species is consumed in North Carolina, Florida, and Texas. As it is a very hardy species, it is an excellent sand sifter and is very beneficial in the reef aquarium.

Strombus gracilior (Sowerby, 1825)

Common name(s): Eastern Pacific fighting conch, or Panama fighting conch.

Global distribution: Gulf of California; West Mexico; Pacific Ocean: Peru.

Habitat: On sand flats and lagoons and offshore to 45 m.

Description: Shell of this species has a high spire which is covered with subsutural spines or pointed nodes on the shoulder of the whorls. Exterior color of the shell is yellowish to yellowish-brown, interrupted in the middle with a lighter band. Aperture and large outer lip is white bordered with orange-brown. Shell is covered with a thin, horn-covered periostracum. Length of the shell varies between 4 and 9.5 cm and width may attain 5 cm.

Strombus pugilis (Linnaeus, 1758)

Common name(s): Fighting conch or West Indian fighting conch.

Global distribution: Western Atlantic.

Habitat: Intertidal; sandy and muddy bottoms; depth range 0–55 m.

Description: This species has a robust, somewhat heavy and solid shell, with a characteristic stromboid notch. It has a well-developed body whorl and a short and pointed spire. There are 8–9 whorls. Each of these whorls is having a single row of subsutural spines which are becoming larger toward the last whorl. These spines, however, may be less conspicuous or even absent in some populations. Its aperture is relatively long and slightly oblique. Posterior angle of the outer lip is distinct, projecting in the posterior direction in an erect fashion. Operculum is sickle shaped and is similar to other Strombus snails. Shell color varies from salmon-pink, cream, or yellow to light or strong orange, and the interior of the aperture is white. Anterior end presents a dark purple stain, which is one of the characteristic features of this species. Maximum shell length is 13 cm.

Biology

Food and feeding: It is a herbivore feeding on plants and algae.

Reproduction: It is a dioecious species. Males are less abundant and smaller than females. Reproductive strategy of this species is copulation and internal fertilization. Development includes a long-lasting, planktotrophic stage. The females deposit egg masses during the first months of the year, in March and April, on muddy bottoms.

Association: Shells of *Strombus pugilis* are often used by the hermit crab *Dardanus insignis*.

Fisheries/aquaculture: Commercial fishery exists for this spices.

Other uses (if any): This species forms an important food source for the people of Southeast Florida, West Indies, and Brazil. It is also a potential indicator species of Co-60 in marine ecosystems as it is known for its fast radiocobalt incorporation (Moraes and Mayr, 1991). It can therefore serve as an important organism in the monitoring procedures around the nuclear power plants. *S. pugilis* is also used as a zootherapeutical product for the treatment of sexual impotence in the traditional Brazilian medicine of the Northeast region of Brazil. The shell is commonly used as a decorative item and is sold in local markets as a souvenir.

3.3 ORDER NEOGASTROPODA

Babyloniidae (sea snails): Shell apex is with numerous narrow whorls which are acute. The columella and the lowest part of the palatal lip of the aperture are on horizontal line in the frontal view. Shell is without a partic-ular microsculpture. Maximum shell height is 9 cm. Species of Babylonia are restricted to the Indo-Pacific ocean.

Babylonia areolata (Link, 1807)

Common name(s): Spotted babylon snail or maculated ivory whelk.

Global distribution: Indo-West Pacific and Southwest Atlantic.

Habitat: Fine sandy bottom; depth range 10–20 m.

Description: Shell of this species is thin, ovate, and light. Spire is tall and apex is pointed. Whorls are rounded. Whorls are with a rather broad but shallow sutural canal, which becomes a shoulder on the last whorl. Outer lip of the aperture is not clearly thickened inside. Thin inner lip has a notch for the umbilicus. Umbilicus is wide open. Upper and left sides are surrounded by a raised fasciole. Inside, the fasciole a comparatively narrow band is seen, which runs to the lower end of the inner lip. Initial whorls are whitish, and the following with reddish-brown spots on a white background. Three widely separated rows of spots are visible on the body whorl. Fasciole and the band around the umbilicus are both white. Periostracum is thin and yellowish, forming a row of irregular little pointed flaps along the outer edge of the sutural canal. Shells are up to 10 cm length, 9 cm high and 5 cm broad.

Biology

Reproduction: Members of the family are mostly gonochoric and broad-cast spawners. Embryos develop into planktonic trocophore larvae and later into juvenile veligers before becoming fully grown adults.

Fisheries/aquaculture: As wild spotted babylon is rare, it is being promoted to culture in commercial farms. There are many cultural snail farms along the gulf and southern part of Thailand.

Other uses (if any): It is an economically important, edible sea-snail. Its taste and texture are acceptable among gourmets.

Babylonia japonica (Reeve, 1842)

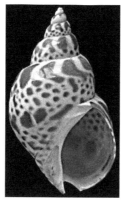

Common name(s): Japanese babylon.

Global distribution: Off Korea, Japan, and Taiwan.

Habitat: Muddy sand.

Description: Dextral shell of this specie has up to about nine whorls and a more or less slender typical buccinoid shape with an acuminate apex.

If the shell is kept in upright position, lowest point of the last whorl and columellar base are situated at the same horizontal line. Apart from the growth-lines and more delicate spiral lines, surface is smooth. Height of adult shells varies from 2 to 9 cm. A sutural canal is seen. Aperture has a small groove for the anus above and a large notch for the siphon below. Exterior shell color is white to orange-yellowish, with orange to brown spots. Dark spots are arranged in four spiral rows. Shell size varies between 4 and 8.5 cm.

Biology

Food and feeding: These snails are carnivores.

Reproduction: The sex ratio in this species is 1:1. Two or three years old adult females with 6–7 cm shell height produce about 10–60 egg capsules with 27–50 eggs each, and about 500–2500 eggs are laid in one season.

Tetrodotoxin poisoning: Due to its tetrodotoxin (a potent neurotoxin, $C_{11}H_{17}N_3O_3$), tetrodotoxin poisoning (TTX) and associated fatal human poisonings has been reported to occur in this species after its consumption. However, these animals are still commercially harvested (Noguchi et al., 2011).

Other uses (if any): This species is eaten in Japan, Taiwan, Korea, and China.

Babylonia spirata (Linnaeus, 1758)

Common name(s): Spiral babylon.

Global distribution: Indo-West Pacific.

Habitat: On mud, sand and shell substrates near seagrasses; depth range 0–60 m.

Description: Shell of this species is thick, heavy, conical, and smooth with distinctive spiral. It has up to more than eight whorls. Aperture is forming more than half the total height. Sutural canal is broad and deep, with a sloping base. Outer lip of the aperture is slightly thickened above. A little anal notch is clearly seen when the shell is seen from above. Inner lip has a heavy callus. Umbilicus is varying from wide open to completely closed. Upper and left sides are surrounded by a slightly raised fasciole. Inside the fasciole, there is a band which is variable in width. Initial whorls are violet, and the following whorls are with orange-brown spots on a whitish background. Operculum thin and flexible is made of a horn-like material. Shell color and pattern varies from plain brown to white with orange or brown spots. Further four bands are usually recognizable: a first (upper) band of large, oblong or moon-shaped spots; a second and a fourth band consisting of a larger number of smaller spots, with in between the larger spots of the third band. Fasciole and band around the umbilicus are whitish with orange-brown spots. Umbilicus and aperture are white. Periostracum is rather thick and felty, usually deciduous in collection material. Body of the animal is pale, with a long muscular foot that is dark with an orange rim, short tentacles, and long siphon. Shell size varies from 4 to 6 cm.

Biology

Reproduction: Sexual maturity occurs in this species at about 2-cm shell length. The male reproductive organs of this snail include the penis, the vas deferent or channel sperrna, and testes, while in the female snails, they are female genital pore or opening, copulatory bursa, gland capsule, fallopian tubes, and ovaries.

Babylonia zeylanica (Bruguière, 1789)

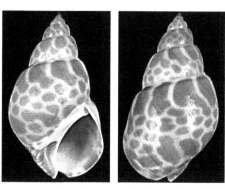

Common name(s): Indian babylon.

Global distribution: Indian Ocean.

Habitat: Intertidal areas; offshore.

Description: Shell of this species is fusiform and less solid with less inflated whorls. Body whorl is narrower than in *Babylonia spirata*. Sutures are not canaliculated. Spire is high ending in dark purple apex. Aperture is dark and outer lip is sharp and smooth, but not flexed at top. Columella is smooth with heavy broad callus posteriorly but narrow anteriorly. A strong parietal ridge is seen almost close to the outer lip. Umbilicus is broadly open with a row of teeth on the outer margin. Fasciole is with a ridge on the inner edge. Anterior canal is broad and deep and posterior canal is not distinct. Surface is smooth and color of shell is white with large brown blotches. Siphonal canal is tinged violet and apex is black or gray. Shell size is up to 7 cm in height.

Biology

Locomotion/behavior: It spends most of the day living under the sand. If and when it smells the food (olfactory stimulant), it erupts from the sand and begin to search for the source of the smell. Then, it moves at a pretty good clip and will consume any meaty food that makes it to the sand bed.

Food and feeding: It is carnivorous feeding on smaller snails. It is also an excellent scavenger.

Fisheries/aquaculture: Commercial fishery exists for this species. In Indian coasts, the trawl fishery of *Babylonia* spp. was unique in that almost 99% of the catch comprised *Babylonia zeylanica* with a very a small quantity of *B. spirata* in 2012.

Other uses (if any): The meat and operculum of the species is commercially important. The meat of this species has been reported to be similar or superior to other marine shellfish, and hence, it could be utilized as an alternative meat source in human diet. It is also an ideal aquarium species. In India, the shell of this species is believed to bring wealth, good luck, and prosperity. Further, the water filled in this shell and sprinkled on the devotees would keep away the evil spirits (Anon. https://www.tradeindia. com/fp3221803/Babylonia-Zeylanica-Medium-Shells.html).

Buccinidae (whelks): Shell of the members of this family is with a fairly high spire and large body whorl. Outer surface is smooth or with sculpture and is without axial varices. Siphonal canal is rather short. Operculum is corneous.

Buccinum undatum (Linnaeus, 1758)

Common name(s): Common whelk, waved buccinum or waved whelk.

Global distribution: Arctic, Northern Atlantic, and the Mediterranean: Canadian Arctic Archipelago and the Atlantic Ocean.

Habitat: On soft sediments including muddy sand and gravel as well as on rocks; areas with sludge; extreme low water mark of the intertidal zone down to depths of 1200 m.

Description: Shell of this species is solid with conical spire which is high. Sometimes, spire is rather flattened. Whorls (7–8) are very convex and suture is markedly impressed. Aperture is ovoidal and outer lip is not thickened or toothed without inner folds. Columellar lip has a fairly striking callus, partly adheres to the ventral surface of the body whorl. Sculpture is of strong crescentic costae, spiral striae, and growth lines. Shell is yellowish white or light hazel in color. Periostracum is light chestnut in color and is not very resistant. Operculum is horny and is smaller than the aperture. Body has large foot. Whole body is cream with blackish patches. Head has a pair of tentacles with an eye at the base of each. Proboscis is long and siphon is well developed. The animal emits a thin and copious slime. These animals prefer moderate or cold sea temperatures and cannot survive at temperatures above 29°C. It does not adapt well to life in the intertidal zone, due to its intolerance for low salinities. Maximum shell height is 11 cm and width is 6 cm.

Biology

Food and feeding: It is a necrophagous, scavenger species feeding on polychaetes, bivalve molluscs, echinoderms, and other smaller crustaceans. It uses the edge of its shell to open bivalve shells and may drill holes into the shell of its prey in order to access the soft tissues. It also scavenges for carrion, which it detects by smell from some distance. When searching for food, these whelks extend a tube known as the "siphon," which is used to funnel water to the gills, and leads to a sensory organ used for smelling the prey (chemoreception).

Reproduction: In this species, the sexes are separate. Sexual maturity occurs after 5–7 years depending on the sex of the individual and other environmental factors. Breeding takes place from October to May, and the eggs are found attached to rocks, shells, and stones in protective capsules. Each capsule may contain as many as 1000 eggs, and the capsules of several females are grouped together in large masses of over 2000. Only a few of these eggs will develop and most eggs are used as a source of food by the growing embryos. There is no planktonic larval stage. Instead, crawling young emerge from the capsules after several months. Common whelks are thought to live for 10 years.

Fisheries/aquaculture: A strong commercial fishery exists for this species in several countries around the North Sea; French and English Channel coast; and Gulf of Maine. Whelks in Maine have traditionally been landed as an incidental by-catch of the lobster fishery. These animals are trapped in pots using dogfish and brown crab as bait.

Other uses (if any): *Buccinum undatum* is eaten widely as food in Europe. In USA, the product is typically shipped live for mostly ethnic (oriental) markets in Boston and New York City. A cottage industry also exists which produces pickled meats and creates shell ornaments from this species.

Cantharus cancellarius (Conrad, 1846)

Common name(s): Crossbarred spindle.

Global distribution: Western Atlantic Ocean: Florida to Mexico.

Habitat: Intertidal zone to deep water; Gulf; bays with higher salinities; jetties.

Description: Shell of this species is spiral in shape. Spire is high and pointed. Shell has 3–5 cords of white-tipped tubercles on each spire whorl and 12–14 on body whorl. Edge of outer lip is wavy. Columellar wall is with ascending spiral fold at bottom. Siphonal canal is short and slightly turned. Aperture is long and its outer margin is crenate (scalloped). Fine dentations are seen on its inner edge. Exterior shell color is yellowish to reddish brown, with white areas. Inside aperture is white. Common shell length is 4 cm.

Neptunea arthritica (Valenciennes, 1858)

Sinistral shell   Dextral shell

Common name(s): Neptune whelk.

Global distribution: Northwest Pacific: Japan.

Habitat: Benthic; depth range 0–100 m.

Description: Shells of this species are medium to shortened length (i.e., length of the shell from apex to end of anterior canal) relative to width. Shells are dextrally coiled and are with distinct patterns of spiral/axial sculpture. Sinistral coiling is rare in this species.

Biology

Reproduction: In this species, the sex ratios of females to males were not significantly different from a 1:1 sex ratio. Females attain sexual maturity when they reach a shell size of 5–6 cm. Spawning occurred between May and August in females and between April and July in males. Spawning peak in females was observed between June and July when the seawater temperature was above 19 °C.

Toxicity: This species has been reported to contain toxic tetramethyl ammonium salts (autonomic gaanglionic agent, tetramine) in its tissues, especially the salivary gland, and has the potential to cause human poisoning. Consumption of this species has caused severe headache, dizziness, seasickness, vomiting, visual disturbances, etc. in Japan (Asano and Itoh, 1960).

Solenosteira pallida (Broderip and Sowerby, 1829) (=*Cantharus pallidus*)

Common name(s): Not assigned.

Global distribution: Peru.

Habitat: Sandy silt and silty clay substrata; depth range 36–83 m.

Description: Shell of this species is with stronger sculpture. It however lacks a clear-cut subsutural groove. This white shell has a densely pilose or velvety periostracum with moderately coarse spiral ribs and low axial undulations. Maximum length of this species is 28 cm.

Other uses (if any): In this species, its muscular foot is used for human consumption in the Mexican Pacific. It has good flavor steamed, in cocktails, fried, or in soups. Further, this species is quite attractive for shell collectors, and it is commonly seen in tourist centers.

Busyconidae (true whelks): The family Busyconidae consists of large sea snails, known as whelks or true whelks. Shell of these animals is large, heavy and pyriform. Body whorl is very large and inflated. Pillar is long and slender. Spiral sculpture is developed.

Busycon carica (Gmelin, 1791)

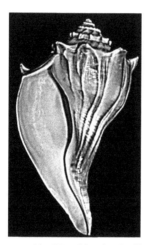

Common name(s): Knobbed whelk.

Global distribution: Western Atlantic Ocean: USA and Mexico.

Habitat: Intertidal to subtidal; sand flats; estuaries, bays and shallow shelf waters; depth range 0–75 m.

Description: Shells of this species are thick and strong with six clockwise coils. These shells are right-handed. Surface is sculpted with fine striations and there is a ring of knob-like projections protruding from the widest part of the coil. Exterior shell color is ivory or pale gray, and the large aperture (the inside of the opening) is orange. Canal inside is wide and the entrance can be closed by a horny oval operculum. Brown streaks are more prominent in juvenile shells. Inner shell ranges in colors from pale yellow to orange and even dark red. Maximum shell length is 30.5 cm.

Biology

Locomotion/behavior: It migrates from deep to shallow waters in times of reproduction and low food supply.

Food and feeding: It is a carnivorous species foraging on intertidal flats and creeks fringed by oyster reef and in subtidal channels. It preys on bivalves such as oysters, scallops, and clams. It uses the lip of its shell to penetrate, break off, and force open the valves of its victim.

Reproduction: Knobbed whelks reach maturity in 3–5 years. They are protandric hermaphrodites initially functioning as males and changing into females as they grow. Adult females are generally larger than adult males of the same age. Mating and egg laying occur during the spring and fall migration. Internally fertilized eggs are surrounded by a transparent mass of albumen (a gel-like material) and are laid in protective flat, rounded egg capsules joined to form a paper-like, thin chain of egg cases, commonly called a "Mermaid's necklace." On an average, each capsule contains a maximum of about 100 eggs, with most strings having a maximum of 160 capsules. After laying their egg cases, female whelk will bury one end of the egg case into the substrate. Fertilized eggs emerge as juvenile knobbed whelks.

Association: The deal shells of this species are host to several species of *Crepidula* and oyster spats and inhabited by hermit crabs.

Fisheries/aquaculture: Commercial whelk trawl fishery exists for this species in the South Carolina coast and Delaware coasts.

Other uses (if any): This species forms an important food source in Caribbean and Bahamas. Its meat is used in such dishes as salads (raw), burgers, fritters, and chowders. Shells of this species are also sold in the tourist trade as ornamentals.

Busycon contrarium (Conrad, 1840)

Common name(s): Lightning whelk.

Global distribution: North American coastline from New Jersey to Texas.

Habitat: Estuaries, creeks, and around oyster bars.

Description: Shell of this large gastropod is very recognizable because it spirals toward the left instead of the right like other gastropod shells. Juvenile shells have lightning bolt shaped stripes on the shell hence its common name. Interior of juvenile shells can be white, yellow, or pale blue. Adults, on the other hand, are gray with a few vertical violet-brown streaks. Aperture of the adult shells can vary from white, pale yellow to orange, or bright red. Maximum size of the shell is 41 cm in length.

Biology

Ecology: Feeding rates and crawling speeds of this species have been reported to decrease as seawater temperatures increase or decrease above or below the intermediate temperatures of 25–28°C. Growth stops in this species when temperatures fall below 20°C.

Locomotion/behavior: This species migrates from deep to shallow waters during reproduction and low food supply.

Food and feeding: These lightning whelks are carnivorous and are preying on bivalves such as clams. These whelks force open the shells of the bivalve with their large foot and hold it open by the edge of their own shell. Once the bivalve is open, the whelk inserts its radula and proboscis inside the clam to scrape and eat the clam meat.

Reproduction: This species has separate sexes. Reproduction is internal and copulation occurs in late autumn to early winter. Females lay long strings of disc-shaped egg capsules in early spring. The string of eggs is anchored to the sand and the capsules break loose when the eggs hatch at the beginning of May. Fertilized eggs of this species develop slowly and hatch in approximately 3–13 months. They emerge as juveniles, which crawl along the bottom.

Association: The dead shells of this species are host to several species of Crepidula and are also inhabited by hermit crabs and also serve as substrata for oyster spat.

Other uses (if any): Members of this genus are used for food and ornaments. The muscular foot of this snail is used in chowders and is served as steaks.

Busycon spiratum (Lamarck, 1816)

Common name(s): Pear whelk or fig whelk.

Global distribution: North Carolina to Florida and Gulf States.

Habitat: Shallow waters.

Description: It is a less common species. Shell of this species is fusiform and pear shaped with four to five whorls. Body whorl is very large. Sutures are distinct, wide and deeply channeled. Shell is thin but sturdy and spire is very short. Outer lip is thin. Aperture is wide, oval-shaped, and prolonged into a straight, open canal. Operculum is horny. Surface sculptures are with weak revolving lines. Flesh is colored with reddish-brown streaks. Maximum shell size is 14 cm length and 12.5 cm height.

Biology

Locomotion/behavior: In this species, feeding rates and crawling speeds are affected by seasonal temperature changes, and these activities are the highest at intermediate temperatures.

Reproduction: This species copulates in late autumn to early winter, lays egg capsules in early spring, with hatching beginning in May. It is a slow-growing species, and growth has been reported to occur only above 20°C.

Busycotypus canaliculatus (Linnaeus, 1758) (=_Busycon canaliculatum_)

Common name(s): Channeled whelk.

Global distribution: Endemic to the eastern coast of the United States; from Cape Cod, Massachusetts to northern Florida; introduced into San Francisco Bay.

Habitat: Subtidal and intertidal zones; buried in or crawling on mud and sand bottoms.

Description: It is a very large predatory sea snail. Shell of this species is smooth and subpyriform (pear shaped), with a large body whorl and a straight siphonal canal. Between the whorls, there is a wide and deep channel at the suture. There are weak knobs at the shoulders of the whorls. Finely sculpted lines start at the siphonal canal and revolve around the shell surface. Shell aperture is located on the right side, that is, the shell of this species is dextral in coiling. Left-handed or sinistral specimens are rare in this species. Color of the shell is buff gray to light tan. Shells reach a size of 12–20 cm in length.

Biology

Locomotion/behavior: It is a nocturnal species. Its operculum does not tightly seal the opening of the shell. Hence, it cannot expose itself to the air as some intertidal animals (like mussels) can. These whelks grow by using their mantle to produce calcium carbonate to extend their shell around a central axis or columella, producing turns, or whorls, as they grow. A whorl is each spiral of the shell. The final whorl (largest) is the body whorl that terminates, providing the aperture into which the animal withdraws.

Food and feeding: It is a carnivore eating mainly the clams. When hunting prey, these whelks use their nose (or proboscis) to find the buried food animals by sensing the stream of water flowing out of the clam's feeding tubes. On locating its prey, the whelk digs down into the shore bottom to capture it. The whelks use their shell's lip to chip and pry the valves of bivalves (clams) apart by holding it with its foot. This slow chipping continues until an opening occurs to allow the whelk to wedge its shell between the clam's valves. Subsequently, it extends its proboscis to begin feeding.

Reproduction: These whelks reproduce through sexual reproduction and internal fertilization. They produce a string of egg capsules that may be about 1 m long, and each capsule has a maximum of 100 eggs inside which hatch into miniature whelks. The egg capsule allows the young whelk embryos to develop and provides protection. Once they have developed, the eggs hatch inside the capsule, and the juvenile whelks leave through an opening.

Predator: The main predator of this species is the blue crab, *C. sapidus*.

Fisheries/aquaculture: It is caught as by-catch in the fisheries industry.

Other uses (if any): It is an important food source for the people of Cape Cod southward to northern Florida and the Gulf of Mexico. The shells of this species are used in the sea shell trade.

Sinistrofulgur perversum (Linnaeus, 1758)

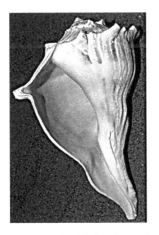

Common name(s): Lightning whelk.

Global distribution: Native to Southeastern North America, south to Florida and the Gulf Coast from North Carolina to Texas.

Habitat: On the bottom of shallow bays in sand or mud near shoal grass or turtle grass meadows.

Description: It is an edible species of very large predatory sea snail or whelk. Shell of this species is off-white to tan or gray with narrow, brown "lightning" streaks from the top to the bottom. Shell is white on the inside. Animal inside the shell is dark brown to black. These whelks are unusual in that they have a counterclockwise shell spiral (left handed). Color of the shell largely depends on light, temperature, and age. Older lightning whelks have pale shells. Maximum shell length is 40 cm.

Biology

Locomotion/behavior: Even on a favorable bottom, the rate of movement of this animal is slow.

Food and feeding: It is a carnivore feeding primarily on the marine bivalve such as oysters, clams, and scallops. It ingests the soft parts of the victim using its proboscis.

Reproduction: The mating season of this species is from late October to early January. Spawning season is from March to April. The female lays eggs in long strings of capsules which are about 80 cm long. Each strand has a maximum of 145 capsules, and each capsule may contain a maximum of 100 eggs. However, only less than 13 of the eggs in each capsule hatch. Juveniles will begin hatching in May and emerge as tiny lightning whelks. There is no free larval stage. The species of Busycon are slow-growing but long-lived individuals. The lifespan of this species is, however, not known.

Association: Hermit crabs make homes of unoccupied lightning whelk shells.

Predators: The main predators of this species are gulls, crabs, and other whelks.

Other uses (if any): For thousands of years, the Native Americans used this species as food. They also used the shells of these animals for tools, ornaments, containers, and to make jewelry, that is, shell gorgets. These sinistral shells were also believed as sacred objects. It is also believed that sailors used the egg cases of this species as sponges for bathing.

Sinistrofulgur sinistrum (Hollister, 1958) (=*Busycon sinistrum*)

Common name(s): Lightning whelk.

Global distribution: Western Atlantic.

Habitat: Estuaries and near shore sandy shallows; depth range 0–73 m.

Description: It is an edible species of large predatory sea snail or whelk. Shell of this species has large body whorls with distinct shoulders bearing a dozen or more knobs. Aperture is wide and is tapering into its long siphon canal. Background color is cream or gray. Living animals possess chocolate-colored body. Size of the shell varies between 20 and 45 cm.

Columbellidae (dove shells): These tiny snails are with small, thick, strong, fusiform, and boldly colored shells. Outer lips of the shells are thickened. Inner lip is not folded. A small, corneous, and elongate operculum is present. Aperture rather long and narrow, with a short siphonal canal. These snails have a diverse diet which includes algae and sessile animals like sponges and sea anemones. Some are scavengers.

Columbella rustica (Linnaeus, 1758)

Common name(s): Rustica dove shell or trumpet.

Global distribution: Northeast Atlantic, Mediterranean Sea, Red Sea and Equatorial Guinea.

Habitat: Coastal zone of rocky bottoms; subtidal, on or under rocks and among eelgrass and Neptune grass; prefers the waters very close to the coast and beat by the waves; benthic and sedentary; depth range 1–12 m.

Description: Shell of this species is solid, conical, spiral, smooth, and shiny. Last spire is much larger, occupying two-thirds of the total shell. Shell opening is long and narrow, as high as the width of the shell, and is pointed up and dug down with a short siphonal channel. Columella is toothed at the base. Outer lip has small teeth. Operculum is small, long, and horny. It does not completely close the shell. Surface is smooth and shiny. Exterior shell color may be red and white with small flames and spots or orange-yellow. Shells are sometimes covered by a thin yellow and periostracum. Siphon, tentacles, and eye are brown mottled white. Foot is stained beige cream. Length of the shell is between 1.2 and 2.5 cm.

Biology

Locomotion/behavior: Breathing in this species is branchial. The palloral siphon of this species is provided with sensory organs which smell the substrate during displacements. The narrow opening of its thick shell allows it to resist predators such as crabs.

Food and feeding: It is very voracious and is a herbivorous microphage feeding on algae and detritus that gnaw on the rocks of the radula.

Reproduction: The breeding season is from October to May. The sexes are separate. Like most marine snails, these animals are dioecious. The breeding season is from October to May. They undergo exclusively sexual reproduction and fertilization takes place internal. The sperm is stored in a seminal receptacle, where fertilization takes place. The females deposit their eggs in ovigerous capsules which adhere directly to the substrate and are joined in groups of 5–30 capsules or are individually attached to the algae or posidonia. Pelagic larvae leave the eggs and after a brief plank-tonic phase, sink to the bottom and are transformed into juveniles.

Other uses (if any): It is potential food source for the people of Mediter-ranean. It is also a Croatian cuisine. The shell of this species is used as ornamentation and elaboration of collars.

Conidae (cones): Shell of the species of this family is cone shaped, with a low spire and a well-developed body whorl tapering toward the narrow anterior end. Aperture is very long and narrow, with a short siphonal canal. Operculum is corneous and is very small or absent. All cone snails are carnivorous, and their prey includes snails, crustaceans, fishes, etc. depending on the species. These cone snails are also venomous. The venom of some species is quite enough to kill humans, and hence, such species should not be handled.

Conus gloriamaris (Chemnitz, 1777)

Common name(s): Glory of the sea cone.

Global distribution: Pacific and Indian Oceans.

Habitat: Deep water; on sand and mud; depth range 5–300 m.

Description: Shell of this species is considered the rarest and the most expensive seashell in the world. Compared with other cones, the shell is large and slender, with a tall spire. Its lip is chipped and heavily filed, and there are breaks on the body whorl and later whorls of the spire. It is finely reticulated with orange-brown lines, enclosing triangular spaces and two or three bands of chestnut hieroglyphic markings across its body. The tan coloration of the shell may vary from a lighter, golden color to a deeper dark brown. Maximum length of the shell is 17 cm.

Biology

Food and feeding: It is a carnivore feeding on other molluscs.

Reproduction: This species has planktotrophic larval development.

Other uses (if any): It is a potential food source for the people of the Philippines. The shell of this species is greatly prized by collectors owing to its great beauty and historical significance. This species is also used in the pharmaceutical industry for development of new drugs.

Conus miles (Linnaeus, 1758)

Common name(s): Soldier cone.

Global distribution: Aldabra, Chagos, Madagascar, Mascarene Basin, Mauritius, Mozambique, the Red Sea and Tanzania; entire Indo-Pacific; off Australia.

Habitat: Bays, intertidal reef, subtidal reef, sand, gravel, reef limestone, beachrock, and lagoons; depth range 0–50 m.

Description: Shell of this species is thick with nodular shoulders of whorls. Spire is obsoletely tuberculate or smooth and rather depressed. Body whorl is bordered by a broad shoulder and is spirally ridged at the base. Exterior color of the shell is yellowish white or pale orange, with close narrow, wavy, thread-like longitudinal chestnut strigations, interrupted by a chocolate, fairly narrow, revolving band above the middle. Base of the shell is stained chocolate, bordered upwards by progressively lighter bands. Aperture is banded, chocolate, and white. Size of an adult shell varies between 5 and 14 cm.

Other uses (if any): It is an edible species. In common with all Conus spp., shells of this species are also traded for the collector market.

Conus planorbis (Born, 1778)

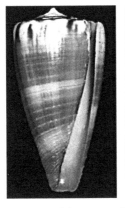 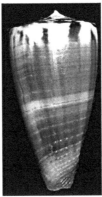

Common name(s): Planorbis cone, ringed cone or calf cone.

Global distribution: Indo-West Pacific.

Habitat: Intertidal; sandy bottoms with lagoon Halimeda spp. algal patches; beneath dead coral on reef rock; depth range 0–20 m.

Description: Whorls of the spire are striate, maculate with chestnut. Body whorl shows beaded striae below. Sometimes, these granular striae cover the entire surface. Exterior shell color is orange-brown or chestnut and is light-banded in the middle. Shell base is darker colored. Size of the shell varies between 3 and 8 cm.

Other uses (if any): It is a potential food source for the people of the Philippines. Shells of this species are traded for the collectors' market.

Conus ventricosus (Gmelin, 1791) (*=Conus mediterraneus*)

Common name(s): Mediterranean cone.

Global distribution: Mediterranean, Eastern Atlantic (Portugal, Canary Islands, and Cape Verde).

Habitat: Shallow waters; sheltered bays, among rocks and rocks covered with brown algae; mattes of the Posidonia meadows; depth range 1–5 m.

Description: Overall shape of the shell of this species resembles the union of two cones of unequal size, one small, one large, joined by their bases. Spire of the shell of this species is elevated, gradate, and maculated. Opening of the shell is long and narrow. Siphonal channel is open and slightly curved. Exterior shell is generally greenish with brown spots and streaks. Two broad transverse white strips are often visible on the shoulder and toward the bottom third of the last lap. Shell is covered with a yellowish and transparent periostracum. Animal is milky white. It is in fact an extremely polymorphic species as it is with more slender shapes and brighter base colors (pink, orange, and red). Rarely sinistral forms may be encountered in this species. Size of the shell varies between 1 and 7.5 cm.

Biology

Locomotion/behavior: It hides in the day in the pockets of sand, under the stones or in the crevices between the rocks.

Food and feeding: This carnivore pricks its prey (mainly worms) by injecting a paralyzing venom. Prey is slowly digested by the enzymes secreted by the cone.

Reproduction: Reproduction in this species is sexual. The sexes are separate and the fertilization is internal. Ovigerous capsules filled with eggs are deposited, by the female under the stones. The hatching of eggs gives rise to benthic larvae which are metamorphosed into juveniles in a few days.

Toxin: A toxic peptide Contryphan-Vn has been extracted from the venom of this species (Massilia et al., 2001).

Other uses (if any): It is a potential food source for the people of Montinegro Coast. It is also a Croatian cuisine. In common with all Conus spp. and other molluscs, the shells of this species are also traded for the Conus shell market.

Costellariidae (Sea snails): Shell of the members of this family is fusiform-ovate, with a predominantly axial sculpture. Aperture is notched by a short siphonal canal. Outer lip is finely lirate inside. Columella is with strong folds and is larger posteriorly. There is no operculum.

Vexillum ebenus (Lamarck, 1811) (*=Pusia ebenus*)

Common name(s): Ebony mitre.

Global distribution: West Europe and Mediterranean Sea.

Habitat: Under stones in a few meters deep.

Description: Shell of this species is solid and slender in shape. Opening on the side columellar has three large teeth and lip is finely striated. Some specimens may have axial ribs which are more or less pronounced. Exterior shell color is from black (ebenus) to brown to greenish gloss. It is yellowish below the shoulder of the turns. Shell size varies between 1 and 3.6 cm.

Other uses (if any): It is a potential food source for the people of Montinegro Coast, Mediterranean and West Europe.

Fasciolariidae (tulip and spindle shells): Shell of the species of this family is fusiform, with a well-developed siphonal canal. Columella is with a few low basal threads. Operculum is corneous. Soft parts of the animal are brilliant scarlet.

Cinctura lilium (Fischer von Waldheim, 1807) (=*Fasciolaria lilium*)

Common name(s): Banded tulip.

Global distribution: Off the coast of South Carolina; Gulf of Mexico: from the Florida coast to the Gulf coast of Texas and Mexico; Caribbean Sea.

Habitat: Sand or muddy sand; depth range 1–50 m.

Description: Shell of this species is large, broadly fusiform and is rather thin. This banded tulip shell does not grow as large as that of the true tulip, *Fasciolaria tulipa*. Shell surface is ornamented with many intact spiral bands. Outer lip is medially convex. Color splotches appear as a redder color. Shell length varies from 6 to 11 cm.

Fasciolaria tulipa (Linnaeus, 1758)

Common name(s): True tulip.

Global distribution: Western Atlantic: from the North Carolina to the Gulf coast of Texas; West Indies and Brazil.

Habitat: Marine and estuarine habitats in tropical and subtropical zones; tidal flats and seagrass beds, especially where the turtle grass (*Thalassia testudinum*) is present, prefer, and/or require warm, saline waters; depth range 0–73 m.

Description: Shell of this species is fusiform with about nine rounded whorls. Surface smooth, except for very fine growth lines. Outer lip thin, with fine denticles on inner edge. Operculum is thick and heavy. Exterior shell is cream, light brown, to reddish orange with irregular blotches of darker brown, white, or cream. Numerous (25–39) interrupted black spiral lines appear across the shell surface. Operculum which is thick and brown to black in color bears concentric growth rings. Body of the living animal is bright orange to red, including the muscular foot. Size of the shell varies from 6 to 25 cm in length.

Biology

Locomotion/behavior: These animals are active in shallow water during high tide and burrow under the sediments when they are exposed during low tide. Like many other snails, this snail reacts to threats by retracting its body into the shell and using its hard operculum to completely seal the aperture, thereby protecting its soft tissue from predators and other environmental problems.

Food and feeding: It is a carnivore feeding on bivalves and other gastropods. The flesh of its victim is scraped out and broken down through its radula. These snails have also been known to scavenge dead or dying animal tissue.

Reproduction: Like other snails of its family, the tulip snail reproduces sexually through copulation. Mating may occur several times in one season. In warm waters, it reproduces year-round and spawning peaks are from October to December. Following a successful mating event, females will produce clusters of egg capsules which are shaped like laterally flattened cones. The cluster is attached to a hard surface such as an empty shell or rock. Only a small portion of the 600–800 eggs laid in a cluster develop normally. Once the snails are fully developed, the hatchlings crawl from the holes formed in the tops of the capsules.

Predators: This species is preyed on by the horse conch, *Pleuroploca gigantea*, sting rays, crabs, sea stars, and bony fishes.

Fisheries/aquaculture: A commercial fishery exists for this species.

Other uses (if any): Although these snails are not commonly found in seafood markets or restaurants, small and localized recreational fisheries exist for this species. Further, cleaned shells and opercula of this species are sold in shell shops.

Fusinus dupetitthouarsi (Kiener, 1846)

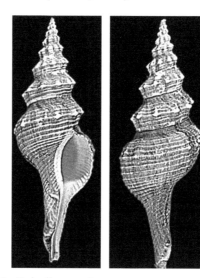

Common name(s): Spindle conch.

Global distribution: Eastern Pacific.

Habitat: Moderately deep waters; sandy silt and silty clay substrata; depth range 50–70 m.

Description: Shell of this species has a large, pyriform shell with a good periostracum of yellowish brown color. Long syphonal canal, long spire, well-delineated sutures, knobs, and vertical folds are seen in the shell. Both spiral ribs and spiral ridges are visible inside the aperture. Some specimens have a left-handed spiral in this species. Normal shell height is 22 cm.

Biology

Carotenoid content: Several caratenoids such as alloxanthin, 3S,3S-astaxanthin- (3S,3'S)-astaxanthin, canthaxanthin, β carotene, diatoxanthin, echinenone, fritschiel-laxanthin, fucoxanthin, isocryptoxanthin, lutein A, mytiloxanthin, 3S-phoenicoxanthin, (3R,3′R)-zeaxanthin have been isolated from this species (Sri Kantha, 1989).

Other uses (if any): The meat of this species is locally used for fish bait but seldom eaten by men. The beautiful shells of this species are most esteemed by shell collectors.

Opeatostoma pseudodon (Burrow, 1815)

Common name(s): Thorn latirus, banded tooth latirus or red footed conch.

Global distribution: Western coast of America; Southern Baja California; from Mexico to Peru; Galápagos.

Habitat: Under rocks; usually partially buried in coral sand.

Description: Shell length of this species may vary from 3 to 7.5. cm. Exterior shells are white or brown with thin brown or black stripes. Body of the snail is reddish. It has the longest apertural teeth developed by any gastropod. Long spines or thorns which are growing from the tip of the snail's opening help the animal anchor in the sandy and coral.

Biology

Locomotion/behavior: These animals are active nocturnally.

Food and feeding: It is a carnivore feeding on small worms and on other gastropods.

Plueroploca trapezium (Linnaeus, 1758)

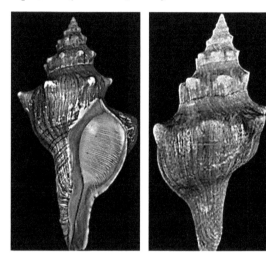

Common name(s): Trapezium horse conch or striped fox horse conch.

Global distribution: East Africa and Indo-West Pacific.

Habitat: Inner reef flats; shallow water near rocky areas; at low tide mark; offshore; depth range 0–40 m.

Description: It is a golden brown shell which is large, elongated, solid, and heavy. Spire is of moderate length. Apex is eroded. Sutures are constricted. Shoulders on the whorls are covered with spiral rows of slightly pointed strong nodules. Surface is covered with fine, brown, incised spiral lines, mainly in pairs. Outer lip is dentate with seven pairs of teeth. Oval aperture is pale with strong ridges internally. Columella is smooth posteriorly. Siphonal canal is extended and short. Fasciole is weak. Its shell size varies between 8.5 and 28.0 cm.

Biology

Food and feeding: These animals are active predators, feeding on tube worms, vermetid, and other molluscs like the spiny cerith, *Cerithium echinatum.*

Fisheries/aquaculture: Commercial fishery exists for this species.

Other uses (if any): This species is collected for food in Southeast Asian countries. Its large and heavy shell is traditionally used as a trumpet when the tip of the spire is cut off. Further, these shells are also sold as handicraft items.

Tarantinaea lignaria (Linnaeus, 1758) (=*Fasciolaria lignaria*)

Common name(s): Tulip snail or tulip shell.

Global distribution: European waters and Mediterranean Sea.

Habitat: Meditoltoral and infralittoral zones.

Description: Shell of this species is smooth and fusiform with high spire. Sculpture is characterized by a coil forming nine turns, each carrying a row of tubers. Shell opening is elliptical and is extended by a siphonal channel. Exterior color of the shell is beige to whitish, sometimes streaked with brown. Interior is brown. When the animal retracts inside the shell, its opening is closed by an operculum (lid) which is horny and brown. Shell length varies from 3 to 7 cm.

Biology

Food and feeding: It consumes other gastropods, bivalves and worms. They are not perforators.

Reproduction: It is a gonochoric species (sexes are separate). Egg laying occurs in summer. The small red eggs are gathered in resistant egg capsules, which are transparent and coriaceous, in the form of horns. Development is intracapsular. The first larvae that hatch feed on the other eggs already present in the capsule, resulting in fewer juveniles per capsule.

Other uses (if any): It is an edible species in Mediterranean and Montinegro coast.

Triplofusus giganteus (Kiener, 1840) (=*P. gigantea*)

Common name(s): Florida horse conch or giant Australian trumpeter.

Global distribution: Western Atlantic: Atlantic coast of the Americas; Southeast United States, in Texas, and Northeast Mexico; Florida.

Habitat: On sand, wood, and mud flats from the low intertidal to shallow subtidal zones; depth range 0–200 m.

Description: It is the largest gastropod in the American waters, and one of the largest univalves in the world. Shell of this species is somewhat fusiform, with a long siphonal canal and 10 whorls. Sculpture is of several spiral cords and axial ribs, some of which may form knobs on the whorls' shoulders. Shell color is bright orange in young individuals. Adult shell is grayish white to salmon-orange with a light tan or dark brown periostracum. Operculum is a leathery brown color and aperture is orange. Animal is brick red in color. Shell length may reach 61 cm. It is also reported that the shell of this species grows up to 91 cm with a weight of 18 kg.

Biology

Food and feeding: The species is a voracious carnivore and it feeds on other large marine gastropods, including the tulip shell (*F. tulipa*), the lightning whelk (*Busycon perversum*), the queen conch (*Eustrombus gigas*), and some Murex species. In aquaria, it eats small hermit crabs of the species *Clibanarius vittatus.*

Reproduction: This species has sexual reproduction. The female attaches egg capsules (containing several dozen eggs in each) to rock or old shell. The young emerge from these egg capsules.

Other uses (if any): It is a potential food source for the people of North Carolina; USA to the Yucatán, Mexico. The US state of Florida declared it as the state seashell in 1969. The shell is popular with shell collectors owing to its great size. The shell has also been used as a bugle or trumpet. In southern Florida, Native Americans used the horse conch to make several types of artifact.

Harpidae (harp shells): Shell of the members of this family is ovate, with an inflated body whorl and a small conical spire. Surface is glossy, with strong axial ribs. Inner lip is covered by a smooth, large callus. Columella is without folds. Siphonal canal is short and wide. Operculum is absent.

Harpa articularis (Lamarck, 1822)

Common name(s): Articulate harp shell or articulate harp.

Global distribution: Indo-West Pacific: from Burma and Indonesia to Fiji Islands; Japan, Queensland and New Caledonia.

Habitat: Depth range 0–250 m.

Description: Shell of this species is thin, ovate, and ventricose. Spire is conical and is indistinctly muricated. Ribs are narrow, distant, and slightly flattened. They are marked by transverse brown lines, articulated and winding. Between these lines, white and violet spots are present. Interstices between the ribs are grayish. Aperture is large and ovate; it is of a violet color upon the edge, and reddish within. Columella is polished and is covered over its whole length by a large brown chestnut-colored spot. Size of the shell varies between 5 and 11 cm.

Fisheries/aquaculture: Commercial fishery exists for this species.

Harpa major (Röding, 1798)

Common name(s): Large harp or major harp.

Global distribution: Indo-Pacific: Mexico.

Habitat: Shallow waters near coral reefs; sandy bottoms; lower intertidal fringe and sublittoral to shelf zones; depth range 36–73 m.

Description: Shells of this species have an ovate body with a heavily calloused spire. Strong axial ribs are seen. A flaring lip on the final whorl dominates the flattened spire. Columella, or the lower portion of the inside coil, has dark brown coloring. Shell has a polished highly patterned appearance. Exterior shell colors range from pinkish gray to bluish white, with square spots of violet in the ribs. It has long eyestalks and a long siphon. Its wide spade-shaped expansion on the front portion of their foot helps for digging. Size of the shell varies between 6 and 13 cm.

Biology

Food and feeding: It uses its very large, broad foot to smother its prey, which are small crabs and shrimps.

Fisheries/aquaculture: Commercial fishery exists for this species.

Other uses (if any): The whole shell of this species is rather polished and is one of the most attractive of the marine shells used in shell trade. It is seldom used for human consumption in Taiwan.

Melongenidae (whelks and false limpets): Shell of the members of this family is pear-shaped to fusiform and it is nodular to spiny on the shoulder. Aperture is anteriorly narrowing into an open siphonal canal. Columella is smooth. Operculum which is drop shaped, oval, and corneous is used as an anchor to drag and move around. Most species feed on clams and oysters.

Melongena corona (Gmelin, 1791)

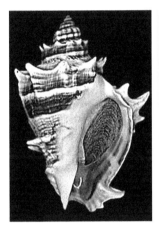

Common name(s): Florida crown conch.

Global distribution: Tropical to subtropical species; Florida peninsula, eastern Alabama, and West Indies south to South America.

Habitat: Intertidal and shallow subtidal (to 2.5 m); seagrass meadows, salt marsh, mangrove marsh, oyster reefs, and tidal mud flats; largest individuals (exceeding 120 mm in length) with oyster reefs and smaller animals in intertidal flats.

Description: Shell of this species is glossy and medium to large with a low spire, large body whorl, and prominent vertical-curved white spines on the shoulder of each whorl. Some shells are smoother than others. There are small spines on the largest whorl of the smoother forms. On the other hand, the most spiny forms have several rows of spines. Aperture of the shell is closed with an operculum. Exterior shell color is brownish-gray to purple with white to yellow-white spiral bands. Columella is thick and white. Maximum length of the shell is 20 cm.

Biology

Ecology: It is a cold-sensitive species. High winter mortalities of this species have been reported with near-freezing temperatures. Though this species is capable of tolerating both low and oceanic salinities, 20–30 ppt are required for favorable reproduction and other life activities. Early stages of this stages are less tolerant of salinity fluctuations and require a narrower range of 25–30 ppt. Developmental abnormalities and increased mortality have been reported in this species at 21.5 ppt and at lower salinities respectively.

Locomotion/behavior: It forages actively day and night. When it is exposed at low tide, it buries itself part-way in the wet sand until the next high tide. Newly hatched crown conchs are believed to bury themselves within the sediments of the upper intertidal and subsist for a time on detrital matter and small bivalves.

Food and feeding: It is an opportunistic predator/scavenger capable of feeding on a variety of live prey items as well as carrion and detrital material. Common food items include bivalves such as *Crassostrea virginica*, *Ensis minor*, *Tagelus divisus*, and scallops; larger gastropods such as *Busycon* spp.; and ascidians. Melongenid species like whelks (*Busycon* spp.) and tulips (*Fasciolaria* spp.), murex snails (Muricidae) and Florida horse conch (*P. gigantea*) are its potential competitors for prey.

Reproduction: In this species, sexes are separate and females are slightly larger than males. It has a direct-development. Like other gastropods of this family, it exhibits sexual reproduction through copulation and internal fertilization. Females lay more than 500 eggs in protective egg capsules which they attach to low intertidal substrata (rocks, shells of both dead and living organisms, mangrove roots, seagrass blades, wood, etc.) in ribbon-like rows of between 6 and 20 capsules. Direct development occurs and there is no free-swimming planktonic larvae in this species. Eggs hatch 20–28 days after deposition. Newly hatched animals measure less than 1 mm.

Predators: Large co-occurring carnivorous gastropods such as the Florida horse conch (*P. gigantea*) and the lace murex (*Chicoreus florifer*).

Impact on ecosystem: As a predatory species, it may potentially impact populations of economically important species such as eastern oysters (*Crassostrea virginica*) and hard clams (*Mercenaria mercenaria*).

Indicator of water quality: This species has been reported to serve as an indicator of poor environmental water quality if it occurred in large numbers in the absence of other species.

Melongena melongena (Linnaeus, 1758)

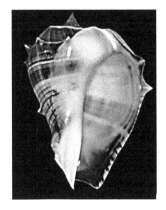

Common name(s): Caribbean crown conch or West Indian crown conch.

Global distribution: Western Central Atlantic.

Habitat: Coastal lagoons, fine-grained sandy bottoms, mangroves, river estuaries, and other low-salinity environments; depth range 0–44 m.

Description: Shell of this species is thick and pear shaped, with large body whorl. Spire is short and last few whorls bear single or double rows of spines. Suture is deeply channeled. Anterior canal is short and broad. Base of shell may have a row of smaller, blunt spines. Exterior shell color is purplish brown, light gray, or white, with bluish, brownish, or grayish bands. Maximum size of the shell is 18 cm.

Biology

Reproduction: It is a dioic species with internal fertilization. The sexual maturity size for the females is about 5-cm shell-length and 5.8-cm shell-length for the males. In the mature stage, the gonad presented a yellowish-orange color which coincided with a great abundance of spermatozoids in the follicular lumen in the males and of oocytes of great length in the females. During spawning, the gonad presented a white color. The spawning times of the females occur in the months of March, July, and December; while in the males, the spawning is observed during April and September and January. No synchronization is observed for this species as regards reproductive timing and gametogenic activity between males and females. This species reproduces through ovicapsules that harbor 300–400 embryos of which 95% are born as larvae with a planktonic life of short duration.

Fisheries/aquaculture: Commercial fishery exists for this species. This species has economic importance within the Golfete de Cuare (zone belonging to the Cuare Wildlife Reserve) where it is exposed to an intense fishing pressure.

Other uses (if any): It is an important food source for the people of Venezuela and West Indies. It is sold in cocktail with other molluscs. The clean and polished shell is used for ornamental purposes.

Volegalea cochlidium (Linnaeus, 1758) (=*Hemifusus pugilinus; Pugilina cochlidium*)

Common name(s): Spiral melongena.

Global distribution: Indo-West Pacific: from Sri Lanka to Papua New Guinea; Philippines; and Queensland.

Habitat: Shallow subtidal zones; brackish water; muddy waters near the shore; near bivalve beds.

Description: Shell of this species is large, solid, heavy, fusiform and longer than wide. Spire is conical and tall, with angulate shoulders and deeply incised sutures. Spire whorls are sculptured with many fine and rough spiral cords, and with broad axial folds. Body whorl is well inflated in its median part, with a distinctly concave shoulder slope. Periostracum is thick and finely wrinkled, becoming somewhat hairy on shoulder slope. Aperture is elongate-ovate but slightly wide. Anterior siphonal canal is broad and moderately long. Outer lip of aperture is angulate at shoulder; strongly sinuated posteriorly; broadly convex anteriorly and with obscure denticulation at inner edge. Inner lip is smooth and calloused. Outside of shell is beige to fawn or purplish brown under a dull, olive brown periostracum, occasionally with obscure darker brown spiral banding. Aperture is polished orange cream. Adult shell size varies between 6 and 15 cm.

Other uses (if any): Frequently appearing in local markets of Indonesia, Malaysia, and the Philippines. In the latter country, the meat is cooked and eaten plain or with a sauce or vegetables. The shell is utilized for making lime.

Mitridae (miter shells): Shell of the members of this family is fusiform-ovate, with a predominantly spiral sculpture. Aperture which is narrow is notched by a short siphonal canal. Outer lip is not lirate inside. Columella is with strong folds and is larger posteriorly. There is no operculum. These species are carnivorous and scavengers.

Mitra zonata (Marryat, 1819)

Common name(s): Zoned mitre.

Global distribution: Mediterranean Sea.

Habitat: Medium deep waters; expanses of sand and mud or debris ones; depth range 20–80 m.

Description: Shell of this species is elongated and slender. Shell opening is elongated and thin with outer lip that ends thinned with five folds and columellari plants. Operculum is absent. All the coils, but particularly the last, are usually marked by longitudinal streaks. Specimens of the Adriatic Sea have the apical part of the shell characteristically curved. Exterior shell is reddish brown with a dark brown spiral band. Interior is white, with a long and narrow mouth. Adult shell length varies from 6 to 10 cm.

Biology

Reproduction: This species has separate sexes. As in many other shellfish, the female produces berried egg capsules (containing eggs) sticking to a substrate. The larvae which appear from these eggs are planktonic for a brief time before returning on the bottom and rise to new mature individuals.

Conservation status: It is quite rare and an endangered marine species.

Other uses (if any): It is a potential food source for the people of Montinegro coast.

Muricidae (murex, rock snails): The muricids are found on rocky shores, or drills as they bore holes on barnacles and other shelled molluscs to feed on them. Shell of the species of this family is thick and strong and is variably shaped. It is with a raised spire and strong sculpture with axial varices, spines, tubercles, or blade-like processes. Periostracum is absent. Aperture is with a well-marked siphonal canal. Operculum is corneous. These animals generally lay masses of yellow egg capsules which turn purple when they hatch. A valuable reddish purple dye had been extracted from various species of murex since ancient time.

Acanthais triangularis (Blainville, 1832) (=*Mancinella triangularis*)

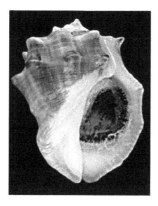

Common name(s): Not assigned.

Global distribution: Cape San Lucas: Baja California.

Habitat: Intertidally on rocks.

Description: Shell of this species is ovate and solid. Siphonal fasciole is seen. This light brown shell lacks the square dots, and the two rows of nodes at and below the shoulder are of equal size. Size of the shell: length 3 cm; diameter 2.6 cm.

Bolinus brandaris (Linnaeus, 1758) (=*Murex brandaris*)

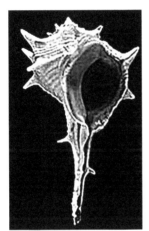

Common name(s): Purple dye murex or spiny dye-murex.

Global distribution: Central and western parts of the Mediterranean Sea, East Atlantic, Indian Ocean and South China Sea.

Habitat: Sandy or muddy beaches, between rocks and sea bed covered with various vegetation; isolated coral atoll beaches; depth range 2–100 m.

Description: Shell of this species has spiny extensions. Shell bulges at one end and the other end is long and straight. Outer shell surface is rough and covered with numerous irregular spiral lines. External shell color varies from yellow to brown. Due to various algae and other organisms that grow on shell, the surface color can vary significantly. Adult shell size of this species is between 6 and 10 cm.

Biology

Reproduction: This species has an annual reproductive cycle, long gonadal activity, and a short resting phase. Spawning occurs mainly between May and July, with a clear spawning peak from June to July. Gonad maturation and spawning seem to be synchronized with the seasonal variation in seawater temperature. The spread of imposex in this species has affected its populations (Rossato et al., 2014).

Secretion: A dye, Tyrian purple (also known as imperial purple, see purple) which is purple-red in color is made from this species.

Fisheries/aquaculture: In Portugal, this species is commercially exploited along the Algarve coast, mainly in the Ria Formosa lagoon, where this artisanal fishery constitutes a locally important activity.

Other uses (if any): It is a potential food source for the people of Mediterranean Sea and Northwest Africa. The meat of this species is quite tasty although it is little bit tougher, especially when compared with fish meat. It is a valuable gastropod in Portugal with great demand and high commercial value in the seafood market. This species was also used by the ancients to produce purple fabric dye (hence Purple dye murex name). Further, it is a bait for many species. As it holds very firmly on the hook, some fishermen just crack the shell and then insert the hook into the snail.

Chicoreus brevifrons (Lamarck, 1822)

Common name(s): West Indian murex.

Global distribution: Western Central Atlantic: from the Caribbean, the Gulf of Mexico, the Antilles to Brazil.

Habitat: Mud flats in protected bays and lagoons; near oyster banks, as well as mangrove areas.

Description: Shell of this species is elongate with a typical muricid outline. Three axial varices are present along its body whorl, and they are ornamented by expanded hollow spines. Flat spiral cords are seen in the interspaces of its surface. Anterior canal is well-developed. Maximum shell length of this species is up to 15 cm.

Other uses (if any): It is of economic importance in the Caribbean. This snail is locally collected for food and is consumed raw or boiled. Its high concentrations of polyunsaturated fatty acids make this species highly suitable for human consumption. The shell is often sold as a souvenir in local markets.

Chicoreus ramosus (Linnaeus, 1758)

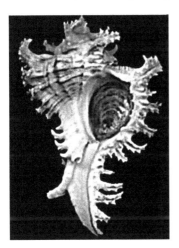

Common name(s): Ramose murex or branched murex.

Global distribution: Indo-Pacific.

Habitat: Sandy and rubble bottoms near coral reefs; depth range 0–50 m.

Description: This species is considered the largest murex in the Indo-Pacific. It has a large, solid, very rugged and heavy shell. It is with a globose outline, a short spire, a slightly inflated body whorl, and a moder-ately long siphonal canal. It has the most striking ornamentations, namely, the conspicuous, leaf-like, recurved hollow digitations. There are three spinose axial varices per whorl, with two elongated nodes between them.

Columella is concave and smooth. Operculum is horny and oval with growth ridges. Exterior shell is colored white to light brown with a white aperture. It is pink toward the inner edge, the outer lip and the columella. Animal has a very powerful muscular foot. Tentacles are short and bear eyes at their base on the outer side. Trunk is long. In the ocean, its shell is often covered in sea moss which camouflages it perfectly in its surroundings. Adult shell size varies from 20 to 33 cm.

Biology

Locomotion/behavior: It is found partially or completely buried in sand.

Food and feeding: It is a carnivore preying on bivalves and other gastropods. This species has been reported to be a pest on pearl oyster beds. Murex are necrophagous and are predators and piercers. Their radula, which has rows of three teeth, is well adapted to drill the shell of other molluscs by forming a circular hole. A gland which is located in the foot, injects an acidic enzyme secretion. This enzyme is believed to dissolve the calcium carbonate of the shell. Subsequently, the murex feeds on the flesh of its prey using its radula.

Reproduction: In this species, sexes are separate and the fertilization is internal. In laboratory conditions, the female lays clusters of egg capsules (about 400 egg capsules) on four frequency times. The capsule length was about 2 cm. The number of eggs per capsule varied from 178 to 214 egg/capsules. The estimated fecundity of this species was relatively high and reached about 79,000 eggs per animal throughout the spawning times. After 70 days of hatching from the egg capsules, the produced larvae reached about 1.1 cm in length and 0.5 cm in width. The average daily growth rate in length was 0.1 and 0.05 mm/day in width. During the first 2 days, the growth rate was about 0.11 mm/day inside the egg capsules (they feed and nurse on the capsule fluid and on the other small eggs), then it increased to 0.14 mm/day by the 70 days of emerging from the egg capsule (Mahmoud et al., 2013). The larvae are pelagic.

Conservation status: This species is listed as 'Endangered' on the Red List of threatened animals of Singapore.

Other uses (if any): It is an economically important species in the Indo-West Pacific, especially in India. All the gastropods in the muricid family produce purple (dye).

Chorus giganteus (Lesson, 1831)

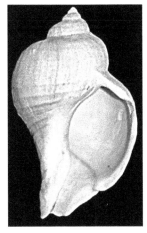 

Common name(s): Not assigned.

Global distribution: It is endemic to the coast of Chile.

Habitat: Rocks in temperate waters; depth range 8–30 m.

Description: Shell of this species is reddish and ventricose, and its girt is with elevated lines. Spire has only two of these lines on each whorl and has a bicarinated appearance. Aperture is wide and outer lip is sinuous. Its tooth is short, broad, and obtuse. Columella is not straight but curved operculum is horny.

Biology

Food and feeding: It is a benthic predator.

Reproduction: These snails develop through a yolk-supported (leci-thotrophic) type of development. The larvae of this species possess sufficient yolk reserves to support development through settlement and metamorphosis. Under laboratory conditions, the maximal duration of intracapsular development was 72 days at 15°C. Before metamorphosing, the hatched veligers were able to swim for about 3–5 days, depending on temperature.

Other uses (if any): This snail is of great commercial value in Chile, and it has been overexploited by local fishermen in much of its range.

Concholepas concholepas (Bruguière, 1789)

Common name(s): Chilean abalone.

Global distribution: Eastern Pacific and Western Atlantic: from Lobos de Afuera Island, Peru to Cape Horn, Chile.

Habitat: Rocky substrates, tidepools, and rock crevices from the lower intertidal to a depth of 40 m.

Description: This species has a thick, slightly oval, and white-brown to purple-gray shell. It is shaped almost like that of an abalone, with a very large aperture compared to other muricids. It has very few whorls. Outer surface of the shell shows strong lamellose ribs which are radial and circular-concentric. It has no operculum; instead, it depends largely on its strong foot to remain in place. Shell is made of calcite with an inner layer of aragonite. Maximum shell length is 15 cm.

Biology

Food and feeding: It is a benthic predator and its diet consists of mytilids (like *Semimytilus algosus* and *Perumytilus purpuratus*) and barnacles (like *Chthamalus scabrosus*).

Reproduction: It is dioecious which means the populations are divided between male and females, with no external evidence of sexual dimorphism. The fertilization is internal. Females lay egg capsules (243–256 egg capsules containing 668–14,250 eggs per capsule) during southern autumn months. These capsules are found attached to a substrate by a stalk. After about a month of development inside the egg capsules, small planktotrophic veliger larvae emerge. These larvae spend the following 3 months at the sea surface until they settle on rocky intertidal and shallow subtidal habitats. Greatest settlement density of this species has been linked to the coupling of upwelling between February and April, associated with the occurrence of the larvae (which hatch from benthic capsules as a late veliger stage)

(Moreno et al., 1998). The normal size at which the snail reaches sexual maturity is between 5 and 7 cm which is reached at about 4 years.

Fisheries/aquaculture: Commercial fishery exists for this species. It is a major product of the aquacultural industry in the Chilean coast.

Other uses (if any): It is a potential food source for the people of Chile and Peru. This species is used in Chilean cuisine and is commercially marketed worldwide as a delicacy. In Chilean cuisine, the meat of the foot of these snails is cooked and eaten with mayonnaise or as a chupe de locos soup in an earthenware bowl. The shells of this species are used as ashtrays in Chile. It has been reported that this species has the ability to provide valuable signals for long-term evolution of the sea surface temperatures in cold seas.

Dicathais orbita (Gmelin, 1791)

Common name(s): Cart-rut whelk, dog winkle or white rock shell.

Global distribution: Indo-West Pacific: Southern Australia and New Zealand.

Habitat: Rocky shores and coral reefs; low tide to immediate subtidal; depths greater than 10 m.

Description: This species has its characteristic feature, namely, deeply grooved shell sculpture (like the marks left by a cart in mud). Adult shell is with a short pointed end (spire). Eastern Australian shell has a strong spiral sculpture. There are seven to nine strong spiral ribs on body whorl, with intervening grooves of about the same width. Ribs and grooves are with secondary spiral riblets. Anterior fasciole is about same size as spiral ribs.

Columella is smooth and outer lip internally reflects external sculpture. Specimens are often partially encrusted by limey worm tubes or barnacles and this may serve as a camouflage from predators for these whelks. Exterior shell color is off-white or gray. Aperture is cream, often pale yellow on the edge. Shell length may be up to 10 cm. Reported age of this species is 20 years.

Biology

Food and feeding: These cart-rut whelks feed on other molluscs, barnacles, and tubeworms by drilling a hole in the shell, through which they extract the soft body.

Reproduction: This species follows an annual reproductive cycle with peaking in early summer (December) in South Australia. Females spawn about 40 egg capsules in a session and each capsule contains about 5000 eggs. Posthatching larval development proceeds through five stages over 41 days. The relatively high fecundity of adults and readiness to spawn in captivity along with the suitability of post hatching larvae for growth under laboratory conditions, makes this species promising for aquaculture.

Fisheries/aquaculture: It is a potential species for aquaculture. Successful aquaculture of this species would provide a sustainable supply of individuals for ongoing development of pharmaceutical leads, as well as for seafood markets.

Other uses (if any): This species is recreationally harvested for food. This species has been reported to produce potent bioactive compounds of interest for development as pharmaceutical leads (Noble, 2014).

Hexaplex brassica (Lamarck, 1822) (=*Chicoreus brassica*)

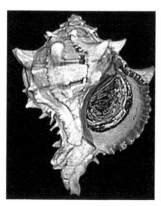

Common name(s): Cabbage murex or rose cabbage murex.

Global distribution: Indigenous to Costa Rica; Southeast Pacific: Peru.

Habitat: Medium sand, sandy silt, and silty clay substrata; from shallow (24 m) to moderately deep waters (83 m).

Description: The species was collected. Some of the shells collected reached a length of 18 cm.

Other uses (if any): It is primarily used as food and sometimes as fish bait. This beautiful shell, commonly known as "caracol chino," has always attracted the attention of collectors.

Hexaplex duplex (Röding, 1798)

Common name(s): Rock murex shell.

Global distribution: The Red Sea Atlantic Ocean: off the Cape Verdes and the Canary Islands; and from Senegal to Angola.

Habitat: Intertidal and subtidal zones in sandy areas.

Description: Shell of this species is medium and large with six to eight varices. It is a spinous species. Spines are somewhat frondose and those on the shoulder of the whorls are usually larger and curved. There are no interstitial ribs. Shell mouth is oval and operculum is round and horny. Color of the shell is light yellowish brown, usually more or less pink banded. Aperture is pink, with three or four darker bands. Length of the shell varies between 3 and 23 cm.

Other uses (if any): It is a potential food source in Senegal.

Hexaplex erythrostomus (Swainson, 1831) (*=Chicoreus erythrostomus*)

Common name(s): Pink murex or pink-mouthed murex.

Global distribution: Central America: Mexico to Colombia and Bermuda.

Habitat: Intertidal and subtidal zones in sandy and rocky areas.

Description: Shell of this species is globose, strong, and ornamented with blunt spines. Aperture is circular. Siphonal canal is narrow, short, and curved with decorations on the outside. Color of the outer part of the shell is white or shades of gray. Interior of the shell and the aperture are deep and glossy. Maximum size of the shell is 9 cm.

Hexaplex fulvescens (Sowerby, 1834) (*=Murex fulvescens*)

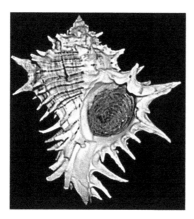 

Common name(s): Giant Eastern murex or giant Atlantic murex or tawny murex.

Global distribution: Western Central Atlantic: from North Carolina to Cape Canaveral, Florida; Gulf of Mexico; from Florida to Texas.

Habitat: Deeper waters; rocky intertidal; gulf; shallow inshore waters; depth range 0–80 m.

Description: This is the largest muricid shell of the Western Atlantic. Shell is massive, spinose, and spiral in shape. It has several straight or bifurcate spines arranged in 6–10 radial rows (varices) with spiraling ridges. Spines may be worn or broken. There are 6–8 indistinct whorls with spiraling ridges. Shell aperture is oval with crenulate edges. Inside aperture is white with some blotching. Siphonal canal is short. Spire is short. Shell surface may be whitish, grayish, or pale brown. Adult shell length of this species may vary from 6 to 22 cm.

Biology

Food and feeding: It is an active carnivore which preys on other molluscs such as mussels, oysters, and clams.

Reproduction: These animals lay their eggs in capsules which are attached under rocks.

Association: *Macrocypraea zebra* and *Hexaplex fulvescens* have been reported to coexist in the same habitat.

Hexaplex nigritus (Philippi, 1845) (=*Muricanthus nigritus*)

Common name(s): Black murex snail, nigrite murex, Northern radix or black-and-white murex.

Global distribution: Gulf of California (Sea of Cortez); Western Mexico.

Habitat: Intertidal and subtidal waters, with sand and gravel substrate; depth range 0–60 m.

Description: Shells of this species are large; and black and white with black or dark brown stripes. Shell surface has short spikes around the body whorl and spire. Aperture is porcelaneous white. Shell length may vary from 7 to 20 cm.

Biology

Food and feeding: It is a carnivore feeding primarily on clams.

Reproduction: This species is capable of reproducing a minimum of five times in its lifetime.

Fisheries/aquaculture: This species is heavily exploited in the Gulf of California and Mexico for the Asian market.

Other uses (if any): It is a potential food species in Gulf of California and Mexico.

Hexaplex princeps (Broderip, 1833) (=*Muricanthus princeps*)

Common name(s): Prince murex.

Global distribution: Southern part of Gulf of California to Peru.

Habitat: Intertidally on rocks at extreme low tide and offshore in shallow waters.

Description: Shell is somewhat biconic in shape with five to eight varices. Shell surface is whitish with ribs and spines which are tinged in brown. Maximum length of the shell is 12.5 cm.

Hexaplex radix (Gmelin, 1791) (=*Chicoreus radix*; *Muricanthus radix*)

Common name(s): Radix murex or root murex.

Global distribution: Western Pacific: Central America (Baja California and from Mexico to Peru).

Habitat: Shallow waters; intertidal rocks.

Description: This species has large, massive, and heavy shells which are globose or pear-shaped and very spiny, with a white surface and blackish-brown foliations and spiral elements. Body whorls have 6–7 varices. Aperture is large, broad, ovate, and porcelaneous white. Outer edges are strongly dentate. Siphonal canal is moderately long. Operculum is dark brown. Shells of this species may reach a size of 5–16 cm.

Biology

Food and feeding: It is a carnivore feeding primarily on clams.

Hexaplex regius (Swainson, 1821) (=*Chicoreus regius*)

Common name(s): Regal murex.

Global distribution: From Western Mexico to Peru.

Habitat: Epifaunal species.

Description: In this species, overall sculpture of the shell is prickly and frond like. Spines are numerous and scaly. In between spine rows, shell is lirate. External colors are white with occasional brownish spiral stripes, which on some specimens may be absent. Color of the aperture, or opening of the shell, is pink at the edges and the glossy shield over the columella. Siphonal canal is a mixture of pink and light brown. Size of the adult shell varies between 6 to 18 cm.

Biology

Food and feeding: It is a carnivore that preys on other invertebrates.

Hexaplex trunculus (Linnaeus, 1758) (=*Phyllonotus trunculatus*)

Common name(s): Banded dye-murex.

Global distribution: Mediterranean Sea; Atlantic coasts of Europe and Africa; Spain, Portugal, Morocco, the Canary Islands, Azores.

Habitat: Shallow, sublittoral waters.

Description: *Hexaplex trunculus* has a broadly conical shell about 4 to 10 cm long. It has a rather high spire with seven angulated whorls. The shell is variable in sculpture and coloring with dark banding, in four varieties. The ribs sometimes develop thickenings or spines and give the shell a rough appearance.

Biology

Reproduction: The spread of imposex in this species has affected its populations (Rossato et al., 2014).

Other uses (if any): This snail is still used as a valuable food source in Portugal, Italy and Spain. This species is also historically important because its hypobranchial gland secretes a mucus that most ancient people of the Mediterranean and classical Greeks used as a distinctive purple-blue indigo dye. This dye's main chemical ingredients is dibromo-indigotin, and it is left in the sun for a few minutes, its color turns to a blue indigo (Wikipedia).

Indothais lacera (Born, 1778) (=*Thais carinifera*)

Common name(s): Carinate rock shell.

Global distribution: Indo-West Pacific: from India to Melanesia; Taiwan Province of China; Indonesia and New Caledonia.

Habitat: Intertidal; subtidal; mangroves; and mudflats.

Description: Shell of this species is medium sized and ovate. Spire is acute. Whorls are angulate. Body whorl is strongly shouldered. Aperture is wide and ovate. Outer lip is prominently crenulated. Lirations are seen within aperture. Columella is smooth and calloused with mild striations. Double rows of tubercles are seen on the upper part of body whorl. Exterior shell color is grayish brown and aperture is light brown. Adult shell size varies between 3 and 5.7 cm.

Other uses (if any): This species is collected for food in various parts of the Indo-West Pacific, particularly in Indonesia, Indo-China, and India.

Murex tribulus (Linnaeus, 1758)

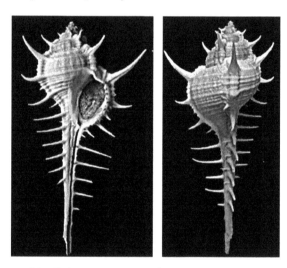

Common name(s): Caltrop murex or thorny murex.

Global distribution: From Central Indian Ocean to Western Pacific Ocean.

Habitat: Shallow water; depth range 1–15 m.

Description: Shell of this species has a very long siphonal canal and numerous very long, fragile, and acute spines which are providing protection to this species against predators. Shell length of this species varies from 6.5 to 16 cm.

Biology

Food and feeding: It is a carnivore feeding on other molluscs.

Fisheries/aquaculture: Commercial fishery exists for this species.

Other uses (if any): This species is used as food locally in the areas of its occurrence and its empty shells are sold for collections.

Neothais harpa (Conrad, 1837)

Common name(s): Harp drupe.

Global distribution: Endemic to Hawaii.

Habitat: Rocks in the intertidal zone.

Description: Maximum size of the shell of this species is 4 cm.

Biology

Locomotion/behavior: These animals are active at night for feeding.

Food and feeding: It is a carnivore feeding on nerites and periwinkles.

Other uses (if any): It is a potential food source for the people of Hawaii.

Phyllonotus margaritensis (Abbott, 1958)

Common name(s): Margarita murex.

Global distribution: Western Atlantic: North Carolina to Florida and Caribbean sea to Brazil; Colombia, Venezuela, and Trinidad.

Habitat: Seagrass beds; rocky areas and reef areas; depth range 6–300 m.

Description: Shell of this species is large and strong with four or five lap varices which are simple, nodular and obtuse. Siphonal canal is wide and curved. Mouth of the shell has colors that may vary from yellow, orange to red. Inner lip is modulated at the base. In this species, females are generally larger than males with greater variation in shell shape of these sexes. Shell size varies from 7 to 11 cm.

Biology

Food and feeding: This muricid is an active predator of bivalves such as sand-dwelling clams.

Other uses (if any): It is a potential food source in Venezuela.

Phyllonotus pomum (Gmelin, 1791) (=*Murex pomum*)

Common name(s): Apple murex.

Global distribution: Caribbean Sea; the Gulf of Mexico and the Lesser Antilles; in the Atlantic Ocean between North Carolina and Northern Brazil.

Habitat: Offshore, in coral reefs and mangroves, particularly on coral, rubble, rock, sand, seagrass, and shell habitats.

Description: Shell of this species is fusiformly oblong, thick, solid and is very rough throughout. It is transversely conspicuously ridged and tuberculated between the varices. Columella and interior of the aperture are ochraceous yellow. Columellar lip is slightly wrinkled; its edge is erected and vividly stained with very black brown. Outer lip is strongly toothed and is ornamented with three black-brown spots. Siphonal canal is rather short, compressed and recurved. Exterior shell is fulvous or reddish brown. Size of an adult shell varies between 4.4 and 13.3 cm.

Biology

Locomotion/behavior: During low tides, these animals bury themselves in the sand.

Food and feeding: These animals are carnivores, and they make their living by drilling holes in oysters and eating the contents.

Other uses (if any): The local people of in Venezuela harvest apple murexes by diving. They eat the shells' contents and sell the more colorful and more ornamented shells to the tourists.

Plicopurpura columellaris (Lamarck, 1816) (=*Purpura columellaris*)

Common name(s): Not assigned.

Global distribution: Southern part of Gulf of California to Chile.

Habitat: Hard substrates in the intertidal areas.

Description: Shell of this species is grayish-brown, thick, and heavy with an orange-brown aperture. Outer lip is studded with teeth. Columella has a couple of raised nodes near its center. Maximum size of the shell: 8 cm length and 6 cm diameter.

Other uses (if any): Though it is an edible species in its areas of its occurrence, caution should be exercised while collecting these specimens in their habitats. It has been reported that eating of this species collected from red tide-infected areas has caused death of Salvadoran girls. Mouse bioassay results indicated that such specimens contained 7 million mouse units/100 g of saxitoxins (Barraza, 2009).

Plicopurpura pansa (Gould, 1853)

Common name(s): Purple snail.

Global distribution: Pacific coast of Mexico.

Habitat: Exposed rocky shores and crevices between high and low tide lines.

Description: This species is a sympatric (two related species or populations existing in the same geographic area) species of *Plicopurpura columellaris*. It is also closely related to other related genera such as Drupa and Thais. This species has a dark gray knobbly shell with pinkish-brown aperture. It can be distinguished from *P. columellaris* by its lack of strong teeth inside the aperture. Maximum shell size is 4.1 cm.

Biology

Food and feeding: These snails are predatory feeding on other molluscs.

Reproduction: In this species, sexes are separate and the female–male ratio has been estimated as 1:0.80. Recruitment in this species has been observed during September and December with peak mating in March. Female lays egg capsules, which hatch into planktotrophic (plankton feeding) which can delay settlement for about 6 months. Lifespan of this species is 11 years.

Imposex: This species has been reported to show abnormalities (imposex; diflaia) in the reproductive system of females. Imposex is a disorder in marine snails caused by the toxic effects of certain marine pollutants. Such pollutants may cause female snails to develop male sex organs such as a penis and a vas deferens. This imposex phenomenon is largely due to the toxic effects of some pollutants that cause damage to the endocrine system. In this species, this phenomenon was found to be due to tributylin chloride, an organotin compound used in the paints of ships, port facilities, and aquaculture artifacts. A female which was exposed to the said toxic compound was found to develop two penises in the laboratory (Domínguez-Ojeda et al., 2013).

Fisheries/aquaculture: Although this species was once extremely abundant, it is becoming less common due to exploitation for dye.

Other uses (if any): This species is edible in the areas of its occurrence. The purple snail resource is considered culturally and economically important because of its dye which is used to color traditional clothing that will not fade or bleach. The strong garlic odor given off by clothes dyed in this method is considered proof of authenticity. To obtain the Tyrian purple dye, the snail is collected at low tide and is poked with a pin until it releases thick liquid. Afterwards, it is released back into the wild, relatively unharmed.

Purpura bufo (Lamarck, 1822) (=*Thais bufo*)

Common name(s): Toad purple.

Global distribution: Indian Ocean along the KwaZulu-Natal coast and Madagascar.

Habitat: Wave exposed rocky shores.

Description: Shell of this species is large sized, solid, heavy, and globose. Spire is short with three or four rounded whorls. Body whorl is large, consisting of 3–4 prominent spiral cords with short blunt spines on the first and last spiral cords, and tubercles on one or two central cords. There are

6–7-m fine spiral cords above the shoulder rib on the subsutural ramp, and three cords between each of the spiral ribs. Aperture is large and ovate and its outer lip is turned outward with crenulated margin. Columella is smooth and calloused, extending posteriorly above shoulder; anal canal is forming a distinct notch in the outer lip; and siphonal canal is short and broadly open. Exterior shell is dark brown with chocolate brown blunt spines and off-white to light orange patches in their interspaces. Aperture is creamy white, and it turns yellow near the margin; outer lip margin is chocolate brown; and columella is creamy yellow. Operculum is 1.5–2.0 cm long, dark brown, and its outer margin is thickened on one side, with a lateral nucleus.

Purpura persica (Linnaeus, 1758) (=*Thais rudolphi*)

Common name(s): Persian purpura.

Global distribution: Indo-Pacific: Southeast and Northeast Asia.

Habitat: Underside of corals and stones; or on rocks exposed to surf action.

Description: Shell of this species is spindle shaped with a broad middle and a short spire. Shell surface is sculptured with spiraled, beaded cords. Last whorl is larger than all the others together. Aperture is very much dilated, oval and is terminated anteriorly by an oblique notch. Columella is flattened, finishing in a point anteriorly; and right lip is sharp-edged, thickened, and furrowed strongly and armed anteriorly with a conical point. Interior of the aperture is white, and the inner edge of the outer lip is banded dark brown with yellow streaks in between. Shell surface shows a chocolate-brown color. Shell size varies between 6 and 11 cm.

Other uses (if any): This species is collected locally for food in Southeast Asia.

Rapana bezoar (Linnaeus, 1767)

Common name(s): Bezoar rapa whelk, small felsic rock snail or little wrinkle whelk.

Global distribution: Indo-West Pacific.

Habitat: Shallow coral reefs; rock bottom; low tide zone.

Description: Shell of this species is white with thick loose ribs. There is a slight shoulder with nodules or short spines.

Other uses (if any): This species is collected for food in China.

Rapana venosa (Valenciennes, 1846)

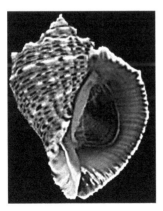

Common name(s): Veined rapana whelk.

Global distribution: Sea of Japan, Yellow Sea and East China Sea; Black Sea; introduced in western Atlantic, and Western France.

Habitat: Subtidal; sandy bottoms, burrows; depth range 12–14 m.

Description: It is a purple-producing mollusk. Shell of this species is globose, with a very large inflated body whorl, a deep umbilicus and low conical spire. Spire whorls have a strong shoulder. Irregular spiral sculpture is seen. Columella is broad, smooth, and slightly concave. Aperture is large, ovate, and slightly expanded, with thin outer lip. Small elongate teeth are present along the edge of the outer lip. Siphonal canal is broad, widely open and is bearing a series of scales. Exterior color of the shell is gray to brownish with irregular darker blotches or flames, which tend to make an interrupted pattern on the cords. Aperture is orange inside. Maximum size of the shell is 18 cm.

Biology

Ecology: *Rapana venosa* is an extremely versatile species tolerating low salinities, water pollution, and oxygen deficient waters. Larvae of this species have been found to tolerate to salinities as low as 15 ppt with minimal mortality.

Locomotion/behavior: It may migrate to warmer, deeper waters in winter, thereby evading cool surface waters. It is a nocturnal species remaining burrowed most of the day, avoiding settlement by epifaunal biota.

Food and feeding: In planktonotrophic veligers of this species, the most important food sources are flagellates, diatoms, organic, and inorganic particles. Young are predators and consume barnacles, mussels, oyster spat, and oysters. Adults are carnivores feeding mainly on bivalves. They can open a clam by smothering the shell and introducing the proboscis between the gaping valve, without any drilling. They may also be scavengers on carrion. This species has become a major pest of oyster beds in the Black Sea.

Reproduction: It is dioecious with separate sexes. Mating occurs for an extended period, mainly during the winter and spring. It reproduces by internal fertilization, after which it lays clusters of egg cases which resemble small mats of white-to-yellow shag carpet, mainly during spring and summer. When laid, the egg cases are white, turning sequentially darker (through lemon to yellow). One adult female can lay multiple egg cases throughout the season, and each cluster may contain 50–500 egg cases, and each egg case may contain 200–1000 eggs. The pelagic veliger larvae then hatch and are persisting in the water column for 14–80 days and are feeding mainly on plankton. These larvae finally settle on the ocean floor where they develop into hard-shelled snails. Pelagic larvae have a long planktonic phase which may last to a maximum of 80 days (Saglam et al., 2009). Veligers larvae settle successfully on a wide range of attached macrofauna including bryozoans and barnacles. They grow quickly on mixed algal diets, reaching shell lengths in excess of 0.5 mm at 21 days. Average lifespan of this species is 10 years. Growth is rapid over the first year of life, and reproduction occurs from the second year onwards. The imposex (the development of a male penis by adult females) phenomenon has been observed in this species.

Association: Bryozoans and barnacles are the most frequent epibionts colonizing the shell of this species. Small sea anemones, chitons, and polychaete tubes are also found occasionally in this species.

Nanobioceramic production: This species could be used as an alternative source for nanobioceramic production. Its shells have been found to be very good candidate materials to produce fine powders for various tissue engineering applications (Ozyegin et al., 2012).

Other uses (if any): In Japan, Bulgaria, Rumania, and Turkey, these animals are sold as seafood in fish markets as whole, cut, fresh, frozen, canned, cured, processed, or smoked products. This species has been reported to manifest the most bioaccumulation capacity of Cd (cadmium) and Ni (nickel), thereby serving as hopeful bioindicators for monitoring metal pollution in waters.

Siratus alabaster (Reeve, 1845)

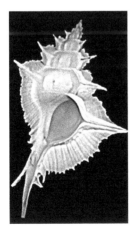 

Common name(s): Alabaster murex.

Global distribution: From Japan to the Philippines.

Habitat: Sand and gravel bottoms; depth range 50–200 m.

Description: Shell of this species is distinctively large, fusiform, and often crisply white or ivory. It possesses paper thin webbed wings. Operculum is complete. Adult shell size may vary from 10 to 22 cm.

Other uses (if any): This species is a potential food source for the people of Philippines and Japan. Shells of this species are beautiful to behold.

Stramonita biserialis (Blainville, 1832) (=*Thais biserialis*)

Common name(s): Rock Thais.

Global distribution: From the West Coast of Mexico to Chile.

Habitat: Intertidal rocks.

Description: It is a purple-producing mollusk. Exterior color pattern is varying from plain to mottle with light and dark brown. Maximum length and diameter of the shell is 7.5 and 5.4 cm, respectively.

Biology

Reproduction: Imposex has been reported in this species collected from Tuticorin, South India. This is possibly due to the contamination of tributyltin (TBT), which could have leached out from antifouling paints of the ships using Tuticorin port (Ramasamy and Murugan, 2002).

Stramonita chocolata (Duclos, 1832) (=*Thaisella chocolata*)

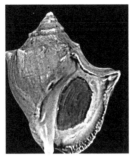 

Common name(s): Chocolate rock shell.

Global distribution: From Ecuador southward to Peru.

Habitat: Marine biome rock sea.

Description: It is a purple-producing mollusk. Columella is tinged with orange. Exterior shell color is uniform brown. Interior is bluish or yellowish.

Biology

Reproduction: As with other muricacean snails, this species also engages in communal egg laying. Females deposit egg capsules in clusters on subtidal rocks. Each cluster of capsules contains a maximum of 150 pedunculate, ampulliform egg capsules, with each capsule containing an average number of 2600 small eggs. Free-swimming veliger larvae are released from these capsules after 49 days incubation. These veliger larvae get metamorphosed into young after finding suitable substrata (Romero et al., 2004).

Other uses (if any): It is a potential food source in Chile.

Stramonita haemastoma (Linnaeus, 1767) (=*Thais haemastoma*)

Common name(s): Red-mouthed rock shell, Florida dog winkle, Florida rocksnail, southern oyster drill, or blood mouth.

Global distribution: Southeast Pacific, Atlantic Ocean, and the Mediterranean.

Habitat: Intertidal to subtidal; muddy and rocky bottoms; up to more than 500 m deep.

Description: Shell of this species is solid and egg shaped with fine spirals and six or seven rows of knuckles. Last whorls are sometimes with nodules on shoulder. Shell opening which is of orange or red color is oval with short siphonal channel and external lip is narrow. Exterior shell color is light gray, yellowish, or tan; usually mottled or checkered with darker brown, grayish, or orange marks. Adult shell size for this species varies between 2.2 and 12 cm.

Biology

Locomotion/behavior: The snails move on rock by continuous, direct, ditaxic, alternate undulations of the foot sole. But on submerged sand, they use slower arrhythmic discontinuous contractions of the foot sole. Most snails burrow into the sand when the rocks become exposed during low tides.

Food and feeding: It is a carnivorous predator and piercer. It feeds mainly on bivalves and other sessile shellfish. It pierces the shells of its victims with its radula, enzymatic secretions, and acid.

Reproduction: The eggs laid by the females of this species are in clusters of hexagonal capsules. The spread of imposex in this species has affected its populations (Rossato et al., 2014).

Bioindicator of organotin contamination: This species has been reported to be a reliable bioindicator of TBT and TPhT contamination in coastal waters (Limaverdea et al., 2007).

Fisheries/aquaculture: Commercial fishery exists for this species.

Other uses (if any): These animals are collected for food in Spain and Mediterranean; and Northwest Africa, Southeast USA, and Brazil. This species is a principal source of Tyrian purple, a highly prized dye which was used in classical times for the clothing of royalty,

Thalessa aculeata (Deshayes, 1844) (=*Thais aculeata*)

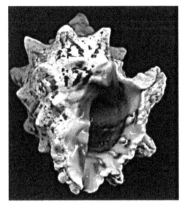

Common name(s): Aculeate rock shell.

Global distribution: Indo-Pacific: East Africa, Northeast and Southeast Asia.

Habitat: Reef-associated and on rocks; common among rock oysters; shallow subtidal.

Description: Maximum shell length of this species is 6 cm.

Other uses (if any): This species is frequently collected for food by coastal inhabitants.

Trophon geversianus (Pallas, 1774)

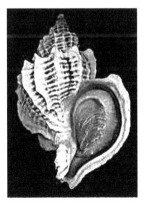

Common name(s): Gevers's trophon.

Global distribution: Southeast Pacific, Southwest Atlantic and Antarctic Indian Ocean. Temperate to polar.

Habitat: On rocky substrates, rock crevices and mussel beds; forests of *M. pyrifera*; sessile; more than 500 m deep.

Description: Shell of this species is large and fusiform with subquadrate profile. Protoconch is of two whorls, smooth, cylindrical, slightly globose and is slightly asymmetrical. Teleoconch is of four shouldered whorls, and its spire is less than one-third of the total shell height. Spire angle is about 50°. Suture is impressed. External surface is covered by concentric, irregular, growth lines. Inner surface attachment area is with three horseshoe-shaped scars. Aperture is ovoid; interior is glossy pinkish; anterior siphonal canal is moderately long (half the height of aperture); and umbilicus is closed or deep. Outer lip is rounded, with reflected edges; and inner lip is curved and adpressed. Operculum is oval and brownish, with terminal nucleus. Exterior shell coloration is varying from creamy white to dark brown. Adult shell size varies between 3 and 11 cm.

Biology

Food and feeding: These animals consume more frequently the available size range of mussels at mid intertidal levels. There is a positive correlation between the size of predator and prey consumption in this species. The drilling rate also decreases with increasing time of aerial exposure.

Reproduction: Sex ratio in the population differed from 1:1 (female biased). The mean shell length was 22 mm for males and 24 mm for females, although the females presented significantly larger maximum sizes. There is no external sexual dimorphism in this species. On the other hand, the female snails differ internally by the presence of the albumin and capsule gland and by gonad color. This species shows a marked reproductive seasonality. Oviposition starts in May and concludes in November, when hatching of crawling embryos is observed up to January. This seasonality coincides with changes in water surface temperature, ambient temperature, and photoperiod. In the aquarium, each female was found to lay an average of 12 egg capsules per oviposition event, and it needed about 25 h to complete attachment of a single egg capsule (Cumplido et al., 2010). Each egg capsule contains a large number of eggs where only a few develop and others serve as nutritional eggs. These are attached to hard substrates by a stalk.

Vasula speciosa (Valenciennes, 1832) (=*Mancinella speciosa*)

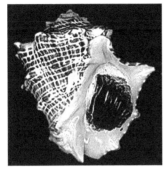 

Common name(s): Not assigned.

Global distribution: Magdalena Bay, Baja California through the Gulf and south to Peru.

Habitat: On rocks between tides.

Description: The white shell of this species is easily recognized by its spiral bands of brown squares. It has a yellowish aperture. Shell size: length 3.6 cm; diameter 3.0 cm.

Nassariidae (dog whelks): Shell of the members of this family is ovately rounded with a long and conical spire. Body whorl is anteriorly bordered by a strong spiral groove. Aperture is rather small and rounded, with a short, recurved siphonal canal. Operculum is corneous and is smaller than the aperture. These nassariids are mostly scavengers, and their siphons allow them to sniff out dead animals. Some dog whelks have small sea anemones growing on their shells.

Buccinanops cochlidium (Dillwyn, 1817)

Common name(s): Gradated bullia.

Global distribution: From Central Brazil to Argentina.

Habitat: Sandy infralittoral.

Description: Shell of this species is ovate-conical, elongated, smooth, and shining. Spire is elongated and is composed of eight turreted whorls which are slightly angular at their upper part and very slightly convex. First whorls are plaited longitudinally. Aperture is ovate, whitish, and strongly emarginated at its base. Outer lip is thin. Columella is smooth and yellowish. Exterior shell is of a reddish yellow color and is scattered over with longitudinal flames of a brown red. A transverse band of the same color surrounds the base of the shell. Size of the adult shell varies between 3 and 11 cm.

Biology

Locomotion/behavior: It remains completely buried.

Food and feeding: It is a carnivorous species, both as a scavenger and as a predator, which feeds mainly on dead crabs and living bivalves. Hatchlings and juveniles feed exclusively on carrion.

Reproduction: Mating in this species is mainly during autumn and winter (March to October). Females carrying egg capsules are found between July and October when water temperature is about 10°C, while development of embryos continues until February (the hatching peak) when water temperature is around 18°C. All egg capsule masses are found attached to the shells of females. The spawn consists of more than 80 egg capsules, which are attached to the apertural callus of the shell. Each egg capsule contains about 3000 eggs. One to 20 embryos complete their development within each egg capsule by ingesting the undeveloped (nurse eggs). During the "veliger" stage, the embryo consumes these nurse eggs and forms a large rounded embryo. After shell development, the embryos hatch as crawling juveniles. Under laboratory conditions complete development is in 4 months.

Association: This species is often found with the sea anemone, *Phlyctenanthus australis*.

Fisheries/aquaculture: This species is fished in the Argentine coast and is traded.

Buccinanops deformis (King, 1832) (=*Buccinanops globulosus*)

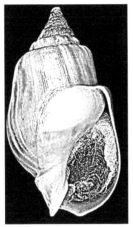

Common name(s): Collared buccinum.

Global distribution: From Uruguay to Argentina; South Pacific Ocean.

Habitat: Soft sediment in intertidal habitats.

Description: Shell of this species is ovate, smooth, and ventricose. Spire is formed of six slightly convex whorls. Body whorl is very large and slightly canaliculated. Base of each whorl of the spire is bordered by a coloration of deep violet and a yellowish ash, with a small white. There exists also at the base of the shell, a large band of a grayish white color. Some specimens may possess longitudinal whitish lines. Aperture is large and ovular. Columella is strongly arched, and upon all its length is seen a callosity of a yellowish color. Outer lip is of a reddish brown internally. Size of the adult shell varies between 2.3 and 7.0 cm.

Biology

Reproduction: Sexual dimorphism in shell morphology (sex-related shape differences in populations) is seen in this species. Further, a relationship between imposex development and alterations on the reproductive output of the females of this species living in adverse environmental conditions (polluted area) has been obtained. Accordingly, higher imposex incidence has been reported in the nongravid females (not carrying egg capsules) of this species.

Tributyltin contamination: Shell-shape variations of this species in TBT contaminated high maritime traffic and low maritime traffic harbor zones have been studied by Primost et al. (2016). Animals from areas of high maritime traffic showed a rounded shell with a shorter spire, and a smaller relative size of the shell aperture, whereas the opposite shape (fusiform shape, elongated-spired shell and bigger relative size of the shell aperture) occurred in gastropod shells from areas of low maritime traffic. Shell variation registered in this study could be useful to detect TBT pollution in populations of *B. globulosus* and another neogastropod species.

Fisheries/aquaculture: Animals of this species are captured and traded in the Argentine coast.

Tritia mutabilis (Linnaeus, 1758) (=*Nassarius mutabilis*)

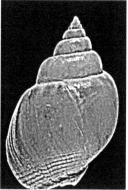

Common name(s): Mutable nassa.

Global distribution: Mediterranean Sea; Black Sea; Atlantic Ocean off Mauritania; and West Africa.

Habitat: From the tidal range up to about 20 m deep.

Description: Shell of this species is smooth, ovate, conical, and slightly ventricose. Spire is composed of seven whorls, which are rounded and swollen at the upper part. Lowest whorl is larger than all the others united. Three upper whorls are finely plaited. Body whorl has a few fine, transverse striae near the base. Aperture is white and ovate; strongly emarginated; and oblique at the base. Depth of the cavity is chestnut colored. Thin outer lip is white and very finely striated internally. Inner lip is thin, white, and shining and partially covers the body of the shell. Columella is arcuated and terminates at the base by a sharp and slightly projecting keel. Exterior of the shell is red or fawn colored and is ornamented with an articulated band of white and violet upon the upper edge of the whorls, with waved longitudinal yellow or red spots. Size of an adult shell varies between 1.4 and 3.0 cm.

Biology

Locomotion/behavior: These animals live somewhat buried beneath the stones, in the sand or mud and allow only their siphon to pass beyond which they create a stream of water to breathe.

Food and feeding: It feeds on dead animals (necrophagous), fish, and crustaceans. It buries itself slightly in the sediment and detects corpses up to about 30 m. A long sensory siphon and its odor sensor (osphradium) serve to search the sediment. The proboscis, equipped with the pink radula, is then inserted into the decaying flesh of the carcass.

Reproduction: This species is gonochoric, that is, the sexes are separate. There is copulation and fertilization is internal. Breeding normally occurs in spring and summer. Small transparent egg capsules (containing eggs) which are flattened and grouped between 5 and 10 are deposited on algae, seagrasses or stones. The hatched out larvae have a long pelagic life.

Predators: The main predators of this species are certain starfish, to which it may, however, escape by successive rocking of the shell and the foot.

Other uses (if any): This species serves as a potential food source in Spain and Italy, Mediterranean; Black Sea; West Africa; and Canaries.

Tritia reticulata (Linnaeus, 1758) (=*Hinia reticulata*)

Common name(s): Netted dog whelk.

Global distribution: Northeast Atlantic; Europe; Atlantic Ocean off the Azores; Canary Islands, Cape Verde Islands and Morocco.

Habitat: Intertidal zone; sandy bottoms, sandy mud, gravel, under stones, and rocks; prefers the sheltered areas; depth range 0–20 m.

Description: Shell of this species is egg shaped, elongated, rounded, and obtuse at its lower extremity, and pointed at the upper extremity. It is moderately thick. Conical spire is composed of 8–9 whorls, which are almost flat, or slightly swollen, but distant from each other. Their surface is deeply checkered by longitudinal folds, crossed by numerous striae. Aperture is moderate, white and ovate. Outer lip is thick and is ornamented within with 7–8 striae, of which those of the middle are generally the largest. Columella is slightly arcuated and is covered with a thin, brilliant plate. Exterior shell color is of a yellowish white, reddish or chestnut color, with a blackish blue band. passing beneath the suture. Length of the adult shell varies between 2 and 3.5 cm.

Biology

Locomotion/behavior: This species burrows itself in the sediment.

Food and feeding: It is a macrophagous and necrophagous species is an excellent carnivorous scavenger also. It can move very quickly and clean a dead or injured prey (fish, crustacean, other mollusk, or any animal). It detects the smell of corpses of animals from far away by means of its siphon barded of chemoreceptors.

Reproduction: It is a gonochoric species, that is, sexes are separate and fertilization is internal. After fertilization, the female deposits nearly 100 egg capsules each containing hundreds of eggs on a solid support. Each egg will give a pelagic larva.

Predators: The main predator of this species is the starfish, from which it can however escape with its sense of smell, and by a series of flip-flops and pirouettes.

Other uses (if any): It is a potential food source for the people of Mediterranean.

Olividae (olive shells): The name "Olive Snails" is largely due to their olive-shaped shells. They are characterized by their smooth and glossy cylindrical shells. Shell is elongate-ovate, with a short spire, a large body whorl and channeled sutures. Surface of the shell is smooth and is highly polished. Aperture is long and narrow with a short siphonal canal. Inner lip is calloused, with oblique grooves anteriorly. Operculum is absent. Olive Snails are carnivorous sand-burrowers, feeding on other small animals or carrion.

Agaronia gibbosa (Born, 1778) (=*Oliva gibbosa*)

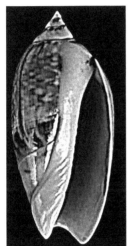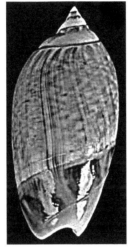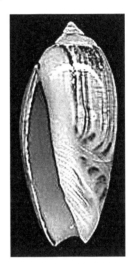

Sinistral   Dextral

Common name(s): Richards seashell.

Global distribution: Western Indo-Pacific.

Habitat: Coastal, sandy bottom.

Description: Shell of this species is moderately large, stout and thick and is fusiformly ovoid. Shell surface is smooth and highly polished. Spire is short, but acuminate; and its apex is pointed. Body whorl is slightly inflated and suture is channeled. Lower part of body whorl is sharply demarcated from the upper by an oblique spiral line. Aperture is rather wide, with a slit-like posterior canal. Columella is with thick, dense, and white callosity extending posteriorly to the penultimate whorl. Fasciole is strong and raised. Aperture is narrow and elongate. Posterior canal is small and slit like. Anterior canal is in the form of a semilunar notch. Exterior shell color is pale yellowish-brown with a prominent yellow band at the base, mottled with black spots. Sometimes it is whitish with zig-zag trans-spiral brownish bands. Spire and columella are yellowish white. Aperture is bluish white. Shell size is up to 6 cm in height.

Americoliva sayana (Ravenel, 1834) (=*Oliva sayana*)

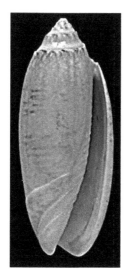

Common name(s): Lettered olive.

Global distribution: Western Atlantic.

Habitat: Gulf and bay (inlets); offshore banks, on soft substrate near-shore waters; shallow sand flats near inlets; depth range 0–130 m.

Description: It is a large predatory sea snail. Shell of this species is smooth, shiny and is cylindrical-shaped with a short spire and five or six whorls. Suture is V-cut and deep. Aperture is narrow, is extending almost the length of shell, and is purplish on the inside. Inner lip of aperture (columellar lip) is with several diagonal folds. Operculum is absent. Exterior shell coloration may vary from cream to a grayish with reddish-brown zigzag markings. The common name of this species is due to its darker surface markings that sometimes resemble letters. Maximum size of the shell is 9.1 cm.

Biology

Locomotion/behavior: Its presence can be detected at very low tides by the trails it leaves when it crawls below the surface on semiexposed sand flats.

Food and feeding: It is a carnivore capturing bivalves and small crustaceans with its foot.

Reproduction: Females lay floating, round egg capsules. Young are free swimming.

Other uses (if any): Both the flesh and shell of this animal are commercially important. Colonists and early Native Americans made jewelry from these shells. This lettered olive is the state shell of South Carolina.

Turbinellidae (Sea snails): Shell of the members of this family is thick and heavy; biconical to fusiform; and often nodulose to spinose on shoulder. Periostracum is conspicuous. Siphonal canal is present. Inner lip is with strong folds. Operculum is corneous.

Syrinx auranus (Linnaeus, 1758)

Common name(s): Australian trumpet shell or false trumpet.

Global distribution: Indo-West Pacific: northern and western Australia; eastern Indonesia and Papua New Guinea.

Habitat: Shallow subtidal sand flats; depth range 0–30 m.

Description: It is the largest extant shelled gastropod species in the world. Shell of this species is spindle shaped. Spire of the shell is high. Whorls usually have a strong keel which may have nodules on it. Shell has a long siphonal canal. There are no folds on the columella. Exterior shell is usually pale apricot in color; however in life, it is covered by thick brown or gray periostracum. Shell color may fade to a creamy yellow. Juvenile shells show a long tower-shaped protoconch or embryonic shell of five whorls, which is usually lost in the adult. Overall height (length) of the adult shell is up to 91 cm and it is 1.8 kg.

Biology

Food and feeding: This carnivorous species is feeding on polychaete worms of the genera *Polyodontes*, *Loimia*, and *Diopatra*. These worms live in tubes. This species can reach their food animals including largest polychaetes (with a length of over 1 m) with its proboscis, which has a length of up to 25 cm.

Other uses (if any): Traditionally, it is fished for its gigantic shell and edible flesh which is sometimes used as bait. The shell is sold for shell collections and is used as a source of lime and as a water carrier.

Turbinella angulata (Lightfoot, 1786)

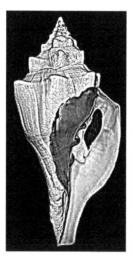 

Common name(s): West Indian chank or lamp shell.

Global distribution: Western Atlantic Ocean.

Habitat: Subtidal and offshore; coral reefs, and mangroves; on mud, rock, sand, and seagrass; depth range 0–81 m.

Description: This is one of the largest gastropods in the Atlantic Ocean. The name "chank" for the shell of this species has been derived from the word "shankha" (the divine conch or sacred conch, *Turbinella pyrum*, a closely related species from the Indian Ocean). Shell of this species is heavy and fusiform, with a sculpture of 8–10 prominent ribs angled at shoulder. Columella has three strong folds. External shell surface is white and the interior is pink or orange. Size of the adult shell may vary from 13 to 50 cm.

Fisheries/aquaculture: Commercial fishery exists for this species.

Other uses (if any): It is a potential food source in Florida, Nicaragua, and North Caribbean.

Turbinella pyrum (Linnaeus, 1767) (=*Xancus pyrum*)

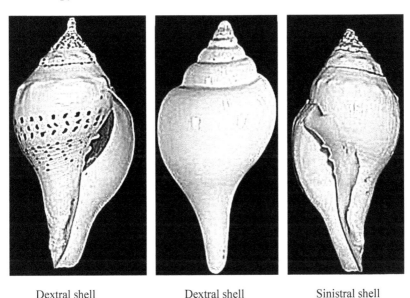

| Dextral shell | Dextral shell | Sinistral shell |

Common name(s): Chank shell, sacred chank, chank or divine conch.

Global distribution: Indian Ocean.

Habitat: Intertidal and reef habitats.

Description: The shell of this species is massive, with three or four prominent columellar plicae. It is normally pure white under a heavy brown or pale apricot periostracum. Sometimes, these shells may be dotted with dark brown. Shells of the normally left-handed *Busycon contrarium* are sometimes sold in imitation of the rare left-handed, sinistral, sacred shells of the *T. pyrum*.

Biology

Reproduction/ growth: The growth of this species has been observed to be high in the first 3 years of age, after which the animal attains maturity.

Seasonal variation has its own impact on the growth of this species as reduced growth was observed during the northeast monsoon months, that is, October and November. The lifespan of this species has been estimated as 13 years (Panda et al., 2011).

Fisheries/aquaculture: This species is exploited (with peak fishing period from October to May) throughout the year along Thoothukudi coast, South India for its flesh and shell. Due to overexploitation, the fishing mortality of this species has been found to more than three times of natural mortality in this coast. In order to conserve this species from endangerment, the minimum size for its capture should be fixed as 14 cm (Panda et al., 2011).

Other uses (if any): The flesh of this species is used as food by the fisher folk of India. The fleshy foot is immersed in boiling water, then cut into slices, which are then dried in the sun. Slices of meat are sold at the bazaar or boiled in oil with spices and served with rice. The shell of this species has considerable significance in certain religions especially Hinduism and Buddhism. The sinistral shell is believed to be very sacred. The shell is modified into a ceremonial trumpet by cutting the tip of its spire. For this purpose, these shells are decorated with metal and semiprecious stones.

Vasum caestus (Broderip, 1833)

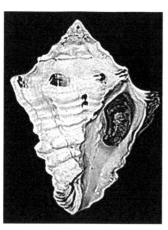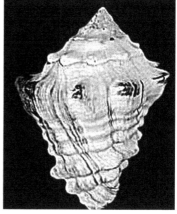

Common name(s): Gauntlet vase, vase snail or vase shell.

Global distribution: Eastern Pacific: Gulf of California to Ecuador.

Habitat: On sand between rocks; depth range 10–12 m.

Description: Shells of this species are somewhat large and are usually very thick and heavy. They are often vase shaped or conical. Shells have a thick periostracum, a low spire, and two, three, or four plaits on the columella. Common length of the shell is 9 cm.

Biology

Food and feeding: mostly rather large predatory sea snails.

Vasum turbinellus (Linnaeus, 1758)

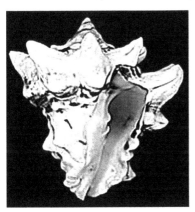

Common name(s): Top vase.

Global distribution: Red sea; Indo-Pacific—from East Africa and the Persian Gulf; Maldives and Thailand to Indonesia; Malaysia, Philippines, South Japan, Papua New Guinea, and Great Barrier Reef.

Habitat: Shallow coral reefs, rocky bottom, and intertidal and shallow subtidal waters.

Description: Shell of this species is reverse conical and forms dense tubular short spines. Short spines of the shoulder are the strongest. Spiral type is low or low-level. Shell mouth is white, narrow, black and white with black and white staggered stains. Shaft lips are with dentition. Coveris horny. Exterior shell is gray with dark spots. Maximum size of the shell varies from 3 to 12 cm.

Other uses (if any): It is a potential food source for the people of Indo-Pacific Philippines. In other areas of its occurrence, this species is collected for subsistence or bait. Shell is used in shellcrafts.

Volutidae (volutes): The volutes are large snails with a short spire. Shell is variable in shape and is often glossy and brightly colored. Aperture is long, with a short siphonal canal. Inner lip is with strong folds which are weaker posteriorly. Operculum is horny when present and is often absent. The large foot is with obvious patterns and colors, and they have a long siphon. These species are carnivorous, and they normally feed on other snails and clams. They can burrow to seek for their prey with their long siphon. Unlike many marine snails, the eggs of the volutes hatch directly into small snails and there is no planktonic larvae. Many of these snails are collected for food and for making decorative items.

Adelomelon ancilla (Lightfoot, 1786)

Common name(s): Ancilla volute.

Global distribution: Southeast Pacific and Southwest Atlantic.

Habitat: Shallow water near the shore; infaunal; mixed gravel and sand bottoms; rocky bottoms with cobbles and shells; depth range 10–600 m.

Description: Shells of this species are smooth and narrower with a smaller aperture, a greater aperture-spiral distance, and a higher spire. Whorls (5) are convex; spiral lines are absent; and there are three columellar folds. Exterior shell is light yellow to orange color; and some may have thin reddish ochre zigzag bands. Color of the foot edge is pinkish brown. These spiral shells reach over 21 cm in maximum length.

Biology

Food and feeding: It is a carnivore feeding mainly on bivalves and secondarily on gastropods. It captures live prey by tightly engulfing it with its foot.

Association: It usually carries on its shell the anemone *Antholoba achates*, which also lives attached to hard substrates in the same area.

Reproduction: Males of this species attain sexual maturity at 7 years with a shell length of 7.4 cm. In the case of females, the values are 9 years and 9.4 cm, respectively. The females lay egg capsules. There are two to eight embryos that develop within the capsule, and it emerges as crawling juveniles. Intracapsular embryonic direct development is seen in the life history of this species. Lifespan of this species is 18 years.

Fisheries/aquaculture: This species has the capacity to serve as source of support for local fisheries at the Argentinian–Uruguayan Common Fishing Zone.

Other uses (if any): It is a potential food source in Argentina.

Adelomelon becki (Broderip, 1836) (=*Voluta fusiformis; Erato voluta*)

Common name(s): Beck's volute.

Global distribution: From Mediterranean to Norway; around British Isles.

Habitat: Sublittoral; associated with ascidians on hard substrata; depth range 20–150 m.

Description: Shell of this species is solid, glossy, and harp shaped with three or four slightly tumid whorls. Spire is short and sutures are shallow; and the last whorl about 80% or more of shell height. Aperture is long and narrow, with parallel sides and short siphonal canal; outer lip is thick, with internal row of teeth; and base of columella is grooved. In the body, cephalic tentacles are rather short, and slender; and mantle is drawn out into long siphon anteriorly and laterally into large tuberculate lobes. Foot is elongate, with double-edged anterior and posterior pedal gland on sole. There is no operculum. Size: Length of the shell varies between 15 and 50 cm.

Other uses (if any): It is a potential food source in Argentina. Caution needs to be exercised while eating this species as it is known to contain saxitoxin obtained through the consumption of bivalves which in turn depend on the paralytic shellfish poisoning (PSP) toxin producing dino-flagellate species, namely, *Alexandrium tamarense* and *Gymnodinium catenatum* for their food (Turner et al., 2014).

Adelomelon brasiliana (Lamarck, 1811) (=*Pachycymbiola brasiliana*)

Common name(s): Caracol negro (black snail).

Global distribution: Brazil: Rio de Janeiro, Sao Paulo, Parana, Santa Catarina, Rio Grande do Sul; Uruguay, Argentina.

Habitat: Sandy-muddy coastal bottoms; infaunal-epifaunal inhabitant, which slides on the substratum half burrowed.

Description: This species characterized by a large solid and oval shell, measuring up to 20 cm in length. Spire is short and conic with a large last whorl which is occupying almost the totality of the length. Shell has a thick, dark brown periostracum. When there is no periostracum, the shell is pale yellowish to pink surface with strong and irregular growth grooves.

Sometimes, it has a sculpture composed by tubercles. Shell opening is large and oval occupying about 5/6 of the length of the shell and is orange-yellowish internally. Side: Columella is short with two to four folds. Siphonal channel is short and wide.

Biology

Food and feeding: It is carnivorous feeding on mussels.

Reproduction: It is a dioic species. Males have a conspicuous penis, and females have a genital pore which is used for copulation and spawning of the egg capsules. Reproduction occurs from September/October to May/June. Egg capsules laid by the females are not attached to the substratum and drift freely on the bottom along a narrow zone close to the shoreline. Twenty percent of the egg capsules. At hatching, each egg capsule may contain 5–15 embryos from which the siblings come out. The average developmental time in this species has been reported as 57 days.

Association: It is usually epibiotized by sea anemones and barnacles.

Fisheries/aquaculture: Extensive commercial fisheries exist in in Uruguay.

Other uses (if any): This species serves as a food source in Argentina.

Adelomelon ferussacii (Donovan, 1824)

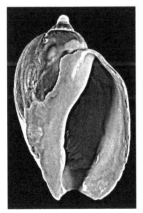

Common name(s): Volute.

Global distribution: From the province of Santa Cruz to Southern Chile.

Habitat: From the intertidal to 70-m-deep water.

Description: Shell of this species is medium in sized, solid, and fusiform. Protoconch is of 1½ smooth whorls and teleoconch has up to four slightly convex whorls. Spire is low and is somewhat upturned with an angle of 80. Suture is well defined. Aperture is semicircular and is dark-brown within. Columella is curved and is orange in color, with 3–6 folds set obliquely to siphonal fasciole. Columellar callus is usually weak, but sometimes thick. Siphonal canal is fairly broad and shallow. Growth lines span the shell surface, sometimes producing irregular costae. Exterior color of the shell is grayish-brown. Shell size is up to 12 cm.

Cymbiola vespertilio (Linnaeus, 1758)

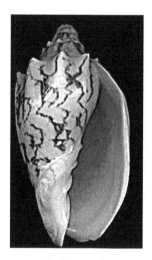 

Common name(s): Bat volute.

Global distribution: Central Indo-Pacific: off the Philippines and Northern Australia.

Habitat: Mud and seagrass in silty habitat; depth range 0–20 m.

Description: This species has a beautiful shell which is with variable in color and pattern. Size of an adult shell varies between 4.5 and 11.5 cm.

Biology

Food and feeding: It is a carnivore feeding on other molluscs. They are amazingly fast during feeding.

Other uses (if any): It is a potential food source in Philippines. Owing to their giant, beautiful shells, this species has commercial value.

Cymbium cymbium (Linnaeus, 1758)

Common name(s): False elephant's snout volute or pig's snout volute.

Global distribution: Eastern Central Atlantic: Mauritania to Guinea.

Habitat: Offshore.

Description: Shell size of this species varies from 10 to 20 cm.

Biology

Food and feeding: It is a predatory snail feeding on bivalves.

Other uses (if any): It is a potential food source in Senegal.

Cymbium glans (Gmelin, 1791)

Common name(s): Elephant's snout volute.

Global distribution: Eastern Atlantic: Senegal to Angola.

Habitat: Offshore.

Description: Shell of this species is large, smooth, shiny, and oblong-oval in shape. A bulged smooth curl covers the whole length of the shell. Shell surface has growth lines. Channel is wide. Mouth of the shell is wide with yellowish-pinkish hues. Inner lip has oblique folds and the outer lip is thin-walled with smooth rounded edge. Exterior shell color is yellowish with pale cream-colored shades. Shell size varies form 10 to 36 cm.

Other uses (if any): It is an edible species in Senegal.

Cymbium marmoratum (Link, 1807)

Common name(s): Marble cymbium volute or mamorate volute.

Global distribution: Eastern Central Atlantic and the Mediterranean: Morocco and Senegal.

Habitat: Depth range 30–40 m.

Description: Common length of the shell is 20 cm.

Other uses (if any): It is an edible species in Senegal.

Cymbium pepo (Lightfoot, 1786)

Common name(s): African neptune volute.

Global distribution: Northwest Coast of Africa (from Morocco to Senegal), Canary Islands.

Habitat: Depth above 30 m.

Description: Shell of this species is wide and oval shaped without any curl. Last turn is greatly advanced and the entire shell is high. Mouth of the shell is wide, with a semicircular deep notch at the bottom. Inner lip carries two distinct elongated prongs at the top. Outer lip is relatively thin, brittle, with a smooth edge, which is gently rounded. Outer lip is widely unfolded out or pressed that determines the shape of the mouth. Surface of the shell is smooth and its clamshell sculpture is represented only by thin growth lines. Exterior shell is ray-brown color and mouth is pink or yellowish and brilliant.

Other uses (if any): It is an edible species in Senegal.

Melo broderipii (Gray in Griffith and Pidgeon, 1833)

Common name(s): Crowned baler.

Global distribution: Western Central Pacific: Philippines.

Habitat: Island; depth range 5–15 m.

Description: It is a large predatory sea snail. It is often referred to as a bailer shell as their shape makes it the perfect canoe bailer. Shell of this species is large to very large and is easily recognized by its vivid yellow coloration and the swollen body whorl, covering most of the spire whorls. Sculpture consists of weak axial threads. Aperture is extremely wide and yellow; and outer lip is thin arcuate and smooth. Parietal wall is yellow and columella is calloused with four strong plaits. There is no operculum. Exterior shell color is yellow with interrupted brown spiral bands (especially in young specimens). Juvenile shells have very large, domed apex, and beautiful color patterns. Animal has a very large, dark brown, and edible foot with thin, yellow lines. Size of the shell varies from 125 to 36.5 cm.

Other uses (if any): Many tribes in New Guinea and other south pacific islands use the various types of Melo to make jewelry and shell money as well as canoe bailers. The large orange pearls of this species found were treasured especially by royalty in Vietnam in the past. Polished shells are highly appreciated by tourists and shell collectors.

Melo melo (Lamarck, 1811)

Common name(s): Indian volute, bailer shell, melo-melo sea snail, zebra sea-snail or bailer volute.

Global distribution: Restricted to Southeast Asia: from Burma, Thailand and Malaysia, to the South China Sea and the Philippines.

Habitat: Littoral and shallow sublittoral zones; muddy bottoms; depth range 0–20 m.

Description: This sea-snail earns its name "bailer volute" because its large shell was often used to bail water out of boats. Shell of this species has a bulbous or nearly oval outline, with a smooth outer surface presenting clear-cut growth lines. Columella has three or four long and easily distinguishable columellar folds. It has a wide aperture, nearly as long as the shell itself. There is no operculum. Shell's spire is completely enclosed by the body whorl, which is inflated and quite large, and has a rounded shoulder with no spines. Apex is of smooth. Exterior shell color is pale orange, sometimes presenting irregular banding of brown spots, while the interior is glossy cream, becoming light yellow near its margin. Maximum shell length is 27.5 cm and common length is 17.5 cm.

Biology

Locomotion/behavior: They are generally nocturnal animals, crawling along the sea floor to find food, and burying themselves in the sand during day.

Food and feeding: It is a specialized predator of other continental shelf-predatory gastropods, particularly *Hemifusus tuba*, *Babylonia lutosa*, and *Strombus canarium*. It is also feeding on other molluscs such as scallops, turban shells, tritons, and even other volutes. To catch the prey, this animal seizes it with its large foot and plugs the prey's aperture, essentially smothering it.

Reproduction: For breeding, the females lay a first layer of eggs onto a rock or another shell and then stack more eggs on top, layer by layer, to form an egg case. The egg case is hollow through the middle and has holes all through it, enabling the water flow around the eggs. An egg case may contain more than 100 juveniles. Like other volutes, Melos sp. are direct developers, which means that juveniles hatch directly from the egg mass and "crawl away," without a free-swimming, (planktonic) larval stage (Anon. http://museum.wa.gov.au/research/collections/aquatic-zoology/baler-shells).

Other uses (if any): This volute is often collected for food by local fishermen. The shells of this species are used in shellcraft. The smooth and cream-colored surface, adorned with attractive orange-brown patterns, makes this shell a prized addition to shell collections. These shells are also traditionally used by the native fishermen to bail out their boats and store water. The nonnacreous pearls of this species are extremely rare (Anon. Melo Melo Pearls—http://www.internetstones.com/melo-melo-pearls.html).

Melo miltonis (Gray in Griffith and Pidgeon, 1833)

Common name(s): Southern bailer or southern baler.

Global distribution: Restricted to Southwest Australia.

Habitat: Shallow seagrass beds, on sand, and around reefs; depth range 0–40 m.

Description: Shells of this species have long been used by the peoples of Australia to carry or remove water, hence the common name "bailer." Shells are with distinctive cream and brown markings. Its large brown foot is with an intricate patterning of cream lines. Maximum length (height) of the shell is up to 45 cm.

Biology

Food and feeding: These animals remain buried under sand during the day and emerge at night to feed. As active predators they glide over the sandy surface with their extended proboscis foraging for other molluscs, including scallops, triton, and turban shellfish which they envelop with their large fleshy foot.

Reproduction: Females of this species lay a large mass of translucent egg capsules (containing over 100 developing shells) which are attached to solid structures such as coral, rock, and other dead shells. They lack a planktonic larval stage, instead developing directly into crawling juveniles.

Other uses (if any): This species also produces the melo pearl in response to irritation, but without nacre. The animal is used for food in Australia, and the shells are sold for a variety of uses. The indigenous people used these shells for "bailing" water out of their canoes. Further, these shells were highly valued by Aborigines for use of storing and holding water.

Voluta ebraea (Linnaeus, 1758)

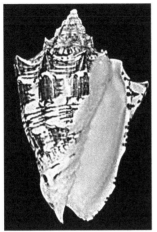 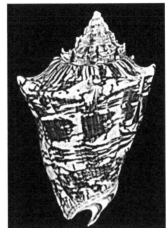

Common name(s): Hebrew volute.

Global distribution: Endemic volutid from north and Northeast Brazil.

Habitat: Sandy bottoms; coral and rocks; from shallow water to depths around 40–70 m.

Description: It has a somewhat robust and solid shell, with a slightly elongate contour. Protoconch is rounded, with two whorls. Shell as a whole has seven slightly convex whorls. These whorls (including the body whorl) are ornamented by several posteriorly oriented sharp spines. Outer lip is thick, and the aperture is relatively long and narrow. Columella has an array of strong oblique columellar folds (also known as plicae) and 9–11 of them are more conspicuous anteriorly. Corneous, claw-like operculum partially covers the shell aperture. Exterior shell is cream colored with a complex series of darker-reddish brown markings and lines which are said to resemble Hebraic figures. Interior of the shell may vary in color from pale to strong orange. Shell size may vary from 10 to 22 cm. This volute has a pale ivory colored body, ornamented by numerous irregular and intertwined thin dark-red to brown colored lines, and several small spots of the same color along the sides of the foot. Some of the distinct external features are its very large foot, and a long siphon.

Biology

Food and feeding: The Hebrew volute is carnivorous and predatory and in the wild, it feeds on the cardiid bivalve *Trachycardium muricatum* during the day, and its prey is largely detected prey through chemoreception. When hunting, these animals move forward toward the prey and then raise the front edge of their muscular feet to cover the bivalve and capture it. In captivity, it has been reported to feed on the sea snails, *S. haemastoma* (a muricid carnivorous gastropod) and *Tegula viridula* (a top snail).

Reproduction: *Voluta ebraea* is dioecious (male or female). It is also sexually dimorphic, which means there is a difference in form between individuals of different sex within this species. In this case, the shells of the males are more elongated with a smoother outer surface, whereas the shells of the females are generally wider and more nodulose. The angle of the spire also differs between males and females. Egg capsules laid by the females are circular and flattened. They are found attached to the alga *Udotia occidentalis* in seagrass beds. An opening through which crawling juveniles emerge is located at the center of the egg capsules with a suture from the base to the central opening (Penchaszadeh et al., 2010).

Imposex: The imposex phenomenon has also been observed in this species. In this phenomenon, females of this species exposed to organic tin compounds, such as TBT, were found to develop masculine sexual organs. Further, this phenomenon has been reported to cause several negative consequences for entire populations of this species, from sterilization of individuals to the complete extinction of those populations. The above-said organic tin compounds are biocide and antifouling agents, which are commonly mixed in paints to prevent marine encrustations on boats and ships. It is not uncommon for high concentrations of such compounds to be present in the sea water near shipyards and docking areas, consequently exposing the nearby marine life to its deleterious effects (Castro et al., 2008).

Predator: The Hebrew volute is the prey of the Bocon toadfish, *Anphichthys cryptocentrus*.

Conservation status: It is an endangered species.

Fisheries/aquaculture: This fished is commonly fished by shrimp trawlers. Recent studies indicate that natural populations of *V. ebraea* may be suffering declines owing to overfishing and overexploitation.

Other uses (if any): The flesh of *V. ebraea* is edible, and it is locally collected for food in many areas. Its shell is also considered a popular and beautiful decorative object and is sold as souvenir in local markets and shell craft stores in several regions of Brazil.

Zidona dufresnei (Donovan, 1823)

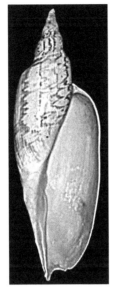

Common name(s): Angulate volute.

Global distribution: Southwest Atlantic and Southeast Pacific: Argentina, Uruguay, Brazil and Chile freshwater.

Habitat: Continental shelf; depth range 15–200 m.

Description: Shell of this species has a slender spire and an angulated body-whorl. Maximum shell length is 27 cm.

Biology

Food and feeding: It is a benthic top predator feeding on a variety of marine molluscs.

Reproduction: In this species, gonadal maturity was found to occur prior to the development of secondary sexual characters, that is, from August to April in females and males were found reproductively active throughout the year. Size at first gonadal maturity in females was found to be 13 cm shell length and in males 12 cm shell length. Females were recorded depositing egg capsules from late August to late April.

Fisheries/aquaculture: Extensive commercial fishery exists for this species in Uruguay. In the Mar del Plata (Argentina) shelf area where it was subjected to unregulated commercial exploitation for more than 20 years.

Other uses (if any): It is a potential food source in Argentina. It is also being exported to Asian countries for human consumption since 1988. Caution needs to be exercised while eating this species as it is known to contain saxitoxin obtained through the consumption of bivalves which in turn depend on the PSP toxin producing dinoflagellate species, namely, *A. tamarense* and *G. catenatum* for their food (Andrew et al., 2016).

3.4   ORDER CAENOGASTROPODA

Bursidae (frog snails or frog shells): The thick, ovate to slightly elongated shells of the members of this family are coarsely sculptured. The intersection of the spiral ribs and the axial sculpture lead to a strong nodulose pattern of more or less round knobs. This warty surface gave these shells the common name—frog shells. The outer varicose lip is dilated and shows a number of labial plicae, resulting in a toothed lip on the inside. The inner lip is calloused, showing transverse plicae. The anterior and posterior canals are well developed. The siphonal canal at the anterior end is usually short. The anal canal at the posterior end is a deep slot. The nucleus of the corneous operculum is situated either at the anterior end or the midinner margin. A periostracum (hairy covering of the outer shell) is usually absent or thin.

Bufonaria crumena (Lamarck, 1816) (=*Bursa crumena*)

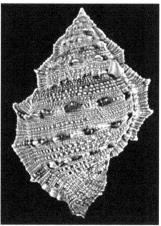

Common name(s): Purse frog shell.

Global distribution: Indo-West Pacific: from East and South Africa to Melanesia; north to the Philippines and south to Queensland.

Habitat: On sand or mud; protected, submerged rocks; sublittoral; depth range 0–50 m.

Description: Shell of this species is moderate sized, broad, and ovate. Apex is pointed. Sculpture is composed of nodulose spiral threads. Body whorl is with rows of short sharp nodes, and the remaining whorls are with single spiral row of tubercles. There are two fin-like varices on both sides; and varices have sharp nodes at regular intervals. Aperture is ovate; outer lip is expanded and supported by a varix; and inside of outer lip is toothed. Columella is denticulate at the base. Columellar callus is well developed. Siphonal canal is short and twisted. Exterior shell color is light brown with dark brown spots close to the nodes. Aperture and lips are white with slight orange tinge. Maximum shell length is 9 cm.

Bufonaria echinata (Link, 1807) (=*Bursa spinosa*)

Common name(s): Spiny frog shell.

Global distribution: Red Sea; off Madagascar Indian Ocean: off the Philippines and China.

Habitat: Intertidal zones.

Description: Shell of this species is elongate shape and large. Shell surface is weakly sculptured. Spire is elevated and is sculptured with fine close set granular spiral ribs. There is a single strong recurved spine adjacent to anterior canal; and inner margin of outer lip is flared and denticulate with irregular shaped tooth on inner margin. Operculum is fan shaped. Exterior shell color is light brown with dark brown markings occasionally. Size of the shell may vary from 5 to 9 cm.

Bufonaria rana (Linnaeus, 1758)

Common name(s): Common frog shell.

Global distribution: Indo-Pacific: from Indonesia to Polynesia; north to Japan and south to southern Queensland.

Habitat: Mud and muddy-sand bottoms; sublittoral and continental shelf; depth range 40–170 m.

Description: Shell of this species is moderately large, solid, and dorso-ventrally compressed. Spire is raised. Varices which are on both sides are fin like with two to three small spines. Shell is sculptured with granulated spiral ribs. Body whorl bears two rows of spiral cords with fine nodes; one row is on penultimate whorl; and rest of the shell surface is with granulose spiral threads and spinous nodes. Aperture is slightly ovate; and outer lip is denticulate. Columella is smooth and denticulate toward base. Siphonal canal is short. Operculum is thin and flexible, made of a horn-like material. Body is pale, with a long muscular foot and short tentacles with a black band at the tips. Exterior shell color is light brown and aperture is white. Maximum shell length is 9 cm.

Biology

Food and feeding: It is a carnivore feeding on fish and tubiculous polychaetes. These animals have an extendible proboscis and large salivary glands, which are probably used to anaesthetize the worms in their tubes. The worms are then sucked out and swallowed whole.

Other uses (if any): It is a potential food source in Hong Kong.

Cassidae (helmet shells): Shell of the members of this family is helmet shaped. It is thick and solid, with a large body whorl and rather small, conical, short spire. Sculpture is variable and axial varices are sometimes present. Aperture is elongate, with a short siphonal canal which is recurved dorsally. Outer lip is thickened. Inner lip is with a shield-like callus. Operculum is quite small and corneous. These species feed on echinoderms, such as sea stars and sea urchins. Some species secrete a paralyzing saliva to prevent the sea urchin prey from releasing its venom. Then, they secrete an acid to dissolve the shell so as to feed on the prey.

Cassis cornuta (Linnaeus, 1758)

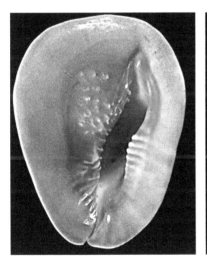 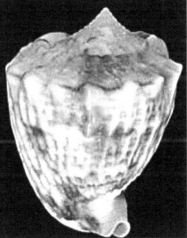

Common name(s): Horned helmet, yellow helmet, or giant helmet.

Global distribution: Indo-Pacific: from East Africa, including Madagascar and the Red Sea to eastern Polynesia; Japan and Hawaii; southern Queensland and New Caledonia.

Habitat: Coral reef areas; on sand and coral rubble bottoms; sublittoral; depth range 2–40 m.

Description: Shell of this species is massive and globose with a wide and flat apertural side. Shell has a short spire and a thick parietal shield that is usually large enough to envelop the apertural side of the body whorl. Body whorl with three or four spiral rows of large tubercles; and those at shoulder are much longer and stouter than the others. Aperture is long and narrow and is heavily calloused. Outer lip is thickened in a broad and flat shelf, with a dorsally recurved outer edge and with 5 to 7 strong teeth on its inner edge. Inner lip is with an extensive callous shield, forming a flange along left side of body whorl, and produced over the spire to join the outer lip at its posterior end. Columella is with irregular spiral ridges. Operculum is elongate-ovate and is about 1/4 the length of aperture. Dorsal side of the shell and spire are grayish white, often somewhat spotted with light brown. Calloused ventral side is glossy cream or orange, with two spiral rows of brown spots in the central region. Teeth and ridges of the aperture are white. Outer lip is with six or seven broad patches of brown on its dorsal side and outer edge. Length of the shell varies between 5 and 41 cm. Horned helmets in sizes greater than 28 cm are extremely rare.

Biology

Locomotion/behavior: These animals which are living in colonies are often seen partly buried in sand (with only the top of the back and a couple of horns showing) when inactive or during feeding.

Food and feeding: It is a carnivore feeding on *Acanthaster planci*, a starfish that feeds on corals. By feeding on *A. planci*, this species helps to maintain the health of coral reefs.

Reproduction: It is gonochoric and a broadcast spawner. Female lays clusters of fertilized egss which are usually attached to macroalgae. Embryos develop into planktonic trochophore larvae and later into juvenile veligers before becoming fully grown adults.

Fisheries/aquaculture: Commercial fishery exists for this species.

Other uses (if any): It is a potential food source for the people of countries of Indo-Pacific. This snail is hunted for its meat, and its shell is a very popular collector's item. The shell is also traditionally used as a decorative item in many parts, or as container for liquids by the natives of the South

Seas. In India, it is normally sold as a souvenir or decoration piece. This species has also been reported to produces non-nacreous pearls.

Cassis flammea (Linnaeus, 1758)

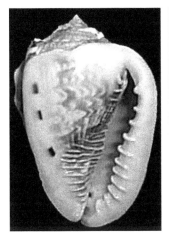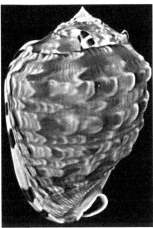

Common name(s): Flame helmet.

Global distribution: Western Atlantic: Florida, the Caribbean, and Bahamas.

Habitat: Shallow subtidal depths; on sand bottoms near seagrass beds and around reefs; depth range 1–12 m.

Description: Shell of this species is large and heavy. Spire is short. Shell surface is smooth, except for knobby projections on body whorl. Parietal shield is large and well defined, oval. Outer lip is with inner tooth-like projections. Exterior shell is brownish cream with large patch of brown at center of parietal shield. Outer lip is entirely cream or cream-white. Shell may be covered by algae or other sessile marine organisms. Maximum recorded shell length is 15 cm.

Biology

Food and feeding: It is a carnivore feeding on sea urchins, which it hunts at night.

Fisheries/aquaculture: Commercial fishery exists for this species.

Other uses (if any): It is a potential food source in Bermuda and Florida. In most countries, it is illegal to bring back these shells from holidays.

Cassis madagascariensis (Lamarck, 1822)

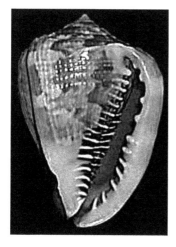

Common name(s): Emperor helmet or giant queen helmet.

Global distribution: Western Atlantic: the Gulf of Mexico and the Caribbean Sea.

Habitat: On sand bottoms near seagrass beds; shallow subtidal areas; depth range 3–180 m.

Description: It is the largest species of the family in the Atlantic Ocean. Shell of this species is very large, heavy. Spire is short. Shell surface is with three rows of large knobs on body whorl. Parietal shield is large and well defined and is triangular. Outer lip is with inner tooth-like projections. Exterior shell has a lighter shade that ranges from creamy white to cream while its inner layer is brown surface. Maximum length of the shell is 41 cm.

Biology

Reproduction: The egg mass laid by the female contains 260 capsules. Individual capsules are slightly opaque, vasiform structures with very smooth walls and are almost without sculpturing.

Fisheries/aquaculture: Commercial fishery exists for this species.

Other uses (if any): It is a potential food source for the people of North Carolina, Bermuda, and West Indies. Shells of this species are used in jewelry to make cameos.

Cassis tuberosa (Linnaeus, 1758)

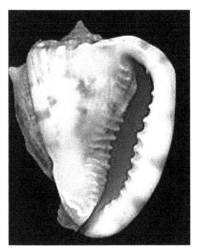 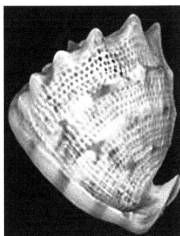

Common name(s): King helmet.

Global distribution: Western Atlantic.

Habitat: On sand bottoms (sometimes buried) near seagrass beds; shallow subtidal depth; offshore coral reefs; depth range 0–27 m.

Description: It is very large sea snail with a solid, heavy shell. Shell surface is with fine reticulated sculpture and knobby projections on body whorl which has three rows of blunt spines. Parietal shield is large and well defined and is triangular. Operculum is small and oblong and brown. Outer lip is with inner tooth-like projections. Exterior shell is brownish cream with large patch of brown at center of parietal shield. Outer lip is entirely cream or cream white with brown between teeth. Maximum recorded shell length is 30 cm.

Biology

Food and feeding: It is a carnivore feeding exclusively on echinoids. This species shows highly predictable feeding behavior.

Fisheries/aquaculture: Commercial fishery exists for this species.

Other uses (if any): It is a potential food source in Brazil, Bermuda, North Carolina, and Cape Verde Islands. Shell of this species has been used for creating cameos.

Galeodea echinophora (Linnaeus, 1758) (=*Cassidaria echinophora*)

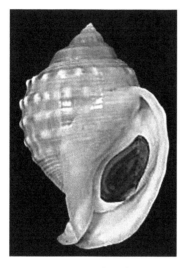 

Common name(s): Spiny bonnet or Mediterranean spiny bonnet.

Global distribution: Eastern Mediterranean Sea; North Atlantic Ocean: Western Africa.

Habitat: Sandy and muddy bottoms over 10 m in depth.

Description: This species has a very flashy shell which is globular or oval and is mainly formed by the body whorl. Shell aperture is wide, with denticulate lips, a curved siphonal canal and a large columellar edge. Tubercles are quite variable, usually not very pronounced and may be entirely absent. Surface of the shell is yellowish-brown. Adult shell length may vary from 5 to 11 cm.

Biology

Food and feeding: These molluscs are carnivorous eating mostly echinoderms, especially *Echinocardium cordatum*. Feeding experiments conducted in the laboratory aquarium showed the following. Every few days, each animal emerged from the sand and foraged for 1–3 h. On detecting the prey, the animal stopped locomotion and attacked the buried prey from the surface of the sand.

Reproduction: In the laboratory, a female laid 120 egg capsules on the side of the aquarium. These capsules hatched after 112–159 days at 13°C.

Within the capsule, veligers were found to feed on the yolk cells of abortive embryos. Juveniles hatched at the crawling stage, and at first secreted strings of mucus acting as drogues, but after 1 day, the juveniles crawled over the substratum. They readily attached themselves and metamorphosed into adults.

Fisheries/aquaculture: It is commercially harvested.

Other uses (if any): The flesh of this species is eaten in Italy and Spain.

Galeodea rugosa (Linnaeus, 1771) (=*Cassidaria tyrrhena*)

Common name(s): Rugose bonnet.

Global distribution: Eastern Atlantic and Mediterranean: Spain, Portugal, Morocco, Italy, and Cyprus.

Habitat: On soft bottoms of several mud volcanoes; depth range 77–1032 m.

Description: Shell size of this species varies between 5 and 14 cm.

Biology

Food and feeding: These animals are carnivores (with a single pair or large salivary glands) feeding on echinoderms.

Fisheries/aquaculture: Commercial fishery exists for this species.

Other uses (if any): It is a potential food source in Spain.

Phalium glaucum (Linnaeus, 1758)

Common name(s): Gray bonnet or glaucus bonnet.

Global distribution: Indo-West Pacific: East Africa to Melanesia; Japan and Queensland.

Habitat: On sandy bottoms with seagrass meadows; intertidal and shallow subtidal areas; depth range 0–10 m.

Description: These shells are helmet shaped with a large body whorl and tiny spires resembling a bonnet. Surface of shell is smooth and uniformly grayish or pale brown. Operculum is bright yellow and fan shaped. Animal has a white body and a large yellowish or whitish foot which is edged in reddish brown. Adult shell length may vary from 6 to 15 cm.

Biology

Locomotion/behavior: This sea snail buries itself in the sandy areas with the long siphon sticking out.

Food and feeding: This species is a carnivore feeding on sea urchins/sand dollars.

Reproduction: In this species, the egg capsules females are forming an irregular mass which is due to several females spawning together.

Fisheries/aquaculture: Commercial fishery exists for this species.

Other uses (if any): This species is collected for food and for the shell trade.

Semicassis granulata undulata (Gmelin, 1791) (=*Phalium undulatum*)

Common name(s): Scotch bonnet.

Global distribution: Mediterranean Sea; Atlantic Ocean: off Northwest Africa and the Macaronesian Islands.

Habitat: This species is found from the intertidal zone to depths of 100 m in rocky and sandy habitats; and also on offshore banks. It is buried in the sand during the day.

Description: It is a medium-sized sea snail, and it shell has six spirals of which, first is very large, more than two-thirds with respect to the rest. It has very strong growth lines. Siphon is small. Maximum size of the shell is 8 cm.

Biology

Food and feeding: It is an epifaunal carnivore.

Other uses (if any): It is a potential food source in Spain, Mediterranean, East Atlantic Islands, and Northwest Africa.

Semicassis saburon (Bruguière, 1792) (=*Phalium saburon*)

Common name(s): Bonnet shell.

Global distribution: Mediterranean Sea: from Bay of Biscay up to Ghana; Azores; and Canaries.

Habitat: Coastal waters.

Description: Shell of this species has a heavily reinforced lip of the aperture. Adult shell size of this species may vary from 3.5 to 7.5 cm.

Other uses (if any): The flesh of this species is eaten in Mediterranean and Spain.

Ficidae (fig shells): The fig snails get their common name as they are resembling the shape of an edible fig. Shell is thin, pear shaped and is drawn out anteriorly into a long, tapered, and curved siphonal canal. They have a very short spire, a wide and long aperture. Operculum is absent. They feed on other invertebrates, such as polychaetes or echinoderms. They are usually found in sandy areas.

Ficus ficus (Linnaeus, 1758) (=*Ficus ficoides*)

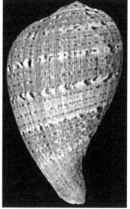

Common name(s): Paper fig shell.

Global distribution: Indian Ocean and West Pacific.

Habitat: Depth range 0–176 m.

Description: This species has an unusual pear-shaped and thin shell with a long narrow aperture and four whorls. Spire is tiny. There is a trellis-like sculpture of fine striations on the pinkish surface. Inside is orange and there is no operculum. Body of the animal has a large foot with two curved flaps near the head and a single long siphon. Shell may reach a maximum size of 14.5 cm in length.

Ficus ventricosa (Sowerby, 1825)

Common name(s): Swollen fig shell.

Global distribution: Gulf of California: off the coast of Mexico; Pacific Ocean: off Peru.

Habitat: Shallow to moderately deep (offshore) waters; medium sand, sandy silt, and silty clay substrata; depth of 35 m.

Description: The shell of this species is moderately large and is strongly inflated with a low spire. Adult shell size varies between 7 and 15 cm. The primary spirals are very strong and are very widely spaced. The secondary spirals are generally of equal width in sets of three. All the spirals are faintly noded by narrow axial threads.

Biology

Reproduction: In this species, sex ratio of female to male was found to range from 0.3 to 2.3 which is largely associated with the egg-laying migration of the females. Normally, a copulated female might migrate to sites with hard substrata to deposit its egg capsules for firm attachment. Laid egg capsules are translucent-white and rectangular. In this species, under laboratory conditions, copulation and egg laying often occur at night to early morning during the period of November to February. It is also reported that this species under stress sheds a certain part of its mantle on the side of the inner lip as a special defensive mechanism. Regeneration of the autonomic tissue occurred in the following week and reached a normal size within 2 m.

Predators: Although predation of this species in the field is unknown, puffer fish and box crabs are known to prey on fig snails in the laboratory.

Other uses (if any): This species is seldom used as food, but the shell is commonly sold in souvenir shops.

Rostellariidae (thumb nails): Shell of the members of this species is thick and glossy with several quite-level whorls. Siphonal canal is long and straight or slightly bent.

Tibia curta (Sowerby, 1842)

Common name(s): Indian tibia.

Global distribution: Persian Gulf to the Bay of Bengal; Yemen to Southern India.

Habitat: Intertidal; mangrove habitat.

Description: Shell of this species is thick, spindle shaped, and very glossy with up to15 quite-level whorls. Last whorl is much shorter than others. There is a broad band of brown color close to the suture of the whorls. Spire has axial ribs. Lower part of the body whorl has subtle spirals. Siphonal canal is straight or slightly bent, about 1/6 of the total length. Outer lip which is irregular has several short projections with the ornamentations. Exterior shell has coffee color. Size of the adult shell varies between 12 and 19 cm.

Tonnidae (tun shells): The Tonnidae is a family of medium-sized to very large marine molluscs, commonly known as "tun shells." The name "tun" refers to the mollusk's shell shape which is resembling wine cask known as "tun." Shell of the members of this family is thin and globose, with a short spire and very inflated body whorl. Sculpture is only spiral. Siphonal canal is short. Operculum is absent. They are mostly seen in all tropical seas, inhabiting sandy areas. During the daytime, they bury themselves in the substrate, emerging at night to feed on echinoderms (particularly sea cucumbers), crustaceans, and bivalves. Larger species may also capture fish, using their expandable prosboces to swallow them whole. Females lay rows of eggs which hatch into free-swimming larvae before settling to the bottom.

Tonna dolium (Linnaeus, 1758)

Common name(s): Spotted tun.

Global distribution: Indo-West Pacific: off Tanzania, Mascarene Basin and off the Philippines.

Habitat: On fine sand and mud bottoms; outer edge of fringing reefs; seagrass meadows; depth range 10–30 m.

Description: Shell of this species is thin, ovate-globose, and ventricose. Spire is generally short. Shell consists of six whorls, which are slightly flattened above. Body whorl is large and very convex. All these whorls are encircled by wide and distant ribs. Surface of the shell is of a white color, slightly grayish, and sometimes rose colored. It is ornamented upon the ribs, with alternate white and red spots. Aperture is very large, colored within of a chestnut tint. Outer lip is thin, notched, and canaliculated within, and its edge is white and undulated. Inner lip is slightly perceptible toward the base, where it forms a part of the umbilicus. Columella is twisted spirally and is with longitudinal ribs externally. Adult snails lack an operculum. Size of the adult shell varies between 10 and 18 cm.

Biology

Locomotion/behavior: The animal has a large foot and can crawl and burrow rapidly.

Food and feeding: It is carnivorous feeding mainly on echinoderms such as sea cucumbers and crustaceans. While eating, the snail first paralyzes its prey with its salivary secretion containing sulfuric acid and then it swallows the prey whole.

Reproduction: The egg mass laid by the female is a wide gelatinous ribbon containing many small, transparent eggs. The hatched out planktonic larvae normally take a very long time to develop.

Fisheries/aquaculture: This species forms an incidental bycatch in fishing nets.

Other uses (if any): It is sometimes collected for food by coastal dwellers, and the shell is used as decorative items.

Tonna galea (Linnaeus, 1758)

Common name(s): Giant tun or fluted tonne.

Global distribution: North Atlantic Ocean: as far as the coast of West Africa; Mediterranean Sea; Caribbean Sea.

Habitat: Seabeds with muddy or sandy with seagrass beds; depth range 0–120 m.

Description: Shell of this species is very large, thin, inflated, and ventricose. In young, the shell is almost diaphanous and its transverse ribs of the surface are only indicated by lines of a slightly deeper tint. Conical spire of the shell is formed of six convex, very distinct whorls which are loaded externally with wide, flat, slightly raised ribs, separated by narrow and superficial furrows. Whorls of the spire are isolated by a deep channeled suture. Body whorl is rounded and very ventricose. Aperture is large and subovate and is colored interiorly with reddish, and marked with transverse ribs. Outer lip is dilated, undulated and is tinged with black, or a deep brown upon the edge. Inner lip is whitish. Umbilicus is open and is rather narrow; and the siphonal canal is short. Columella is smooth and polished. Operculum is absent. External surface of the shell is of an uniform reddish fawn color. Ribs are marked with wide spots or irregular brown and white blotches. Soft parts of the animal are white, black spotted. A broad foot, sensory tentacles (with the small eyes at the base) and a long and very extensible siphon, are also seen. Proboscis may be observed at the time of capture and ingestion of prey. Average height of the shell is 15 cm.

Biology

Locomotion/behavior: It is a nocturnal species, remaining buried in the sand during the day. The highly extensible siphon helps to explore its immediate surroundings. For this purpose, it brings water to the osphradium, in the pallal cavity, near the gills. This osphradium, in which the chemoreceptors are located, exerts the olfactory function in locating its prey.

Food and feeding: It is carnivorous feeding on sea cucumber, echinoderms (such as goats and starfish) fish, bivalves, and crustaceans. To help in feeding, it is equipped with a proboscis, which is capable of enveloping its prey. Assisted by special hooks, the animal can swallow it in a few minutes, without anesthetizing it. The fluted ton has also enormous salivary glands that produce a secretion containing 2–5% sulfuric acid, which uses to kill its prey and initiate digestion.

Reproduction: Reproduction of *Tonna galea* is gendered. After internal fertilization, the female lays eggs forming a long gelatinous ribbon of pale pink color. This ribbon, measuring several centimeters, consists of several thousand capsules. Each capsule contains about 100 pink embryos. Hatching occurs about 1 month after spawning, directly at the veliger stage. After a planktonic life of about 8 months, the larva will settle on the bottom to complete its metamorphosis.

Bioluminescence: The fluted tonne is a bioluminescent species. The animal produces a white-greenish light when it moves.

Other uses (if any): It is a potential food source for the people of Caribbean, Atlantic, and Mediterranean.

Tonna perdix (Linnaeus, 1758)

Common name(s): Partridge tun, Pacific partridge tun.

Global distribution: Indo-Pacific; Red Sea.

Habitat: Intertidal to subtidal; on fine sandy bottoms and on coral reefs; prefers deeper and rapid waters; depth range 0–50 m.

Description: Shell of this species is pretty thin, ovate-oblong, and ventricose. Conical spire is slightly projecting and, pointed. It is composed of five to six whorls which are furnished with numerous ribs which are widened, feebly convex and are separated by furrows. Suture is very distinct and is slightly channeled toward the body whorl. Aperture is large, subovate and is marked by transverse and slightly projecting bands. Interior of this cavity is of a fawn color. Outer lip is thin, everted, a little undulated, and adorned with a white band; and the whole length of its interior is deep brown colored. Inner lip is spread out over the body of the shell. It is very thin, transparent, and terminated below by a projecting plate which covers the umbilicus. The latter is open and deep. Columella is smooth, polished, and twisted. Operculum is absent. Exterior shell is of a reddish-brown color, with white spots in transverse series. Foot is colored in the same way. Tentacles are doubly ringed with brown. Size of an adult shell varies between 7 and 22 cm.

Biology

Locomotion/behavior: It is a nocturnal species, remaining buried in the sand during the day. It is emerging from hiding at night to hunt for sea cucumbers.

Food and feeding: It is a highly specialized carnivorous predator, preying on sea cucumbers, sea urchins, crustaceans and small fish. Mode of feeding of this species is very much similar to that of *T. galea.*

Reproduction: Reproduction in this species is sexual. After internal fertilization, the female lays eggs forming a long gelatinous ribbon. Each fertilized egg produces a ciliated larva, called a trochophore, which then becomes a umbilical larva.

Other uses (if any): The flesh of this species is used as food in countries of the Indo-West Pacific.

Tonna variegata (Lamarck, 1822)

Common name(s): Variegated tun.

Global distribution: Off South and Southeast Africa; off New Zealand and off Australia (Western Australia).

Habitat: Continental shelf; sand and mud substrate.

Description: Shell of this species is thin, ovate-globose, and ventricose. Spire is composed of six convex whorls, which are slightly separated by a shallow suture, and loaded with transverse rounded ribs. Body whorl composes almost the whole of the shell. There are 20–26 transverse ribs are seen on its surface. All the ribs are separated by furrows which are not of the same size. Aperture is wide, large and ovate. Its lips are white, and the interior is reddish. Outer lip is traversed by a canal of slight depth. Inner lip is white, thin, applied to the body of the body whorl and forms a part of the umbilicus. Columella is twisted spirally. Exterior shell is whitish, varied with red, and covered, upon the transverse ribs, with irregular spots. These spots may form longitudinal or zigzag bands of a deeper color. Periostracum is thin and reddish. Size of the adult shell varies between 7 and 20 cm.

Other uses (if any): It is a potential food source in Indo Pacific; south coast New South Wales (NSW); West Australia; and New Zealand.

3.5 UNRANKED

Calliotropidae (sea snails): The shell of the members of this family has a conical shape with a high spire. It is carinated and umbilicated. The base of the shell is inflated. The shell is covered with a thin, smooth, and fibrous epidermis which swells up and becomes pustulated in water. The axis of the shell is perforated, and the columella is thin is reverted.

Bathybembix bairdii (Dall, 1889)

Common name(s): Baird's top shell.

Global distribution: Pacific Ocean: Bering Sea to Chile.

Habitat: Deep water species; depth range 300–1500 m.

Description: Shell of this species is large and is covered with a yellow-green periostracum. Shell size varies between 3.8 and 5 cm.

Biology

Food and feeding: It is a megafaunal surface-deposit feeder on the giant kelp, *M. pyrifera.*

Other uses (if any): The flesh of this species is eaten by the people of Alaska to Chile.

Colloniidae (dwarf turbans): The members of this family are mostly small sea snails with globose shell which is very solid and imperforate. The small, globose shell is very solid and imperforate. The spire is conical and more or less depressed. There are several whorls. Operculum is often oval and calcareous.

Homalopoma luridum (Dall, 1885) (*=Homalopoma carpenter*)

Common name(s): Dall's dwarf turban or dark dwarf-turban.

Global distribution: Northeast Pacific: from Sitka, Alaska to Northern Baja California; Mexico.

Habitat: Under rocks at low tide.

Description: Shell of this species is small, globose, very solid, and imperforate. Spire is conic and more or less depressed. Suture is moderately impressed. There are five whorls which are slightly convex. Last whorl is deflected toward the aperture and is encircled by about 15 subequal spiral lirae. Oblique aperture which measures about half the length of shell is pearly white within. Columella is arcuate. Base of the shell is uni- bi-, or tridentate. Rounded oval operculum is nearly smooth and slightly concave. Exterior shell color is red, or purple. Height of the adult shell varies between 0.6 and 1.0 cm.

Association: The empty shells of this species are often occupied by the hermit crab, *Pagurus hartae* (Sato and Jensen, http://people.oregonstate. edu/~satomei/MeiSato/Publications_files/SatoJensen2005.pdf).

Haminoeidae (Atys bubble shells): Shells of this family are rather fragile with large comma-shaped aperture. Operculum is absent. These molluscs are colorful. Generally, they have a rounded head shield, and fleshy wing-like outgrowths (parapodia) which partially or completely cover the shell, thus providing protection while burrowing in soft sediments in search of food.

Bullacta exarata (Philippi, 1849)

Common name(s): Korean mud snail.

Global distribution: Western Pacific: China and Korea.

Habitat: Intertidal flats including the supratidal zone and subtidal zone.

Description: Shell of this species is bullate, fairly thick, white, and spirally striate, with a well-developed periostracum. Surface of the shell is smooth and the ribs are blurred throughout. There is no spire and no umbilicus. Columella is smooth and simple. Aperture extends for the whole length of the shell and is narrower above than below. Apertural lip extends upwards beyond the apex of the shell. Exterior shell color is pale yellow, or white. Height and width of the shell are 0.8 and 0.6 cm, respectively.

Biology

Food and feeding: This species feeds mainly on diatoms. It is an important consumer in the tidal flat ecosystem.

Reproduction: *Bullacta exarata* is a hermaphroditic species.

Fisheries/aquaculture: Commercial fishery exists for this species. This species is being cultivated in mariculture, especially in Zhejiang Province and Cixi City.

Other uses (if any): This commercially important mollusk is used as a food item in eastern China. This delicious *B. exarata* goes well with assorted dishes nicely. This species has been reported to possess high nutrition value. There is good amount of omega-3 fatty acids (600 mg of eicosapentaenoic acid in 100 g of meat) in the canned meat of this species. It is exported as a food source to Hong Kong, Macau, Taiwan, and to Southeast Asia.

Nacellidae (true limpets): Nacellidae is a family of sea snails commonly called as true limpets. (limpet is a common name for any marine mollusk with a flattened conical shell). Interior shells of these nacellid limpets are having a mother-of-pearl iridescence which may also be seen on the exterior surface in well-eroded shells. Also, they have folds on the mantle edges acting as secondary gill leaflets, and a long radula (tongue-like structure for feeding), these snails are much longer than that of their shells.

Cellana exarata (Reeve, 1854)

Common name(s): Black-foot opihi or Hawaiian blackfoot.

Global distribution: Endemic to the islands of Hawaii.

Habitat: Splash zone; intertidal rocks and reefs; high shore where thermal and desiccation stress is severe.

Description: This high-shore limpet is characterized by its low, domed shell, and short mantle tentacles. Shell of this species has close-set ribs with narrower ones between, not extending far beyond shell margin. Both the shell and the foot of the animal are black in color. Interior is dark gray. Maximum size of the shell is 5 cm.

Biology

Locomotion/behavior: This species is well adapted to its surf-punded habitat. Its ridged shell has been reported to deflect the force of the water and its single foot can cling so tightly to the rock that about 40-kg pull is necessary to break its shell lose.

Food and feeding: These animals graze on algae and most of them may creep about to graze, but return to their "home scar" after feeding.

Reproduction: Gametes are shed into the water where fertilization is external. Veligers have a short planktonic life. Spawning occurs mainly in December and January.

Fisheries/aquaculture: It is heavily fished in Hawaiian Islands. Localized heavy fishing pressure is the most significant threat to this species. It is, however, protected by fishing regulations.

Other uses (if any): It is used as a food item in Hawaiian Island. However, it is less popular for consumption as the flesh is soft. Native Hawaiians used the shells of this species as scrapers and tools.

Cellana melanostoma (Pilsbry, 1891)

Common name(s): Green-foot opihi.

Global distribution: Indo-Pacific.

Habitat: Splash zone; higher up in the intertidal zone.

Description: This species is characteristically by its steep-sided shell. Maximum shell diameter is 4.3 cm.

Biology

Food and feeding: These animals graze on algae and most of them may creep about to graze, but return to their "home scar" after feeding.

Fisheries/aquaculture: It is heavily fished in Hawaiian Islands. Localized heavy fishing pressure is the most significant threat to this species. It is, however, protected by fishing regulations.

Other uses (if any): It is used as a food item in Hawaiian Island. Native Hawaiians used the shells as scrapers and tools.

Cellana radiata (Born, 1778)

Common name(s): Rayed wheel limpet.

Global distribution: Indo-Pacific, mainly Australia.

Habitat: Rocky bottoms in intertidal and midlittoral reef-associated coastal waters.

Description: Shell of this species is moderately large sized, flat, and roundly ovate in outline. Apex is slightly out of the middle and is often worn out. Shell surface is sculptured with fine granular radially arranged ribs and riblets. Exterior shell color is pale yellowish brown, often encrusted by algae. Rays are purplish brown. Interior is pale white and iridescent. Markings of the muscles contrast with a dark gray color. Height of the shell may vary from 1.3 to 4.5 cm.

Biology

Locomotion/behavior: This species has been found to stick itself on the vertical rocks of the pools and the creeks. This helps the animal to move along the upcoming tidal water for food from water.

Food and feeding: It is a herbivore feeding mainly on minute algae.

Reproduction: In this a species, the sexes are uniformly distributed throughout all size groups suggesting the absence of sex reversal in the limpets. The gonads develop from February to May and spawning begins in June and extends to February or March. Peak spawning periods are from June to August and December to February.

Fisheries/aquaculture: It is caught for human consumption.

Other uses (if any): It is edible in the areas of its occurrence.

Cellana sandwicensis (Pease, 1861)

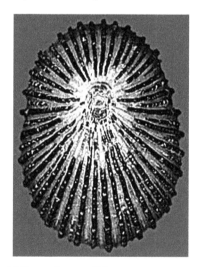

Common name(s): Yellow-foot opihi.

Global distribution: Eastern Central Pacific: Hawaii.

(Its occurrence in Taiwanese coastal waters is largely due to the spread of its larvae transported by ship ballast water.)

Habitat: Water's edge at low tide; on wave-washed coralline algae in the lower intertidal zone; on the mid shore where hydrodynamic forces are severe.

Description: Shell of this species is suboval, with slightly broader posterior end. Apex is subcentral and low. Radiating ribs are strong and subcarinate, extending beyond margin of shell. Shell is dark green on outer surface and shiny silvery white internally. Animal mantle margin is with black-pigmented circumpallial tentacles. A wide black circular band is present on skirt of mantle. Shell muscle is horseshoe-shaped. Head is short, stout, and white. There is a pair of nonpapillate cephalic tentacles which are dark pigmented dorsally. Foot is large, gray on ventral margin, and whitish. Shell is often are covered with seaweed. Maximum diameter of the shell is 3.2 cm and height is 6.5 cm.

Biology

Locomotion/behavior: This species is well adapted to its surf-punded habitat because its low, domed, ridged shell can deflect the force of the water and its single foot can cling so hardly to the rock that about 40-kg pull is necessary to break its shell lose.

Food and feeding: These animals graze on algae and most of them may creep about to graze, but return to their "home scar" after feeding.

Reproduction: In this species, the gametes are shed into the water where fertilization is external. Veligers have a short planktonic life. Spawning occurs mainly in December and January. Supplementation of arachidonic acid and eicosapentaenoic acid in the broodstock diet, fed to the animals during the natural final maturation and spawning season were found to have potential for inducing final maturation in limpets (Hua and Ako, 2014).

Fisheries/aquaculture: It is heavily fished in Hawaiian Islands. Localized heavy fishing pressure is the most significant threat to this species. It is, however, protected by fishing regulations.

Other uses (if any): This is the most popular species for human consumption in Hawaii. It is considered a higher quality food than the black-foot opihi. Native Hawaiians used the shells of this species as scrapers and tools.

Cellana talcosa (Gould, 1846)

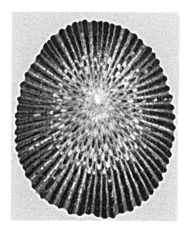

Common name(s): Talc limpet or turtle limpet.

Global distribution: Endemic to the Hawaiian Islands.

Habitat: Low-shallow subtidal shore.

Description: This species is characterized by a flat shell that is thin in juveniles and disproportionately massive in adults. Its mantle tentacles are of varying lengths. Shells are often covered with seaweed. Maximum diameter of the shell is 10 cm.

Locally common on boulders exposed to heavy surf as deep as 20 feet. Shell dome shaped with fine radiating ribs, interior is white. The animal's foot is yellow. Usually cooked to soften the tough muscular foot. Attains 4 in. Rarely encountered west of Moloka'i. Endemic to Hawaii.

Biology

Food and feeding: These animals graze on algae and most of them may creep about to graze, but return to their "home scar" after feeding.

Reproduction: Gametes are shed into the water where fertilization is external. Veligers have a short planktonic life. Spawning information is unknown for the green foot *Cellana talcosa*.

Fisheries/aquaculture: It is heavily fished in Hawaiian Islands. Localized heavy fishing pressure is the most significant threat to this species. Populations in the wild have decreased greatly, and this can impact their reproductive success protected by fishing regulations.

Other uses (if any): Besides eating them, native Hawaiians used the shells as scrapers and tools—Hawaiian Islands; Indo Pacific.

Planaxidae (clusterwinks): Shell of the members of this family is ovate-conical, smooth, or spirally grooved. Aperture is with a very short, distinct siphonal canal. Outer lip is grooved within. Operculum is corneous.

Planaxis sulcatus (Born, 1778)

Common name(s): Tropical periwinkle, sulcate planaxis, or furrowed clusterwink.

Global distribution: Indo-West Pacific: Red Sea; off Mozambique, Kenya, Madagascar, Tanzania and off Mauritius, Chagos, Aldabra, and the Mascarene Basin; off the Pakistani coast.

Habitat: Sheltered rocky intertidal areas. Subtidal, on rubble; in groups on large boulders and seawalls; depth range 0–1 m.

Description: Shell of this large "groovy" snail is conical with strong squarish spiral cords. Shell opening is wide and inner surface is white sometimes with dark purple grooves. Operculum is thin, horny, and dark colored. Exterior shell color is blackish to cream sometimes with white or yellowish spots. Body is pale and its small foot is with a pale underside and dark mottled pattern above. Long tentacles are with dark bands. Females are generally larger than the males. These animals are usually very heavily infected by one or more species of trematodes. Adult shell size varies between 1 and 3.5 cm.

Biology

Locomotion/behavior: These animals are usually active at low tide. They show different avoidance behavior.

Food and feeding: *These animals are* herbivorous, feeding primarily on microalgae. They are active crawling foragers, emerging from the rocks at the incoming tide to graze on these algae. During low tides, it withdraws into its shell behind the operculum and attaches itself to the substratum.

Reproduction: It is a gonochoristic and oviviparous species. Females are generally larger and dominant in this species. It displays poecilogonony as its reproductive strategy. That is the females have a special brood pouch in the foot where the embryos are reared before they are released into the water. Fertilization is internal via copulation between a male–female pairs during summer. Pairing of these snails normally occurs from June through August when their gonads are well developed. The females release veligers. Parthenogenetic development and rearing of embryos up to crawling juveniles have been reported for the populations of the Arabian Sea.

Ranellidae (triton shells or tritons): Shell of the members of this family is thick and ovate-fusiform, with a strong sculpture and axial varices. Periostracum is well developed and hairy. Aperture is with a long siphonal canal. Larger species were used as trumpets in olden days by blowing into the siphonal canal. Operculum is corneous.

Charonia lampas (Linnaeus, 1758)

Common name(s): Knobbed triton or trumpet shell.

Global distribution: Atlantic, Mediterranean, and Indo-West Pacific.

Habitat: Rocky substrata; depth range 0–200 m.

Description: Shell of this species large, solid, and glossy, with a tall pointed spire and an angulated profile. It has 7–8 whorls meeting at shallow sutures. Last whorl is large occupying about two-thirds of shell height. Entire surface is covered with small spiral ridges, but up to ten on the last whorl and two on each whorl in the spire are much bigger than the others and are also nodose. There are several varices and each varix consists of a prominent swelling across the whorl. Aperture occupies a little more than half of shell height. It is a broad oval, pointed above and below and siphonal canal is short and open. Inner lip spreads at its base over an umbilical groove. Columella is fluted with ridges. Outer lip is with paired short ridges internally. Shell is white with brown blotches, especially near sutures, on the inner folds of the outer lip, and at the base of the columella. Flesh is reddish with many scattered brown spots. Each tentacle has two longitudinal black lines. This species harbors epibionts including bivalves and gastropods. Maximum recorded shell length is 40 cm.

Fisheries/aquaculture: Commercial fishery exists for this species.

Other uses (if any): During the neolithic period, the shells of this species were used in necklaces.

Charonis tritonis (Linnaeus, 1758)

Common name(s): Pacific triton, Atlantic trumpet triton, or giant triton.

Global distribution: Throughout the Indo-Pacific Oceans; Red Sea.

Habitat: Among shallow coral and sand.

Description: This species is one of the biggest molluscs in the coral reef. It has a distinctive shell, with a pointed spire and a large body whorl. Exterior shell is creamy with darker brown dashes and chevrons. Aperture is large and orange colored with banded lip. Adult shell size varies from 9.5 to 60 cm.

Biology

Food and feeding: It is a carnivore feeding on echinoderms, especially *A. planci*, a starfish that feeds on corals.

Sterols: The sterols of this species are mainly composed of cholesterol (25.9%), 24-methylcholesterol (16.7%), and 24-methylcholest-7-enol (15.5%). Since the triton feeds on the starfish, *A. planci*, containing Δ7-sterols as principal sterols, the Δ7-sterols of this snail are assumed to have been derived probably from the said starfish species (Teshima et al., 1979).

Other uses (if any): Flesh of this species is eaten in the countries of Indo-West Pacific. Shells of this species are widely sold as decorative items. They are also used as trumpets in some places. In India, a single shell of this species may cost as much as 30 US dollars (Rs. 2000) or more. By feeding on *A. planci* that feeds on corals, these snail populations help maintain the health of coral reefs. Heavy exploitation and excessive trade of these snails may have harmful effects on coral reefs.

Charonia variegata (Lamarck, 1816)

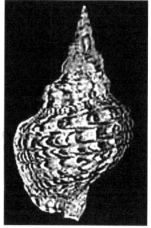

Common name(s): Atlantic triton or Atlantic triton's trumpet.

Global distribution: European waters; Mediterranean Sea; Atlantic Ocean: Cape Verde, off the Canary Islands, Northwest Africa, and Tanzania; Caribbean Sea and Gulf of Mexico; and from North Carolina to eastern Brazil.

Habitat: Shallow subtidal and sandy offshore bottoms around reefs; on algae, boulder, coral, rock, sand and shell habitats; depth range 1–380 m.

Description: Shell of this species is elongate, large, and heavy. Spire is long, conical, and sharply pointed without any knobs. Anterior canal is very short. Varices are present on last whorls. Parietal region is with narrow dark brown inner lip covered by regularly spaced, spirally oriented, white, cord-like plicae. Outer lip is internally with pairs of fine white teeth superimposed on square blotches of dark brown color. Exterior shell color is cream white with brown markings, usually crescent shaped. Inside of the large aperture is orange pink, and the interior is white. Tentacles are yellow with black bands, and the color of the living animal is mottled in shades of yellow to brown. Maximum shell length is 38 cm.

Biology

Locomotion/behavior: During the day it hides in crevices.

Reproduction: The veliger larvae have a period of pelagic development of more than 3 months. The larval shell reaches about 5 mm when fully developed.

Fisheries/aquaculture: Commercial fishery exists for this species.

Other uses (if any): In most countries, it is illegal to bring back the shells of this species from holidays.

Ranella olearium (Linnaeus, 1758) (=*Argobuccinum olearium*)

Common name(s): Wandering triton, little frog triton or olive trumpet.

Global distribution: European waters; Mediterranean Sea; Central and South Atlantic Ocean: Cape Verde and West Africa; Indian Ocean: Mozambique and South Africa; New Zealand, Caribbean Sea (along Colombia) and Southwestern Pacific.

Habitat: Sandy or muddy bottoms of outer shelf-upper bathyal regions; depth range 100–280 m.

Description: It is a highly variable species. Shells are large, elongated, thick, and sturdy, with rounded whorls and with tubercles more or less developed on some sutures. Mouth is large and has a rounded section. Siphonal channel is moderately long and lip bears many teeth, often double, arranged along the polished, white or brown edge. External surface of the shell is brown ocher, with clearer tubercles and other protruding parts. Inner surface and columella are white. In the living individual, the shell is often covered with an outer velvety layer. Shell size of this species varies between 9 and 24 cm.

Tegulidae (top shells): Shell of the members of this family is often large, heavy, conical, with rounded shoulders. Umbilicus is deep and round. Operculum is frequently circular and is well developed.

Cittarium pica (Linnaeus, 1758)

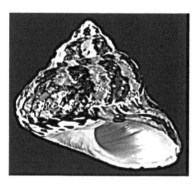 

Common name(s): West Indian top shell or magpie shell.

Global distribution: Western central Atlantic.

Habitat: Shallow subtidal, on rocks and shell rubble; intertidal and offshore coral reefs, particularly on algae, boulder, and coral habitats; depth range 0–7 m.

Description: Shell of this species is large, heavy, and conical. Shell is between trochiform and turbiniform in shape with rounded shoulders.

Spire is conoidal. It contains about six convex whorls. Large body whorl is depressed globose. Outer lip is simple. Columella is arcuate. Umbilicus is deep and round. Operculum is multispiral and circular. Exterior shell color is purple-black on a whitish background. Aperture is white and internally it is nacreous. Operculum is iridescent brown. Maximum shell length is 13.7 cm.

Biology

Food and feeding: It is a nocturnal and herbivore, feeding on a large variety of algae, and sometimes also on detritus. It actively scrapes the algal growths off rocks, and this tends to cause erosion over time. Feeding commonly occurs during night, when these snails are most active.

Reproduction: *Cittarium pica* is dioecious (i.e., sexes are separate) and fertilization is external. During the reproductive season, which normally occurs in the field from June to November during lunar period. Males release their sperm into the water, as females simultaneously release their green colored unfertilized eggs. The encounter of those gametes produces yolky fertilized eggs, which will further develop into lecitotrophic (yolk feeding) larvae. The latter emerge from the egg capsules as shell-cap-bearing trocophore larvae which do not spend much time in the plankton. Settlement occurs immediately after about 4 days. Individuals attain sexual maturity at shell lengths of about 3 cm. The lifespan of this species which is still unknown may be about 30 years.

Association: A small limpet, *Lottia leucopleura*, often lives on the underside of the shell of this species. The crab *Pinnotheres barbatus* is a commensal. The sessile, ringed wormsnail, *Dendropoma corrodens* and the tube dwelling polychaete *Spirorbis* may live attached to the shell of this species along with several species of algae. In the wild, the shell of this species is used extensively by the large land hermit crab species *Coenobita clypeatus*.

Fisheries/aquaculture: Commercial fishery exists for this species. Stocks have been locally extirpated or are diminishing due to over exploitation. Nowadays, *C. pica* is a legally protected species in Bermuda, where its collection is forbidden. Because of their popularity as a food item and overfishing, in the US Virgin Islands, there are territorial regulations to protect these snails: collection is not allowed during reproductive season, and prescribing minimum harvest size. A majority of whelk fishers have observed negative changes in size and abundance of whelks over the last decade, and made suggestions for management action (Nelson et al., 2012).

Other uses (if any): Flesh of this species is a popular food item in Venezuela, Guadeloupe, and in the English-speaking Caribbean islands of the West Indies. These large sea snails are boiled and eaten in a variety of different local recipes. *C. pica* is considered the third most economically important invertebrate species in the Caribbean, after the spiny lobster (*Panulirus argus*) and the queen conch (*E. gigas*).

Omphalius rusticus (Gmelin, 1791)

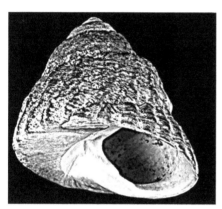 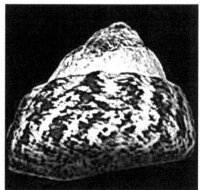

Common name(s): Wild turban snail.

Global distribution: Off Japan.

Habitat: Sublittoral.

Description: Shell of this species is umbilicate, heavy, and solid with a conical shape. Conical spire is more or less elevated. Suture is distinctly impressed. There are six to seven whorls which are moderately convex or nearly flat, and either smooth or longitudinally plicate. Folds are usually obsolescent, and visible only for a short distance below the sutures. Shell is spirally obsoletely striate. Body whorl is obtusely angular at the periphery. Base of the shell is nearly flat. Aperture is very oblique. Columella shows one or two teeth below. Umbilicus is narrow and deep circular. Exterior shell is chocolate colored or brownish-olivaceous. Height and diameter of the shell attain 3 cm.

Biology

Reproduction: *Omphalius rusticus* is dioecious and oviparous. The ovary and the testis are composed of a number of oogenic follicles and several spermatogenic follicles, respectively. First sexual maturity of female and male snails is at about 1 cm in shell height. Reproductive cycle of this species can be categorized into five successive stages: in females, early active (October to April), late active (December to June), ripe (April to September), spawning (July to September) and recovery (September to January); in males, early active (November to March), late active (December to June), ripe (April to September), spawning (July to September) and recovery (September to December). Gonadal development, gameto-genesis, reproductive cycle, and spawning are closely related to the seawater temperature (Lee, 2001).

Tectus niloticus (Linnaeus, 1767) (=*Trochus niloticus*)

Common name(s): Commercial top shell.

Global distribution: Indo-Pacific (Indian Ocean, New Ireland, New Caledonia, North Australia, French Polynesia, etc.).

Habitat: Atoll reefs along the reef crest or on reef slopes at depths of 0–20 m; larger specimens in deeper water, but small ones under stones on the intertidal areas; juveniles prefer shallow areas on intertidal reef flats.

Description: It has a large conical shaped and subperforate shell. It is covered by a corneous striate, brown or yellowish cuticle which is lost on the upper whorls. Its color beneath the cuticle is white, longitudinally striped with crimson, violet, or reddish brown. Base of the shell is maculate or radiately strigate with a lighter shade of the same. Spire is strictly conical. Apex is acute and usually eroded. Shell contains 8–10 whorls. Upper whorls are tuberculate at the sutures, and spirally beaded and the following whorls are flat on their outer surfaces, smooth, separated by linear suture. Body whorl is expanded, dilated, and compressed at the obtuse periphery, more or less convex below, and indented at the axis. Umbilical tract is covered by a spiral pearly deeply entering callus. Aperture is transverse and very oblique. Columella is oblique, terminating in a denticle below, and with a strong spiral fold above, deeply inserted into the axis. Operculum is circular, thin, corneous, orange-brown, and composed of about 10 whorls. Body pale, large foot pale on the underside, mottled on the upper side. A pair of long tentacles at the head which is brown with three white circles. Length of the shell varies between 5 and 17 cm. and its diameter between 10 and 12 cm.

Biology

Food and feeding: It feeds on very small plants and filamentous algae.

Reproduction: This species has separate sexes and are able to reproduce at about 2 years of age when they reach a base diameter of 5–7 cm. Spawning occurs throughout the year in warmer areas and during the warmer months in cooler areas. During spawning, females release more than 1 million eggs that are fertilized by sperm released by males. The fertilized eggs hatch to very small larval stages which drift with currents for up to 5 days before settling on a rocky surface. After 2 or more years they may become adults. They can live for up to 15 years.

Circadian behavior: This species has been reported to undertake circadian behavior. Its activity began at dusk and gradually stopped during the night, before sunrise. This nocturnal behavior was characterized by short movements and was associated with foraging behavior. We assumed that activity ceased once the animal was satiated. Generally smaller specimens displayed greater activity. This nocturnal behavior is mainly associated with attempting to avoid visual predators whilst feeding (Jolivet et al., 2015).

Conservation status: This species is now listed as "Vulnerable" on the Red List of threatened animals of Singapore. Like other creatures of the intertidal zone, it is affected by human activities such as reclamation and pollution. Overcollection of shells may also have an impact on local populations.

Other uses (if any): This species is the most economically important in the tropical West Pacific as an important traditional food and a leading export item as the source of mother-of-pearl buttons and jewelry. As a result of overfishing, aquaculture trials are underway. It is also used in large aquaria.

Tectus pyramis (Born, 1778)

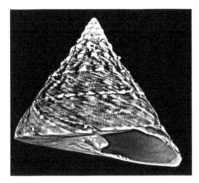 

Common name(s): Pyram top shell, pyramid top or green top shell.

Global distribution: Indo-Pacific: from Madagascar, India, Sri Lanka, Andaman and Nicobar Islands, and Christmas Island, to Southeast Asia and Australia; American Samoa and Japan and China.

Habitat: Rocky shore habitats; shallow subtidal zones; depth range 0–10 m.

Description: Shell of this species is imperforate, solid, and thick with a strictly conical shape. Spire is more or less attenuated above. Apex is acute. Shell contains 12–14 whorls. Upper whorls are slightly extended outwards and plicate, tuberculate or undulating at the sutures. Folds or tubercles are obsolete on the lower whorls. Body whorl is beaded, but smoother than the preceding, or radiately finely wrinkled. Base of the shell is flat, concentrically lirate and the ribs are smooth, wide and are, separated by shallow grooves, obsolete toward the outer margin. Aperture is transverse, very oblique, and subtriangular, and the outer wall is grooved within. Columella is very short, with a very strong acutely carinated spiral fold. Exterior color of the shell is yellowish or grayish and more or less mottled and marbled with green or brown. Base is white, green, or brown. Size of the adult shell varies between 4.5 and 15 cm.

Biology

Arsenic content: This species has been reported to contain two major arsenic compounds arsenic constituents, namely, arsenobetaine and tetramethylarsonium ion (Francesconi et al., 1998).

Fisheries/aquaculture: Commercial fishery exists for this species.

Other uses (if any): Though it is edible, caution needs to be exercised while collecting this species from different habitats.

Trochus radiatus (Gmelin, 1791)

Common name(s): Radiate top shell, redstripe trochus snail or banded trochus.

Global distribution: Indian Ocean off Madagascar; Western Pacific.

Habitat: Intertidal rock boulders; mangroves.

Description: Shell of this species is thick, rather solid and trochoidal with a moderately elevated spire. Spire has nearly straight outlines. Apex is acute, generally eroded and orange colored. There are about seven whorls which are planulate, sometimes a little concave in the middle. Body whorl has a sharply angled periphery. Shell upper surface is circled by irregularly beaded bands which are five or six on each whorl, uneven in size. Base of the shell is nearly flat, concentrically lirate. Lirae are granulose, rather coarse, with broad interspaces, which are frequently occupied by revolving lirulae or striae. Oblique columella is strongly plicate above and its edge is nearly smooth with blunt teeth. Large aperture is subrhomboidal, lirate within, and grooved. Basal lip is thickened and crenate. Umbilicus is wide and deep. Umbilical tract is funnel shaped, rather broad, with a central rib. Color of the shell is yellowish whitish, tinged with green, and radiately striped with broad or narrow uninterrupted, axial, crimson flames. Base of the shell is white or pink, radiately marked or minutely speckled with red. Aperture, columella, and umbilical area are pearl white. Adult and large shells are normally encrusted with algae. Shell length varies between 1.7 and 4.0 cm.

Biology

Food and feeding: Members of this family browse on detritus and algae. They may also undertake filter-feeding.

Other uses (if any): Flesh of this species is an important food item in many countries. It is also an ornamental shell for use in shellcraft.

Tegula funebralis (Adams, 1855)

Common name(s): Black turban snail or black tegula.

Global distribution: Northeast Pacific: British Columbia, Canada to Mexico.

Habitat: High to middle intertidal zones on rocky surfaces or in pools not covered in algae.

Description: This species is similar to *Tegula gallina* in form and characters of the aperture. Shell is lusterless, purple or black. Apex is usually eroded and orange colored. Teeth of the columella are white. Whorls are spirally lirate. Surface is margined below by an impressed line, and by elevated, foliaceous incremental lamellae. Empty shells of this species are very often used by hermit crabs, especially *Pagurus samuelis*. Adult shells have a maximum length of 3 cm.

Biology

Locomotion/behavior: When fleeing a predator on a sloping substrate, the snail may simply detach itself and thus it will roll or drop away.

Food and feeding: These snails graze on many species of micro and macroscopic algae.

Reproduction: It is sexually dimorphic and not hermaphroditic. Larvae in laboratory metamorphose at about 6–7 days. These snails may live up to 25 years.

Predators: Predators of this species include sea otters, and predatory starfish such as *Pisaster ochraceous*.

Fisheries/aquaculture: It is harvested in the early period by the Native American peoples. Overharvesting for food in the early 1900s, mainly by Southern European immigrants, caused a drastic decline in populations. They are still harvested by both Americans of Southern European and Asian ancestries. Therefore, heavy regulations have been promulgated by the Department of Fish and Game in order to recover the populations of this species.

Other uses (if any): It serves as a food in areas of its occurrence.

Turbinidae (turbans): Shell of the members of this family is thick and turbinate to conical. Outer sculpture is often spiral to nodular. Aperture, which is without a siphonal canal, is rounded and nacreous within. Operculum is thick and strongly calcified. Opercula of different species are in different colors and patterns and are sometimes called "cat's eye." They are used in shell jewelries. Turban snails feed on algae.

Astralium calcar (Linnaeus, 1758)

 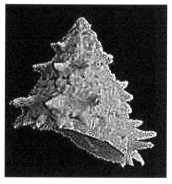

Common name(s): Spurred turban or Ninja star snail.

Global distribution: Indo-West Pacific: off the Philippines, Indo-Malaysia and Queensland, Australia.

Habitat: On rocky shores and reefs; shallow subtidal waters; depth range 3–12 m.

Description: Shell of this species is flattened and wheel shaped and is more or less depressed at the apex. There are six whorls which are flattened above, and radiately plicate, and the folds are rather unequal and irregular. Periphery is carinated spinose, bearing 12 radiating more or less foliated spines upon the body whorl. Body whorl descends deeply toward the aperture. Convex base is concentrically more or less densely squamosely lirate. Outer lirae are generally prominent and subspinose, sometimes making the periphery to appear bicarinate. Aperture is transversely oval, very oblique, white, smooth, and pearly. Operculum is small, chalky, and hemispherical with a smooth glossy surface. Body is pale with fine black stripes and long tentacles with fine black bars. Shell is usually encrusted and thus well camouflaged on the rocks. Shell's color pattern is grayish greenish or brownish cinereous. Maximum shell length is 6.0 cm.

Other uses (if any): This species is sometimes gathered for food by coastal dwellers. It is occasionally appearing on local markets of the northern Philippines.

Bolma rugosa (Linnaeus, 1767)

Common name(s): Turbo cap or Eye of Saint Lucia.

Global distribution: Mediterranean and near Atlantic.

Habitat: On the rocky and muddy bottoms rich in brown algae; Posidonia meadows and coralligenous zones; depth range 3–100 m.

Description: Shell of this species is solid and imperforate with a conic shape. Suture is canaliculate and is bordered below by a series of curved radiating tubercles. There are six to seven whorls which are obliquely lamellose striate. Upper whorls are carinate and tuberculate or spinose at the periphery. The body whorl descends rounded or bicarinate and is spirally lirate. Base of the shell is conspicuously radiately striate. Aperture is obliquely, transversely oval, and pearly within. Columella is arched, white, and pearly. Operculum is short-oval and brown within; and it is with four whorls and the nucleus is situated one-third the distance across the face. Exterior shell is bright orange and is polished, with a spiral callous ridge. Size of the shell varies between 2.5 and 8.0 cm.

Biology

Food and feeding: It feeds mainly on the most tendered parts of algae with its large-toothed radula.

Reproduction: The sexes are separate in this species and its spawning season is between March and July.

Association: The shell of adult specimens is almost entirely covered with brown algae and small worms with calcareous tubes. Further, the shell is often colonized by bernard-giant hermit (*Dardanus arrosor*).

Fisheries/aquaculture: It has commercial fishery and is fished artisanally with gill nets.

Other uses (if any): This gastropod is edible for humans. It is found on the stalls of fishermen mainly in Sicily and France on the Vieux Port in Marseille.

Cookia sulcata (Gmelin, 1791)

 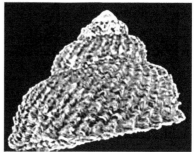

Common name(s): Cook's turban.

Global distribution: Off New Zealand.

Habitat: Ledges or seaweeds; under rocks or along beaches on the open coast at low tide.

Description: This species has a shell which is heavy, solid and strong. Shell is spirally coiled forming a protoconch with up to four whorls and has a rounded to subangulated periphery with a strong and arcuated shell surface. Spiral structure inclines forward and is overridden by sharp spiral cords. Entire shell is crossed by fine lamellose growth striate. These fine striae are rarely visible on adult shells as layers of calcified material form a coating up to about 6 mm thick on the top and sides of the shell. Spire of the shell is often encrusted with white, calcareous algae, which gives the shell a red color. Surface of the shell is often worn off, showing the silver layer underneath. Interior is white and is sealed off by an operculum. The large oval shelly operculum is an oblong plate with several spiral ridges. Size of an adult shell varies from 6.5 to 8.0 cm in height and from 6.5 to 9.0 cm in width.

Biology

Biocontrol agent: This species has been reported to be a potential biocontrol agent to mitigate biofouling on marine structures. On pontoons, this species was found to reduce established biofouling cover by >55% and largely prevented the accumulation of new biofouling over 3 months. On wharf piles it removed 65% of biofouling biomass and reduced its cover by 73% (Atalah et al., 2014).

Food and feeding: It is a herbivore feeding on algae and seaweed.

Other uses (if any): In New Zealand, Maori people consume the meat of this species and use the shells for making tiny detailed wood-carving and for making fish-hooks. The shells of this snail also have a commercial value after it is polished and carved into exquisite spoons or mounted in gold or silver, set with jewels.

Lunella cinerea (Born, 1778) (=*Turbo cinereus*)

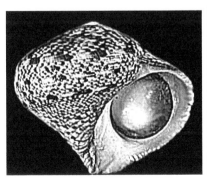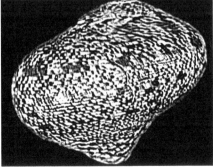

Common name(s): Smooth moon turban.

Global distribution: Eastern Indian Ocean and tropical West Pacific: from eastern India to Melanesia; Japan, Queensland, and New Caledonia.

Habitat: Intertidal; among rocks or gravel.

Description: Shell of this species is solid and umbilicate with a depressed-globose shape and a strong spiral sculpture. Spire is obtuse. Suture is slightly undulating. There are five whorls which are spirally lirate and with lirulae in the interstices. Umbilicus is a narrow, but very deep. Mouth has a rounded shape. Inner lip in its lower part is widened out. Outer lip is thickened. Exterior shell is reddish brown with small greenish spots. Mouth is pearlescent. Size of the shell varies between 2 and 5 cm.

Lunella coronata (Gmelin, 1791) (=*Turbo coronatus*)

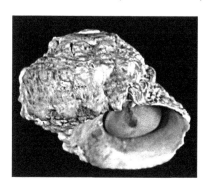 

Common name(s): Crowned turban shell, coronate moon turban or horned turban.

Global distribution: Red Sea, off Southeast Africa, the Mascarene Basin and in the Indo-Pacific.

Habitat: Coastal rock or rocky tidal intertidal zone; on both boulder and sandy shores.

Description: Shell of this species is solid and imperforate with a depressed-turbinate shape. It is covered with an irregular spiral series of nodules and granules. Spire is depressed and dome-shaped with an apex that is frequently eroded and red. Shell contains four to five whorls, of which the last is very large. Aperture is large, round, and iridescent within. Wide columella is flattened and excavated; deflexed recurved and somewhat channeled at its base. Inside of the operculum is flat, greenish and golden, iridescent, with about 5–6 whorls and a subcentral nucleus.

Its outside is convex, greenish, and sparsely granulate all over. Exterior shell coloration varies from brown to reddish, and to white for the heavily encrusted specimens. Interior is nacreous. Shell grows to a maximum length of 4 cm.

Other uses (if any): In Indonesia, these snails are eaten and their shells are used for craftsmanship.

Lunella torquata (Gmelin, 1791)

Common name(s): Twisted necklace.

Global distribution: From New South Wales to Western Australia and off New Zealand.

Habitat: Shallow waters of rocky reefs where they shelter in crevices and under rocks; living amongst algae.

Description: Shell of this species is large, solid, and umbilicate with an orbiculate, conic shape. This species varies much in degree of elevation and carination. There are six whorls which show dense lamellose incremental striae and coarse spiral lirae. Upper whorls are carinated and the carinas are becoming obsolete on the body whorl. Sutures are canaliculated and are bordered below by a row of nodules. Round aperture is oblique and white within. White columella is perforated by the wide and deep umbilicus. Oval operculum is flat within and is with concentric groove. Its nucleus is situated one-third the distance across the face. Its outside is white and is with two strong spiral ribs. Exterior shell is whitish, mottled, and strigate with dark brown. It grows to about 11 cm wide and 10 cm high.

Biology

Food and feeding: It is herbivorous, feeding on algae.

Lunella undulata (Lightfoot, 1786)

Common name(s): Common warrener or lightning turbo.

Global distribution: Endemic to Australia; off New South Wales, South Australia, Tasmania, Victoria and Western Australia.

Habitat: Exposed and semiprotected rocky shores; crevices and under rocks in the shallow intertidal and subtidal zones.

Description: Shell of this species is solid and umbilicate with a depressed-globose shape. Obtuse spire is dome shaped or low-conic and contains five whorls. Upper whorls are angulate and spirally lirate. Last whorl descends and is somewhat concave below the suture. Oval aperture is white within. Columella has a very wide white flattened callus which extends over the umbilical tract. Umbilicus is wide and deep. Operculum is rather circular and chalky-white. Exterior shell is bright green and is longitudinally strigate with white under a brown epidermis. Size of the shell varies between 3 and 7.5 cm.

Biology

Food and feeding: It is a herbivore feeding on algae.

Fisheries/aquaculture: A Tasmanian fishery exists for this species.

Megastraea undosa (Wood, 1828) (=*Astraea undosa*)

Common name(s): Wavy turban snail.

Global distribution: Eastern Pacific: USA to Mexico.

Habitat: On rocks and in kelp beds; from low intertidal to a depth of more than 80 m.

Description: Shells of this species are heavy and sculptured with a turbinate-conical shape and without an umbilicus. Like other shells of the family turbinidae, it is composed of a thick inner nacreous layer, covered by a thinner porcellanous layer. Periphery of the shell forms a twisted ridge at the outer edge of each whorl. Each whorl also has regular, coarsely sculpted rows of fine knobs and folds. Base is marked with several spiral cords concentric to the arcuated columella which has a pearly groove. Operculum has four strong ridges on its outer side decorated with hard shelly bristles that radiate in a curvilinear fashion from its pointed edge. Exterior shell has a light brown or tan color. Periostracum is often covered with coralline algae and other epiphytes. Shell size varies from 4 to 15 cm.

Biology

Locomotion/behavior: To escape predation within kelp forests wavy turban snails crawl or migrate up into the canopy of the giant kelp plants each night.

Food and feeding: The wavy turban snail is a herbivorous generalist, and individuals have been observed feeding on kelp and coralline algae.

Reproduction: These snails are slow growing. Growth rates are higher for smaller sized snails and progressively slower as size increases. Sexual differences in growth rate have been observed with females growing more slowly than males. Reproductive activity is year-round with major peaks in the spring and fall. Fully mature gonads are seen in females with shell diameter greater than 9 cm and males greater than 7.5 cm. There are three reproductive phases occur during the year. Gonad growth and maturity take place during the spring and early summer, followed by spawning in late summer. Somatic growth occurs during the fall and winter. Recruitment of new juveniles has been observed from January to April.

Predators: Predators of this snail are likely the sea stars and the Kellet's whelks based on demonstrated escape responses in laboratory experiments. Other predators include octopuses, lobsters, and fishes.

Other uses (if any): Flesh of this species is eaten in California and West Mexico. Today, turban snails are of commercial value in southern California and Baja California, Mexico. Current market demand for this species is for the foot, which is processed and sold to restaurants as an abalone-like product called wavalone. Other potential markets for this species occur in Mexico.

Turbo argyrostomus (Linnaeus, 1758)

Common name(s): Silver-mouthed turban.

Global distribution: Indo-Pacific.

Habitat: Intertidal; moderately exposed habitats; coral reef and rocky areas; lagoons and atolls; depth range 0–30 m.

Description: Shells of this species is solid and large with an ovate-pointed shape. It is a very variable species. Apex is almost pink. There are six whorls which are convex and separated by subcanaliculate sutures. Upper two whorls are smooth, and the lower whorls are spirally lirate and radiately more or less squamose striate. Lirae are subequal and nearly smooth. Main body whorl has 20 distinct spiral ribs which are mostly flat topped and some are with fluted scales. Body whorl contains 13 lirae, which are wider than their interstices. Penultimate and last whorl bear numerous elevated vaulted scales upon the lirae. Aperture is pearly white or brownish tinted within, about half the length of the shell. It is round-ovate, angled above, dilated, and subchanneled below. Columella is thickened, slightly flattened, and grooved below the narrow deeply perforating umbilicus. Operculum is flat inside with five whorls. Outer surface is convex, with coarse obtuse granules, which are largest upon the higher part, nearly surrounded by a marginal series of fine oblique wrinkles. Exterior shell color is white, more or less tinged with flesh color upon the outer half, and with a narrow marginal orange line. Maximum shell length is 10 cm.

Fisheries/aquaculture: Commercial fishery exists for this species.

Other uses (if any): It is one of the most frequently collected species in the Southwest Pacific. Flesh of this species is used as food and its shell is used for making buttons.

Turbo bruneus (Röding, 1798)

Common name(s): Brown (Pacific) dwarf turban, brown turban snail, little burnt turbo or brown pacific turban.

Global distribution: Indo-West Pacific: from Madagascar and India to eastern Indonesia; north to the Philippines and south to northern Australia.

Habitat: Rocky shores; shallow intertidal and subtidal waters.

Description: Shell of this species is thick and heavy with numerous spiral ridges running along the whorl. Middle ridge is largest ending at the margin of the outer lip as distinct tooth. Narrow and deep umbilicus with a thick keel around it. Chalky operculum is hemispherical with many tiny bumps and is dark green with grayish and white margins. As this structure adapts and protects the opening of the shell, it is often called as the "eye of cat." Body is with brown mottles and a pair of slender tentacles. Exterior shell color is dark greenish-brown irregularly marked with yellow blotches. Inner lip is shiny. They have a lid limestone thick enough. Length of the adult shell varies between 2 and 6 cm.

Biology

Ecology: These snails are very tolerant with the physicochemical parameters of the water of the reef aquaria.

Locomotion/behavior: It shows no danger either for its own congeners than for other occupants of an aquarium.

Food and feeding: It is herbivorous feeding on small epibenthic algae and vegetable detritus.

Reproduction: Sexes are separate in this species. Life expectancy does not exceed 4 years in wild, but it can reach 5 years in aquarium.

Other uses (if any): In several countries, this species is collected for food, and the eye of the cat (operculum) is used in jewelry. It is a very suitable species for reef aquaria.

Turbo chrysostomus (Linnaeus, 1758)

Common name(s): Gold-mouthed turban.

Global distribution: Indo-West Pacific: off Madagascar and Mascarene Basin; Philippines, New Caledonia, Samoa and off Australia (Northern Territory, Queensland and Western Australia).

Habitat: Coral reef and rocky areas; intertidal and shallow sublittoral zones; depth range 0–20 m.

Description: Shell of this species is moderately large, solid and heavy, and turbinate in shape. Spire is well developed and pointed. Whorls are strongly convex with angular shoulders. Outer sculpture is variable, with rounded, unequal spiral cords, and many fine scaly axial threads. Body whorl is biangulate, with stronger spiral cords at shoulder and periphery, each bearing a row of short open spines or nodules. Aperture is roughly rounded-ovate, forming a blunt angle at the slightly flaring anterior end of smooth columella. Outer lip is marginally serrate and smoothish inside. Umbilicus is closed or reduced to a slight chink. Operculum is almost circular in outline, with a subcentral nucleus. Exterior of operculum is very convex and smooth, with fine oblique grooves on its outer margin.

Outside of shell is brownish or cream colored, often marbled with irregular axial stripes of darker brown and/or green. Aperture is bright orange to golden yellow with white margins. Exterior of operculum is brown or dark green, becoming whitish at periphery. Length of the adult shell varies between 3.5 and 8 cm.

Enzyme: A new enzyme, arylsulfatase (EC 3.1.6.1) has been isolated from the liver of this species. This enzyme is capable of catalyzing the hydrolysis of potassium p-nitrophenylsulfate. The inactivation half-life of this enzyme was found to be 15 min at 55°C. The bioactive role of this enzyme is yet to be known (Pesentseva et al., 2012).

Fisheries/aquaculture: This species is heavily exploited in Fiji Islands.

Other uses (if any): Flesh of this species is used as food by coastal people. Shells of this species are used in shellcraft.

Turbo cornutus (Lightfoot, 1786) (=*Batillus cornutus*)

Common name(s): Horned turban or Japanese turban shell.

Global distribution: Northwest Pacific: Japan, South Korea and Hong Kong.

Habitat: Shallow coastal waters; on rocks; depth range 0–40 m.

Description: Shell of this species is hard, spiny, and imperforate. It has a large, thick, green-gray shell with irregular incremental striae and spiral lirae. Shell has about 5–6 whorls, which turn clockwise and have horny protuberances. Body whorl is ventricose, slightly bicarinate and is armed about the middle with two spiral series of erect tubular spines. Sutures are deeply impressed. Oblique aperture is rounded and is green or red-brown. Thin inside lip of the shell is not smooth, but rough and granular. Broad columella is flattened and somewhat grooved, produced and channeled at its base. Operculum is calcareous, concave and green or red brown. Length of the adult shell varies between 6.5 and 14 cm.

Biology

Food and feeding: It feeds on various kinds of algae including larger seaweed, *Sargassum hemiphyllum*, *Padina* spp., *Corallina* spp., and encrusting brown algae. Young horned turban shells eat red-turf algae, *Gelidium pusillum*, and *Polysiphonia* spp.

Reproduction: *Turbo cornutus* spawns from August to September. Lecithotrophic larvae have a very short period (3–5 days) as free-floating plankton, after which they settle down and metamorphose into adults.

Fisheries/aquaculture: Commercial fishery exists for this species. This species is a commercially important shellfish in Japan where artificial enhancement of its production is attempted by stocking of open-sea areas with a large number of artificially bred juveniles.

Other uses (if any): The flesh of this species is enjoyed as a delicacy in Japan and China.

Turbo crassus (Wood, 1828)

Common name(s): Crass turban, heavy turban or thick turban.

Global distribution: Western Pacific and Polynesia; off Australia (Queensland).

Habitat: Among seaweeds and under stones sublittorally; sandy bottoms; near reefs in shallow subtidal waters; depth range 1–50 m.

Description: Shell of this species is large, heavy, solid and, imperforate with an ovate-conic shape. Its shell color pattern is dirty white, or greenish, maculated with angular, alternating blackish or brown and light patches on the broad flat spiral ribs. Interstices are narrow, superficial, and whitish. There are six whorls which are convex and more or less prominently shouldered above. Ribs are obsolete around the axis. Aperture is white within. It is ovate and angled posteriorly. Its margin is more or less green tinged. Operculum is subcircular, concave internally, with a nucleus. Its outer surface is very convex and the center is dark-brown. Size of the adult shell varies between 5 and 8.5 cm.

Fisheries/aquaculture: This species has subsistence fisheries.

Other uses (if any): The flesh of this species is also uses as bait occasionally.

Turbo intercostalis (Menke, 1846)

Common name(s): Ribbed turban.

Global distribution: Western and Southwestern Australia; in the Indian Ocean off Chagos, the Mascarene Basin and the Hawaiian Islands.

Habitat: Marine; epifaunal.

Description: Shell of this species is solid, smooth, and perforate with an ovate-conic shape. There are six whorls which are convex and sometimes subangulate above. They contain several unequal revolving lirae and obsolescent incremental striae. Aperture is round, and its upper angle is sometimes separated from the body whorl and projecting. Base of the shell is rounded. Columella is excavated at the umbilicus. Chalky operculum is hemispherical and smooth. Its dark green center is with yellowish and white margins. Exterior shell color pattern is green or gray and is radiately flammulated with black, green, or brown. Length of the adult shell varies between 2.5 and 8 cm.

Turbo marmoratus (Linnaeus, 1758)

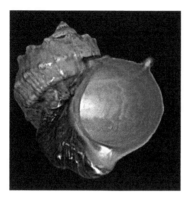 

Common name(s): Green turban, marbled turban or great green turban.

Global distribution: Indo-Pacific: East Africa, to Melanesia; Japan and Queensland and Fiji Islands; introduced in French Polynesia.

Habitat: Subtidal, coral reef areas open to a constant flow of clean oceanic water; depth range 1–30 m.

Description: Shell of this species is large, imperforate, solid, ventricose, and as broad as long. There are 6–7 whorls which are flattened or concave above. Its large, circular aperture has a golden, pearly shine. Base of the shell is produced. Columellar region is more or less excavated. Subcircular operculum is somewhat concave within. Its outer surface is closely tuberculate and whitish. Exterior shell color pattern is green, marbled with white and rich brown. Maximum shell length is 22 cm.

Biology

Locomotion/behavior: The animal is active at night (nocturnal) and prefers dead coral beds where micro- and macroalgae grow abundantly.

Food and feeding: It is primarily herbivorous feeding on algae including microalgae at night. It may also few small animals that are associated with the assemblage of algae.

Reproduction: Spawning occurs in this species in evening. Male releases white clouds (sperm) directly into the water. The sperm are normally expelled through its siphon by alternate contraction-relaxation movement of the soft part of the body. Further, the sperm are intermittently released every 5 to 10 min. The spawning in the female is very short (for about 30 min) but intense. The total number of eggs spawned from the female has been estimated at 1.8 million. Fourteen hours after fertilization, the eggs are hatched into free-swimming trochophore larvae, and after about 22 h (i.e., 36 h after fertilization), most of the larvae metamorphose into veliger larvae. At 60 h old, these larvae reach the pediveliger stage and move on the substrate before finally metamorphosing into benthic living juveniles.

Conservation status: This species is believed to be endangered in India.

Fisheries/aquaculture: This green snail is harvested for its protein-rich flesh and valuable shell, which has a very high mother of pearl content. According to the Food and Agriculture Organization of the United Nations, world production of green snail shell was estimated at 800 t and 1000 t for 1986 and 1987, respectively. Intensive fishing for the mother-of-pearl trade has drastically reduced many turban populations in the recent years. In order to increase the possibilities of long-term exploitations of this species, several attempts are under way in the area. These include juvenile production, reintroduction, translocation, and commercial legislation.

Other uses (if any): Flesh of this species is an important part of the diet of fishermen and local communities throughout the Indo-West Pacific. Its nacreous shell is used in the manufacture of buttons, and as inlay material for lacquerware, furniture, and jewelry. The shells of marbled turbans are used as a source of nacre and are sold as decorative items. The large opercula of *Turbo marmoratus* are sold as paperweights or door stops. The pearly shell of green snail is also used in the ornamental, handicraft, paint, and cosmetic industries.

Turbo militaris (Reeve, 1848)

Common name(s): Military turban.

Global distribution: Off Australia from North Queensland to New South Wales, Australia.

Habitat: On rocky shores, from low tide down to about 5 m.

Description: Shell of this species is large and solid with rounded whorls. It is variable in its external morphology, due to the presence or absence of spines. There are both smooth and spiny forms in which two rows of open-fronted spines are seen on the body whorl. There are also forms with a morphology between these two categories. All these forms differ in the presence of the anterior canal, which is almost nonexistent in the smooth forms, but prominent in the spiny forms. Aperture is subcircular and pearly white within. Outer lip is simple and rather thin. Columella is smooth with a white callus with green edges. Subcircular operculum is calcareous. Its outer surface is white with a slight green. Exterior shell color pattern is formed by the spiral bands of brown or green over a fawn background. Length of the adult shell varies between 6 and 10 cm.

Turbo petholatus (Linnaeus, 1758)

Common name(s): Tapestry turban.

Global distribution: Indo-Pacific.

Habitat: Intertidal shallow coral reefs and rocky shores in protected habitats; sublittoral; depth range 0–40 m.

Description: Shell of this species is moderately large, thick, heavy, and turbinate in shape with length equal to or slightly greater than width. Spire is well developed and pointed. Whorls are strongly convex with a rounded outline becoming somewhat flattened beneath the impressed sutures. Outer surface of shell is smooth and highly polished. Aperture is rounded-ovate and is extending on about half the total length of shell. Outer lip is thin and smooth inside. Columella is smooth, without an umbilicus. Operculum is nearly circular in outline, with a subcentral nucleus and a convex, smooth external surface. Outside of shell is variable in color and pattern, usually brown, red, orange, or greenish and is often ornamented with dark spiral bands and/or thin, chevron-shaped, pale-colored axial stripes. Aperture is silvery white inside, often suffused with yellow, orange, or green on margins, especially on inner lip margin. Exterior of operculum is shiny and bluish-green in the center, becoming brown toward the margins. Maximum shell length is 8.5 cm.

Fisheries/aquaculture: Commercial fishery exists for this species.

Other uses (if any): This species is collected for food and for its highly polished, colorful shell. Its operculum is well known in shell jewelry under the name "cat's eye."

Turbo sarmaticus (Linnaeus, 1758)

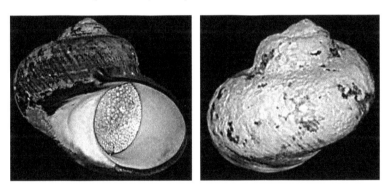

Common name(s): South African turban, giant periwrinkle or African turbo shell.

Global distribution: Southeast Atlantic: off the south coast of South Africa.

Habitat: Depth range 0–8 m.

Description: Shell of this species is imperforate with a globose-depressed shape. Its color pattern is dull brownish; above flammulated and below it is more or less banded or maculate with white. Spire is very short and conic. There are five to six whorls which are convex but concave above. Upper whorls contain revolving lirae. They are frequently carinated, and the last is traversed by several rows of nodules. Large, orbicular aperture is very oblique and is beautifully nacreous within. Outer lip is thin and is margined with intense black within. Arcuate columella is wide and is slightly produced below and is broadly excavated above. Operculum is flat within. It contains 5–6 whorls and a submedian nucleus. Its outer surface is convex, whitish and is covered with calcareous pustules on the exterior side. Length of the adult shell varies between 4 and 12 cm.

Fisheries/aquaculture: Commercial fishery exists for this species in South Africa.

Other uses (if any): Flesh of this species is an important food source in South Africa. The stunningly beautiful polished shells of this species are much prized for their distinctive markings. They are often used as bathroom décor or as a striking center piece in glass vases. Its flesh is occasionally used as bait by the fishermen.

Turbo setosus (Gmelin, 1791)

Common name(s): Rough turban.

Global distribution: Indo-Pacific: off Madagascar, Mascarene Basin and Mauritius; off Australia.

Habitat: Sublittoral zone, in shallow water; exposed areas of coral reefs; intertidal; depth range 0–5 m.

Description: Shell of this species is moderately large, thick, and heavy and turbinate in shape with length usually greater than width. Spire is pointed and is well developed but relatively short. Whorls are strongly convex and are rounded, with moderately impressed sutures. Outer sculpture is of strong, unequal spiral cords, with groove-like interstices and very fine axial striae. Aperture is large and oval and is slightly flaring at the anterior end of the smooth columella. Outer lip is crenulated at margin and spirally fluted inside. Umbilicus is almost closed. Operculum is almost circular with a subcentral nucleus. Exterior of operculum is smoothly convex, finely granulated in the center and striated at the outer margin. Outside of shell is whitish or light fawn, irregularly maculated with dark brown or grayish green along the spiral cords. Aperture is silvery white inside and outer lip is with a green hue. Columella is glossy white. Operculum is whitish externally. Length of the adult shell varies between 2.2 and 8 cm.

Biology

Food and feeding: It is herbivorous and detritus feeder.

Fisheries/aquaculture: Commercial fishery exists for this species.

Other uses (if any): It is one of the most frequently collected Turbinidae in the area, mainly for food. Shell is used as material for making buttons.

Vermetidae (worm snails or worm shells): Shell of the members of this family is composed of three layers and is irregularly coiled or even disjunct, resembling a worm tube. The interior of these tubes, however, is cleaner and glossier, due to a layer of mother-of-pearl (nacre). Aperture is without siphonal canal. Operculum is horny and spiral, and sometimes it may be absent. When these animals are young, they are free-living, but settle down and cement themselves to hard substrates as they grow. While feeding, they produce mucus strands to stick and trap their food organisms, namely, plankton. Some species of this family may come out of their tubes to feed on tiny animals.

Ceraesignum maximum (Sowerby, 1825) (=*Dendropoma maximum*)

Common name(s): Great worm shell, operculate worm shell or great coral worm snail.

Global distribution: Indo-Pacific.

Habitat: Sessile tube-dwelling snail; outer parts of coral reefs; dead and live coral.

Description: It is the largest vermetid species with the maximum shell length of 45 cm.

Biology

Impact on corals: This species has the potential to drive dramatic changes in coral reef ecosystems. It can significantly reduce the growth and survival of corals, change coral morphology, and shift coral species composition.

Food and feeding: It is a ciliary feeding species and consumes zooplankton and phytoplankton via mucus-net feeding.

Reproduction: In this species, the sex ratio is dominated by females with increasing body size, which is suggestive of protandric hermaphroditism. The number and size of brooded egg capsules, and the number of embryos per capsule have been found to be positively related to female size. Females release veliger larvae which settle down and metamorphose into juveniles.

Other uses (if any): It is traditionally collected for food in eastern Polynesia.

KEYWORDS

- profile of edible gastropod species
- biology
- fisheries
- aquaculture

CHAPTER 4

NUTRITIONAL VALUE OF EDIBLE MARINE GASTROPOD MOLLUSCS

CONTENTS

ABSTRACT

The nutritional values of edible marine gastropod molluscs, namely, carbohydrate, protein, lipid, ash, minerals, vitamins, etc., are dealt with in this chapter.

Acmaea sp.

Proximate composition (% FW)				
Moisture	Protein	Fat	Carbohydrate	Lipid
75.7	17.3	1.9	4.2	0.9

Source: Miller and Boxt (http://www.pcas.org/documents/BoxtandMillerweb.pdf).

Babylonia areolata

Proximate composition (%)			
Ash	Crude fat	Crude protein	Crude fiber
5.0	0.8–1.0	65.6–68.9	0.1

Source: Aziz (2016).

Babylonia japonica

Vitamin B12	(μg/100 g)
Meat	Viscera
27.2	92.8

Source: Teng et al. (2015).

Babylonia spirata

Proximate composition (%)		
Protein	Carbohydrate	Lipid
53.9	16.9	9.3

Amino acid composition	
Amino acids	**(mg/g)**
Isoleucine	3.08
Valine	2.07
Lysine	1.01
Phenylalanine	1.01
Leucine	0.89
Methionine	0.77
Proline	0.67
Tryptophan	0.22
Glutamic acid	0.11
Alanine	0.09

Source: Periyasamy et al. (2011).

Proximate composition (%)			
Protein	**Carbohydrate**	**Lipid**	**Ash**
17.4	2.7	0.3	1.2
Fatty acids(%)			
Lauric acid	0.6		
Linoleic acid	1.0		
Arachidonic acid	5.2		

Source: Abdullah et al. (2016).

Proximate composition (%)				
Protein	**Carbohydrate**	**Lipid**	**Ash**	**Moisture**
41.2	17.5	6.6	0.9	80.0

Source: Govindarajalu et al. (2016).

Proximate composition (%)				
Water	**Protein**	**Carbohydrate**	**Fat**	**Ash**
78.4	17.4	2.7	0.3	1.2
Vitamin (µg/100 g) and minerals (ppm) (= mg/kg)				
Vitamin B12	16.6			

Macrominerals	
Ca	764.8
K	1894.2
Mg	1886.4
Na	2481.3
P	677.8
Microminerals	
Fe	13.1
Zn	24.2
Cu	11.5
Se	Undetected

Source: Nurjanah et al. (2015).

Babylonia zeylanica

Proximate composition (% DW)			
Protein	**Carbohydrate**	**Fat**	**Ash**
40.3–42.1	6.8–8.2	1.0–2.5	0.7–2.4

Vitamins	**(mg/g)**
Vitamin A	6.5
Vitamin B2	4.1
Vitamin C	2.1
Vitamin E	2.2
Vitamin B6	0.9
Minerals	**(mg/g)**
Calcium	5.1
Chloride	1.6
Magnesium	1.0

Source: Jayalakshmi (2016).

Babylonia zeylonica

Proximate composition (%)				
Protein	Carbohydrate	Lipid	Ash	Moisture
40.8	16.6	6.1	1.2	78.1

Source: Govindarajalu et al. (2016).

Batillus cornutus

Water	Protein	Lipid	Retinol	Carotene (per 100 g edible portion)
(g)	(g)	(g)	(µg)	(µg)
76.7	19.9	0.4	0	140

Source: Sri Kantha (1989).

Bolinus brandaris

Nutritional facts	(g/100 g edible portion)
Protein	14.5
Lipid	0.9
Total minerals	2.5
mg/100 g edible portion	
Vitamin E	0.9
Iron	1.1

Source: http://www.fisheat.it/murex-bolinus-brandaris/.

Bufonaria echinata

Proximate composition of foot (%)				
Protein	Carbohydrate	Lipid	Water	Ash
22.1	4.4	2.8	50.7	1.0

Amino acid composition (%)	
Essential amino acids	
Phenylalanine	0.7
Valine	0.2
Threonine	11.3
Histidine	4.5
Isoleucine	1.5
Methionine	14.5
Leucine	10.8
Lysine	6.8
Nonessential amino acids	
Alanine	0.8
Arginine	10.7
Asparagine	1.6
Glutamic acid	10.9
Glutamine	10.9
Glycine	10.6
Tyrosine	1.5

Fatty acid composition (% of total fatty acids)	
Saturated fatty acids	
Lauric	0.03
Myristic	9.4
Palmitic	22.4
Stearic	2.2
Arachidonic	3.0
Behanic	1.8
Monounsaturated fatty acids	
Oleic	4.3
Polyunsaturated fatty acids	
Linoleic	11.4
Linolenic	0.4
Eicosapentaenoic acid	16.4
Docosahexaenoic acid	8.1

Source: Babu et al. (2010).

Bullacta exaracta

Nutritional facts (per 100 g)	
Protein	33.4 g
Carbohydrates	2.2 g
Fat	1.5 g
Saturated fat	0.5 g
Monosaturated fat	0.4 g
Polyunsaturated fat	0.3 g
Cholesterol	82.6 mg

Minerals and Vitamins	
Minerals	194.3 mg
Sodium	
Potassium	207.0 mg
Calcium	124.5 mg
Iron	1.8 mg
Magnesium	302.3 mg
Phosphorus	275.6 mg
Zinc	2.2 mg
Copper	0.6 mg
Selenium	51.2 µg
Vitamins	
Vitamin A	8.9 IU
Vitamin B6	0.1 mg
Vitamin B12	6.7 µg
Vitamin E	8 mg
Vitamin K	0.3 µg

Source: Anon. Daily Diet Guide (2006), Santhanam (2015).

Cellana exarata

Nutritional facts (for a serving size of 53 g)	
Macronutrients	
Protein	9 mg
Carbohydrate	1 mg
Fat	0
Cholesterol	117 mg
Mineral	
Sodium	350 mg

Source: Anon. Dietfacts.com., Santhanam (2015).

Cerithidea obtusa

Proximate composition (%)		
Protein	Carbohydrate	Lipid
20.4	0.8	19.5

Source: Chakravarty et al. (2015).

Nutritional facts Proximate composition (%)			
Water	**Protein**	**Fat**	**Ash**
77.5	13.8	2.8	4.5
Minerals (mg/100 g)			
Sodium	283.5		
Calcium	39.8		
Selenium	39.3		
Copper	0.3		
Amino acids (%)			
Isoleucine	4.8		
Arginine	1.0		
Glutamic acid	12.1		
Cysteine	0.8		

Source: Purwaningsih (ISSN 0853-7291, e-ISSN: 2406-7598).

Fatty acids (% of total lipids)	
14:0	8.8
14:1	4.7
15:0	0.3
16:0	27.8
16:1	1.3
17:0	0.5
18:0	5.5
18:1	5.0
18:2ω6	4.2
20:0	0.7
18:3ω3	13.6
20:1	1.5
18:4ω3	0.5
20:2ω9	0.8
20:3ω9	2.5
22:0	0.4
22:1	0.5
20:4ω6	4.5
20:5ω3	10.7
22:4ω6	1.0
22:5ω6	0.9
22:5ω3	0.7
22:6ω3	3.7

Source: Misra et al. (1986).

Chicoreus brevifrons

Fatty acids (% of total lipids)	
C14:0	1.8
C16:0	4.8
C18:0	4.8
C20:0	2.7
C21:0	1.8
C24:0	4.5
Total saturated	20.3
C16:1	1.1
C18:1 (ω-7+ω-9)	2.5
C20:1	6.7
C22:1 ω-9	3.3
C22:1 ω-11	3.1
C24:1 T.I	7.6
Total monounsaturated	24.3
C16:2	1.0
C18:2	1.8
C18:3	2.0
C18:4	4.3
C20:2	4.1
C20:3	2.2
C20:5 ω-3	1.5
C22:6 ω-3	28.9
Total polyunsaturated	45.8
Total unsaturated	70.1
Total unidentified	9.6

Source: D'Armas et al. (2010).

Chicoreus ramosus

Proximate composition (%)				
Protein	**Carbohydrate**	**Lipid**	**Ash**	**Moisture**
37.3	15.2	2.0	1.2	65.4

Source: Govindarajalu et al. (2016).

Proximate composition (%)		
Protein	Carbohydrate	Lipid
24.4	10.5	1.2

Source: Giftson et al. (2015).

Cookia sulcata

Proximate composition (% FW)				
Moisture	Protein	Fat	Ash	Carbohydrate
78.0	17.6	0.7	1.8	1.9

Amino acids (g/100 g protein)	
Asp	6.9
Glu	14.1
Ser	3.8
His[a]	1.4
Gly	9.6
Thr[a]	3.5
Arg	10.2
Ala	4.4
Tau	9.3
Tyr	2.3
Val[a]	2.5
Phe[a]	2.4
Ile[a]	2.5
Lys[a]	5.7
Leu[a]	6.5
Pro	4.8
Cys	0.3
Met[a]	0.3
Trp[a]	5.3

[a]Essential amino acids.

Fatty acids (% Fatty acids)	
SFA	42.2
MUFA	10.8
PUFA	36.3
n3	15.4
n6	15.8
Minerals	
Macroelements (mg/g FW)	
Na	4.1
K	2.8
Ca	0.7
Mg	3.1
P	0.9
S	5.5
Trace elements (mg/kg FW)	
Fe	74.1
Cu	7.8
Zn	38.2
Vitamin E (mg/100 g FW): 3.71	
Cholesterol (mg/g FW): 1.32	
FW, fresh weight	

Source: Shi (https://researcharchive.lincoln.ac.nz/bitstream/handle/10182/2863/Shi_MApplSci.pdf;jsessionid=8CEB19C444DA715A3BE8BE8640F37503?sequence=5).

Crepidula fornicata

Lipid content (% dry matter) and lipid class composition (% total lipids)			
Season	**Total lipids**	**Glycolipids**	**Phospholipids**
Winter	5.3	5.5	69
Spring	2.7	14.7	62
Summer	3.3	13	56
Autumn	3.1	11	61
Beneficial lipids (% dry weight)			
Lipid	5.3		
Phospholipids (PLs)	69		
20:5n-3	9.4		
22:6n-3	7.3		
Cholesterol	31.3		

Source: Dagorn et al. (2014).

Crepidula onyx

Proximate composition (% FW)				
Moisture	Protein	Fat	Carbohydrate	Lipid
83.0	8.4	0.6	5.0	3.1

FW, fresh weight

Source: Miller and Boxt (http://www.pcas.org/documents/BoxtandMillerweb.pdf).

Cymbium glans

Proximate composition (%)				
Moisture	Carbohydrate	Protein	Fat	Ash
76.1–77.8	6.7–22.5	59.1–61.3	4.3–4.8	6.1–7.8

Source: Udotong and Ukot (http://www.journalcra.com/article/microbiological-and- nutritional-quality-cymbium-glans-qua-iboe-river-estuary-nigeria).

Dicathais orbita

Proximate composition (fed with different diets) (g/g)					
Moisture	Ash	Crude lipid	Crude protein	Glycogen	Energy[a]
0.70–0.84	0.03–0.08	0.07–0.10	0.01–0.13	0.02–0.05	18.5–22.9

[a]kJ/g.

Source: Woodcock and Benkendorff (2008).

Ficus ficus

Proximate composition (%)				
Protein	Carbohydrate	Lipid	Ash	Moisture
31.2	13.5	2.2	0.9	67.1

Source: Govindarajalu et al. (2016).

Proximate composition of foot of male and female (%)		
	Male	Female
Protein	25.2	23.9
Carbohydrate	18.3	13.3
Lipid	23.9	21.1

Source: Selvi and Jeevanandham (2016).

Haliotis cracherodii

Proximate composition (%)				
Moisture	Protein	Fat	Ash	Carbohydrate
68–72	18–23	0.8–3.0	3.4	1.5–7.5

Source: Krzynowek and Murphy (1987).

Haliotis discus hannai

Proximate composition (%)				
Moisture	Protein	Fat	Ash	Carbohydrate
72–78	7.5–12.5	1.0–1.5	1.2–2.5	0.1–0.5

Source: Krzynowek and Murphy (1987).

Nitrogen distribution in meat (g/100 g of sample)	
Total-N	2.84
Arginine-N	0.55
Histidine-N	0.19
Lystine-N	0.23
Cystine-N	0.02
Mono amino acid total-N	1.62
Mono amino acid amino-N	1.46
Mono amino acid nonamino-N	0.16

Source: Tanikawa and Yamashita (1961).

Proximate composition (%) of meat (cultured)				
Moisture	Crude protein	Carbohydrate	Crude lipid	Crude ash
78.6	15.6	3.4	0.3	2.2

Source: Jang et al. (2010).

Amino acid (% to total amino acid)	
Aspartic acid	9.5
Threonine	4.0
Serine	5.0
Glutamic acid	15.2
Proline	4.9
Glycine	10.4
Alanine	5.8
Cystine	1.0
Valine	4.0
Methionine	2.2
Isoleucine	3.4
Leucine	6.4
Tyrosine	2.9
Phenylalanine	3.0
Histidine	1.5
Lysine	5.6
Arginine	10.1
EAA[a]	40.6

[a]*EAA*, essential amino acids.

Source: Jang et al. (2010).

Fatty acid (%)	
14:0	6.9
16:0	23.3
18:0	4.4
SFA	34.6
16:1n-7	2.1
18:1n-9	17.1
MUFA	19.2
18:2n-6	1.8
18:3n-3	1.5
20:2n-6	4.6
20:3n-3	14.6
20:5n-3	11.1
22:5n-3	4.1
22:6n-3	8.6
PUFA	46.3

SFA, saturated fatty acid; *MUFA*, monounsaturated fatty acid; *PUFA*, polyunsaturated fatty acid.

Source: Jang et al. (2010).

Haliotis gigantea

Nutritional facts (%)			
Moisture	Protein	Fat	Ash
76	0.4	1.5	2.3

Source: Krzynowek and Murphy (1987).

Proximate composition (%)				
Moisture	Crude protein	Carbohydrate	Crude lipid	Crude ash
82.5	11.1	4.0	0.3	2.2

Source: Jang et al. (2010).

Amino acid (% to total amino acid)	
Aspartic acid	10.4
Threonine	4.1
Serine	5.2
Glutamic acid	15.5
Proline	0.9
Glycine	11.1
Alanine	6.2
Cystine	0.9
Valine	4.1
Methionine	2.3
Isoleucine	3.5
Leucine	6.6
Tyrosine	2.9
Phenylalanine	3.1
Histidine	1.5
Lysine	6.1
Arginine	10.1
EAA[a]	41.3

[a]*EAA*, essential amino acids.

Source: Jang et al. (2010).

Fatty acid (%)	
14:0	6.0
16:0	21.2
18:0	4.8
SFA	32.0
16:1n-7	1.1
18:1n-9	16.3
MUFA	17.4
18:2n-6	1.2
18:3n-3	1.4
20:2n-6	4.7
20:3n-3	14.9
20:5n-3	13.6
22:5n-3	3.7
22:6n-3	11.1
PUFA	50.6

SFA, saturated fatty acid; *MUFA*, monounsaturated fatty acid; *PUFA*, polyunsaturated fatty acid.

Source: Jang et al. (2010).

Haliotis fulgens

Proximate composition of freeze–dried muscle of juveniles (mg/g)		
Protein	Carbohydrate	Lipid
64	16.6	3.3

Source: Perez-Estrada et al. (2011).

Fatty acid profile (mg/g dry weight) of muscle of juveniles	
Fatty acid	
14:0	0.6
16:0	9.9
17:0	0.6
18:0	6.6
Σ SFA	17.7
16:1n-7	0.6
18:1n-9	4.6
18:1n-7	2.7
20:1n-9	2.0
Σ MFA	10.0
16:2n-6	4.8
16:2n-4	0.5
16:4n-1	2.8
18:2n-6	1.5
18:3n-3	1.3
Σ PUFA	16.5
20:3n-9	0.2
20:4n-6	1.8
20:5n-3	3.2
21:5n-3	0.9
22:4n-6	0.5
22:5n-3	3.5
22:6n-3	1.1
Σ LC-PUFA	11.3

SFA, saturated; *MFA*, monounsaturated; *PUFA*, polyunsaturated; *LC-PUFA*, longchain polyunsaturated.

Source: Durazo and Viana (2013).

Haliotis rubra

Nutritional facts	
Protein content:	12.4 mg/g (whole body)
Lipid content (per 100 g of raw product)	
Total fat (oil)	0.8 g
Saturated fat	31% (of total fat)
Monounsaturated fat	22% (of total fat)
Polyunsaturated	47% (of total fat)
Omega-3, EPA	48 mg
Omega-3, DHA	2 mg
Omega-6, AA	100 mg

Source: Freeman (2001).

Haliotis rubra conicopora

Nutritional facts information (per 100 g)	
Fat (total)	0.8 g
Saturated fat	31% of total fat
Monounsaturated fat	22% of total fat
Polyunsaturated fat	47% of total fat
Alpha-linolenic acid	100 mg
Docosahexaenoic acid	2 mg
Eicosapentaenoic acid	48 mg

Source: Anon. Fishflies (http://www.fishfiles.com.au/knowing/species/molluscs/abalo-nes/Pages/Brownlip-Abalone.aspx).

Haliotis tuberculata

Proximate composition (per 100 g edible portion)					
Moisture	Carbohydrate	Protein	Lipid	Ash	Food energy
75.8%	3.4 g	18.7 g	0.5 g	1.6 g	98 cal.

Minerals (mg/100 g)		
Ca	**P**	**Fe**
37	191	2.4
Vitamins (mg/100 g)		
Thiamine	Riboflavin	
0.18	0.14	
Microelements of muscle (ppm dry tissue)		
Co	0.1	
Cu	12.1	
Fe	30.3	
Zn	38.0	

Source: Mgaya (1995).

Haliotis tuberculata tuberculata (=*Haliotis japonica*)

Proximate composition (%)				
Moisture	**Protein**	**Fat**	**Ash**	**Carbohydrate**
76	10.2	0.3	1.4	7.0

Source: Krzynowek and Murphy (1987).

Haliotis wallalensis

Proximate composition (% fresh weight)				
Moisture	**Protein**	**Fat**	**Carbohydrate**	**Lipid**
75.0	18.9	0.9	2.6	2.6

Source: Miller and Boxt (http://www.pcas.org/documents/BoxtandMillerweb.pdf).

Harpa articularis

Proximate composition (%)				
Protein	**Carbohydrate**	**Lipid**	**Ash**	**Moisture**
35.2	14.3	4.7	0.8	72.0

Source: Govindarajalu et al. (2016).

Hexaplex trunculus

Proximate composition (% wet weight)		
Protein	Carbohydrate	Lipid
15.9–28.0	1.5–3.3	1.1–3.5

Source: Gharsallah et al. (2010).

Hexaplex trunculus

Proximate composition (%)		
Protein	Lipids	Ash
47.8	27.3	15.2

Amino acids (mg/g protein) dry weight	
Aspartic acid	79.0
Threonine[a]	19.3
Serine	43.6
Glutamic acid	12.3
Glycine	63.5
Alanine	35.6
Valine[a]	47.2
Methionine[a]	19.9
Isoleucine[a]	48.1
Leucine[a]	69.6
Tyrosine	31.7
Phenylalanine[a]	41.9
Histidine[a]	39.2
Lysine[a]	83.5
Arginine	74.8
Cystine	3.6
Proline	30.1
Total essential amino acids	369
Total nonessential amino acids	375

[a]Essential amino acids.

Fatty acid composition (% of total fatty acids)	
Neutral lipid	**Polar lipid**
77.0 (%)	23.0 (%)
SFA	33.4
MUFA	9.8
PUFA	58.4
UFA	68.2
Total n-3	34.1
Total n-6	20.2

SFA, saturated fatty acid; *MUFA*, monounsaturated; *PUFA*, polyunsaturated; *UFA*, unsaturated.

Mineral contents (wet basis)	
Br (mg/kg)	2.6
Ca (mg/100g)	674.4
Cl (mg/100 g)	495.2
Cu (mg/kg)	31.0
Fe (mg/kg)	81.0
K (mg/100 g)	224.8
Mg (mg/100 g)	178.7
Mn (mg/kg)	6.9
Na (mg/100 g)	196.1
Ni (mg/kg)	10.0
P (mg/100 g)	95.2
Zn (mg/100 g)	112.8

Source: Zarai et al. (2011).

Homalopoma luriudm

Proximate composition (% fresh weight)				
Moisture	Protein	Fat	Carbohydrate	Lipid
74.1	20.3	1.0	2.4	0.5

Source: Miller and Boxt (http://www.pcas.org/documents/BoxtandMillerweb.pdf).

Lambis lambis

Proximate composition (%)		
Protein	Carbohydrate	Lipid
5.2	0.5	1.0

Littorina littorea

Proximate composition (%)		
Water	Protein	Fat
80	15	1.4

Source: Anon. USDA National Nutrient Database.

Lunella coronata

Fatty acid composition (%)		
DHA	PUFA	SFA
0.2	40.9	49.3

Source: Freije and Awadh (http://dx.doi.org/10.1108/00070701011080195).

Lunella torquata

Proximate composition (% fresh weight)				
Moisture	Protein	Carbohydrate	Lipid	Ash
68.5	18.0	2.9	8.5	2.1

Fatty acids (% of total fatty acids) (based on fresh weight)	
Myristic	0.6
Pentadecanoic	1.2
Palmitic	23.0
Margaric	2.4
Stearic	5.7
Lignoceric	7.1
Palmitoleic	3.2
Oleic	8.2
Eicosenoic	2.8
Erucic	0.2
Linoleic	1.6
α-Linoleic	0.9
11,13-Eicosadienoic	0.1
Eicosatrienoic	0.2
Arachidonic	14.9
Eicosapentaenoic	5.3
5,13-Docosadienoic	6.6
Docosahexaenoic	0.8
Docosapentaenoic	15.3

Trace elements	
Macroelements (mg/g fresh weight)	
Na	3.0
K	3.1
Ca	2.4
Mg	0.7
P	1.5
S	11.2
Microelements (mg/kg fresh weight)	
Fe	32.4
Zn	14.0
Cu	1.1
Mo	0.2
Co	0
Se	0.2

Source: Lah et al. (2016).

Lunella undulata

Proximate composition (% fresh weight)				
Moisture	Protein	Carbohydrate	Lipid	Ash
70.8	18.5	3.5	5.2	2.0

Fatty acids (% of total fatty acids) (based on fresh weight)	
Myristic	1.0
Pentadecanoic	1.1
Palmitic	21.6
Margaric	1.9
Stearic	6.6
Lignoceric	6.5
Palmitoleic	2.4
Oleic	8.4
Eicosenoic	3.6
Erucic	0.5
Linoleic	2.8
α-Linoleic	2.3
11,13-Eicosadienoic	0.2
Eicosatrienoic	0.4
Arachidonic	16.0
Eicosapentaenoic	4.6
5,13-Docosadienoic	5.8
Docosahexaenoic	0.5
Docosapentaenoic	13.8

Trace elements	
Macroelements (mg/g fresh weight)	
Na	2.7
K	3.3
Ca	0.4
Mg	0.7
P	1.6
S	12.3

Microelements (mg/kg fresh weight)	
Fe	41.1
Zn	15.2
Cu	0.6
Mo	0.1
Co	0.1
Se	0.1

Source: Lah et al. (2016).

Melo melo

Proximate composition of body tissue (%)			
Protein	Carbohydrate	Lipid	Water
20.9	2.6–5.1	2.7	76.6–83.5

Fatty acids of body tissues (%)	
Saturated fatty acids (SFA)	
12:0	4.2
14:0	2.8
16:0	7.4
17:0	9.8
18:0	8.4
Monounsaturated fatty acids (MUFA)	
16:1	31.3
18:1 w9c	9.8
Polyunsaturated fatty acids (PUFA)	
20:4ω6,9	22.6[a]

[a]Foot

Source: Palpandi et al. (2010).

Monodonta labio

Proximate composition (%)				
Water	Protein	Carbohydrates	Fat	Ash
76.7	17.3	1.0	1.6	2.9

Amino acids (% of total amino acids)		
Essential amino acids	Nonessential amino acids	
35.2	82.5	
Fatty acids (% of total fatty acids)		
Saturated	Monounsaturated	Polyunsaturated
38.94	26.4	34.7

Source: Zhang et al. (2011).

Nerita albicilla

Proximate composition of dried powder (% dry weight)					
Moisture	Protein	Carbohydrate	Crude fiber	Fat	Ash
12.5	62.1	4.2	6.6	5.6	9.2

Amino acids of dried powder (mg/g protein)	
Essential	
Histidine	30.4
Arginine	30.9
Threonine	44.9
Valine	47.4
Methionine	38.9
Isoleucine	75.9
Leucine	64.5
Phenylalanine	36.6
Lysine	53.8
Tyrosine	44.2
Cysteine	19.7
Nonessential	
Aspartate	83.7
Glutamate	143.2
Serine	17.3
Glycine	37.4
Alanine	36.3
Proline	27.0

Source: Hardjito et al. (2012).

Nerita balteata

Nutritional facts (per 100 g)	
Total carbohydrate	7.4 g
Dietary fiber	1.0 g
Sugars	1.0 g
Protein	15.9 g
Total fat	1.7 g
Saturated fat	1.0 g
Trans fat	1.0 g
Cholesterol	1.0 mg
Minerals	
Sodium	1225 mg
Calcium	528 mg
Potassium	531 mg
Phosphorus	1.0 mg

Source: Anon (https://au.nutrihand.com/Nutrihand/pctools/showFoodFacts1.do;jsessionid
=3727F0B3FBF3EAF76514434892C9992B?nodeID=&customerUserID=&mealTypeID
=&mealPlanID=&styleID=&inFramed=&foodLogMiscID=&courseID=&spicinessID=&
source=&recipeID=&foodID=1120441&recordType=).

Patella vulgata

Proximate composition (% fresh weight)		
Moisture	Lipid	Ash
75.0–84.0	10.5–53.3	8.6–34.7

Source: Anon (https://researcharchive.lincoln.ac.nz/bitstream/handle/10182/2863/Shi_
MApplSci.pdf?sequence=5).

Proximate composition (%)			
Moisture	Protein	Lipid	Ash
77.3	13.0	2.5	2.4

Amino acids (%)	
Aspartic acid	9.2
Glutamic acid	11.7
Serine	5.5
Glycine	10.4
Histidine	1.1
Arginine	9.1
Threonine	5.6
Alanine	7.3
Proline	8.4
Valine	4.8
Methionine	1.2
Cysteine	0.3
Isoleucine	1.5
Leucine	7.6
Phenylalanine	2.9
Tyrosine	2.8
Tryptophan	0.5
Lysine	10.0

Fatty acids (%)	
Saturated fatty acids	28.1
Monounsaturated fatty acids	24.3
Polyunsaturated fatty acids	47.6
Omega-3 fatty acids	
DHA	0.6
EPA	13.1

Source: Rambli et al. (https://www.researchgate.net/).

Phalium glaucum

Proximate composition (%)				
Moisture	Protein	Carbohydrate	Lipid	Ash
83.7	35.1	14.2	3.9	1.0

Source: Govindarajalu et al. (2016).

Amino acids (%)	
Total amino acids	17.3
Total essential	8.1
Isoleucine	1.2
Methionine	1.1
Phenylalanine	1.1
Lysine	0.5
Proline	0.5
Total nonessential	9.3
Alanine	1.2
Arginine	1.2
Glutamic acid	1.1
Serine	0
Asparagine	0.5

Fatty acids (%)	
SFA	
Palmitic	0.8
Margaric	0.5
Stearic acid	0.9
MUFA	
Oleic acid	1.0
PUFA	
Linoleic acid	1.8
Linolenic acid	1.1
Morotic	0.3

SFA, saturated; *MUFA*, monounsaturated; *PUFA*, polyunsaturated.

Source: Babu et al. (2011).

Phorcus turbinatus

Proximate composition (% fresh weight)				
Protein	Lipids	Moisture	Ash	Carbohydrate
51.2[a]	2.9[a]	82.8	N/A	N/A

N/A, not available.

[a]The percentage on dry matter basis.

Source: Anon. (https://researcharchive.lincoln.ac.nz/bitstream/handle/10182/2863/Shi_MApplSci.pdf?sequence=5&isAllowed=y).

Phyllonotus pomum

Fatty acids (% of total lipids)	
C9:0	0.1
C10:0	0.1
C11:0	1.3
C14:0	2.3
C16:0	4.1
C18:0	5.7
C21:0	2.1
C24:0	6.5
Total saturated	22.1
C16:1	1.6
C18:1 (ω-7+ω-9)	5.3
C20:1	10.4
C22:1 ω-9	7.3
C22:1 ω-11	7.9
C24:1 T.I	6.5
Total monounsaturated	39.0
C16: 2	0.5
C18: 2	2.0
C18: 3	3.3
C18: 4	2.3
C20: 2	1.4
C20: 3	8.1
C20:5 ω-3	1.9
C22:6 ω-3	7.4
Total polyunsaturated	26.3
Total unsaturated	65.1
Total unidentified	12.7

Source: D'Armas et al. (2010).

Pinaxia coronata

Mineral content (mg/100 g)	
Copper	10.2
Iron	12.4
Manganese	1.8
Calcium	122.9
Sodium	17.4

Source: Davies and Jamabo (2016).

Pleuroploca trapezium

Proximate composition (%)		
Protein	Carbohydrate	Lipid
10.35	4.31	1.74

Vitamins/100 g

Vitamin B1	Vitamin B2	Vitamin B6	Vitamin B12	Vitamin C	Niacinamide
Trace	1.01 mg	1.01 mg	0.25 µg	0.19 mg	0.01 mg

Minerals (mg/100 g)				
Calcium	Magnesium	Sodium	Potassium	Phosphorus
8.9	3.1	120	78.5	3.0

Trace elements (mg/100 g)

Iron	Zinc	Copper
0.025	0.01	0

Fatty acids

Fatty acid (mg/100 mg of lipid)

Saturated fatty acids

Caproic acid	0.02
Caprolyic acid	0.17
Capric acid	0.05
Lauric acid	0.09
Myristic acid	0.60
Palmitic acid	0.09
Stearic acid	0.05
Arachidic acid	Trace
Behenic acid	Trace
Lignoceric acid	Trace

Monounsaturated fatty acids (MUFA)	
Myristoleic acid	0.04
Oleic acid	Trace
Polyunsaturated fatty acids (PUFA)	
Linoleic acid	0.64
Linolenic acid	0.01
Eicosatrienoic acid	Trace
Arachidonic acid	Trace
Eicosapentaenoic acid (EPA)—Trace	

Source: Anand et al. (2010).

Proximate composition (%)		
Protein	**Carbohydrate**	**Lipid**
14.1	12.2	3.0

Source: Anand et al. (2013).

Purpura bufo

Proximate composition (%)		
Protein	**Carbohydrate**	**Lipid**
22.3	19.3	4.6
Amino acids (%)		
Histidine	Methionine	Isoleucine
0.4	0.3	0.1

Fatty acids (%)		
PUFA	**SFA**	**MSFA**
7.7	6.7	4.5

PUFA, polyunsaturated; *SFA*, saturated; *MUFA*, monosaturated.

Source: Margret and Jansi (2013).

Rapana venosa

Proximate composition (%)				
Moisture	Protein	Lipids	Carbohydrate	Ash
67.5	64.7	1.9	24.1	9.4

Source: Celik et al. (2014).

Proximate composition (%)			
Protein	Carbohydrate	Lipid	Ash
55.9	23.4	5.6	15.1

Source: Anon (http://nopr.niscair.res.in/handle/123456789/28650).

Strombus gracilior

	Proximate composition (g%)					
Sex	Moisturea	Protein	Carbohydrate	Crude fiber	Ash	Lipid
Male	72.6	19.1	1.9	0.4	2.3–3.7	0.9
Female	70.9	26.9	1.4	0.4	2.3–3.7	0.9

a%.

	Minerals (µg/g)	
Sex	Na	K
Male	1440.9	1898.3
Female	1377.0	2393.1

Source: Jiménez-Arce (1993).

Telescopium telescopium

Proximate composition (%)		
Protein	Carbohydrate	Lipid
21.0	0.8	18.9

Source: Chakravarty et al. (2015).

Lipids	
Lipid classes (mg/g of wet tissue (foot))	**(% of total lipid)**
Total lipid	11.7
Total cholesterol	3.0
Phospholipid	6.8
Triglycerides	1.2
Others (hydrocarbons, sterol esters, free fatty acids, etc.)	0.4

Fatty acids (% of total lipids) of foot	
Saturated	
14:0	3.8
15:0	0.7
16:0	21.1
17:0	1.4
20:0	10.6
Monounsaturated	
14:1	1.2
16:1	4.9
18:1	13.6
20:1	0.6
22:1	0.9
Polyunsaturated	
16:2	2.5
18:2	4.9
18:3	10.7
20: 2	6.8
20:3	1.6
20:4	3.1
20:5 (ω3)	8.1
22:4 (ω6)	1.8

Source: Rakshit et al. (1997).

Tibia curta

Amino acid composition (%)	
Essential	
Phenyl alanine	0.1
Lysine	2.0
Methionine	1.7
Isoleusine	Trace
Arginine	5.1
Leusine	0.7
Nonessential	
Glutamic acid	0.5
Serine	2.3
Glycine	1.7
Proline	0.9
Tyrosine	0.4

Fatty acids (%)	
Saturated	
C14:0	5.0
C15:0	1.8
C16:0	22.7
C17:0	2.4
C18:0	6.7
Monounsaturated	
C16:1n3	14.8
C18:1n9	4.2
C20:1n5	7.4
C22:1n9	0.3
Polyunsaturated	
C22:2n6	1.4
C20:5n3	12.4
C22:6n6	4.5
C22:6n3	7.7
C22:5n3	2.5

Source: Ragi et al. (2016).

Tonna dolium

Proximate composition (%)				
Moisture	Protein	Carbohydrate	Lipid	Ash
60.2	33.6	12.3	1.9	0.8

Source: Govindarajalu et al. (2016).

Amino acids (%)	
Total amino acids	19.1
Essential	
Isoleucine	1.2
Methionine	1.4
Phenylalanine	1.3
Leucine	1.2
Lysine	0.5
Proline	0.5
Total nonessential	9.1
Alanine	1.4
Arginine	1.3
Glycine	1.2
Serine	0
Asparagine	0.5

Fatty acids (%)	
SFA	
Palmitic	1.1
Margaric	0.8
Stearic acid	1.2
MUFA	
Oleic acid	1.1
PUFA	
Linoleic acid	2.0
Linolenic acid	1.2
Morotic	0.4

SFA, saturated; *MUFA*, monounsaturated; *PUFA*, polyunsaturated.

Source: Babu et al. (2011).

Turbo sarmaticus

Proximate composition (% dry weight)			
Protein	Carbohydrate	Lipids	Ash
62.7–74.8	15.9–23.1	3.2–6.7	4.7–7.6

Source: Anon (https://researcharchive.lincoln.ac.nz/bitstream/handle/10182/2863/Shi_MApplSci.pdf?sequence=5&isAllowed=y).

Turbinella pyrum

Proximate composition (%)				
Moisture	Protein	Carbohydrate	Lipid	Ash
81.2	37.2	14.5	4.3	0.9

Source: Govindarajalu et al. (2016).

Turbo cornutus

Vitamin B12 (µg/100 g)	
Meat	Viscera
3.0	15.1

Source: Teng et al. (2015).

Fatty acids (%)	
14:0	6.0
16:0	16.7
16:1	1.5
16:2	3.8
17:0	2.4
18:0	14.6
18:1	5.0
18:2	3.4
20:0	1.2
20:1	16.3
21:0	4.1
23:0	15.5

Source: Hayashi et al. (1969).

Turbo militaris

Proximate contents (%) (based on fresh weight)				
Moisture	Protein	Carbohydrate	Lipid	Ash
73.1	16.2	3.0	5.6	2.1

Fatty acids (% of total fatty acids) (based on fresh weight)	
Myristic	0.1
Pentadecanoic	1.5
Palmitic	22.1
Margaric	3.0
Stearic	5.5
Lignoceric	7.7
Palmitoleic	3.0
Oleic	7.9
Eicosenoic	2.9
Erucic	0.3
Linoleic	2.9
α-Linoleic	1.9
11,13-Eicosadienoic	0.1
Eicosatrienoic	0.3
Arachidonic	15.1
Eicosapentaenoic	3.7
5,13-Docosadienoic	7.5
Docosahexaenoic	0.4
Docosapentaenoic	13.3

Trace elements	
Macroelements (mg/g FW)	
Na	4.0
K	2.7
Ca	0.6
Mg	0.8
P	1.2
S	8.3

Microelements (mg/kg FW)	
Fe	19.3
Zn	12.2
Cu	2.8
Mo	0.1
Co	0
Se	0.177

Source: Lah et al. (2016).

Turbo setosus

Proximate composition (%)				
Moisture	Carbohydrate	Protein	Fat	Ash
74.8	6.8	16.0	0.02	0.8[a]
10.2	10.1	70.3	2.2	6.9[b]

[a]Fresh weight.

[b]Dry weight.

Macro and microminerals in meat (ppm)						
K	Ca	Mg	Fe	Zn	Cu	Se
724.7	228.0	448.2	14.7	10.0	0.5	<0.002[a]
8225.3	4056.7	1987.3	98.7	48.2	4.4	<0.002[b]

[a]Fresh weight.

[b]Dry weight.

Vitamins (fresh meat) (µg/100 g)	
Vitamin A:	90.1
Vitamin:	2.7
Vitamin E:	2.4
Vitamins (dried meat) (µg/100 g)	
Vitamin A:	70.3
Vitamin B12:	0.5
Vitamin E:	9.7

Amino acids (%)	
Essential amino acids:	7.4
Nonessential amino acids:	8.0
Fatty acids (%)	
Saturated fatty acids:	5.8
Unsaturated fatty acids:	3.7
Other fats	
EPA of fresh meat:	2970 mg/100 g
Cholesterol of fresh meat:	96.2 mg/100 g
Cholesterol of dried meat:	60.3 mg/100 g

Source: Merdekawati (http://repository.ipb.ac.id/bitstream/handle/123456789/66859/2013 dme.pdf?sequence=1&isAllowed=y).

Volegalea cochlidium

Proximate composition (%)				
Moisture	Protein	Carbohydrate	Lipid	Ash
71.3	38.9	16.9	1.1	1.1

Source: Govindarajalu et al. (2016).

KEYWORDS

- nutritional profile
- proximate composition
- amino acids
- fatty acids
- minerals
- vitamins

CHAPTER 5

PHARMACEUTICAL VALUE OF EDIBLE MARINE GASTROPOD MOLLUSCS

CONTENTS

ABSTRACT

The pharmaceutical compounds produced by the edible species of marine gastropod molluscs with activities such as anticancer, antiulcer, antioxidant, anticoagulant, antibacterial, antifungal, anti-HSV, antiviral, etc. are dealt within this chapter.

There is a continual need for new therapeutic agents, especially to treat a large variety of diseases for which there are no effective therapies. Many forms of cancer and neurodegenerative diseases cannot be treated successfully. In this context, the biological diversity of seas and oceans offer great promise as a source of new drugs for the future. In this regard, it is worth-mentioning here that the marine gastropods have largely been used in traditional medicines. The shells of these animals were used for the treatment of skin diseases, wounds in the stomach, arthritis, and eye and ear diseases. It was also used to regulate the menstrual cycle in women and as a purgative and an emetic. Similarly, the operculum of these gastropods were used for treating stomach, skin diseases, teeth problems, eye diseases, uterus diseases, epilepsy, paralysis, tumors, and rheumatism (Lev and Amar, 2008). The potential of marine gastropods as sources of biologically active products is hitherto largely unexplored. Since the natural products of these molluscs are becoming increasingly attractive due to their potential applications in the pharmaceutical industries, there is an urgent need to undertake intensive research on this vital aspect. This chapter deals with the pharmaceutically important edible marine gastropods, their bioactive compounds, and their therapeutic activities.

Babylonia japonica

Antinicotinic activity: The toxins viz. surugatoxin and neosurugatoxin isolated from this species, possess antinicotinic activity.However, the surugatoxin has been reported to be more potent (about 100 times) than neosurugatoxin (Datta1 et al., 2015).

Babylonia spirata

Anticoagulant activity: The bioactive compound glycosaminoglycans have been isolated from this species at 8.7 g/kg. This species showed anticoagulant activity at 134 USP units/mg (Periasamy et al., 2013).

Antimicrobial activity: The maximum inhibition zone (12 mm) was observed against *Pseudomonas aeruginosa* in the crude ethanol extract of this species, and the minimum inhibition zone (2 mm) was noticed against

Staphylococcus aureus in the crude methanol extract. Water extract of *B. spirata* showed the highest activity against *Vibrio parahaemolyticus*, *S. aureus*, and *Candida albicans*. Ethanol, acetone, methanol, chloroform, and water extracts of this species showed antimicrobial activity against almost all the bacteria and fungus (Periyasamy et al., 2012). Sri Kumaran et al. (2011) reported that the butanol extracts of this species showed high zone of inhibition in the bacterial pathogen *Proteus mirabilis* and fungal pathogen *C. albicans*, and the values were 7.0 mm and 8.1 mm, respectively.

Babylonia zeylanica

Antioxidant activity: The methanolic extract of this species exhibited total antioxidant activity (78.6 at 10 mg/mL) (Velayutham et al., 2014).

Antimicrobial activity: The human bacterial pathogen *Klebsiella pneumoniae* was found to show maximum susceptibility (10.1-mm inhibition zone) against the ethanol extracts of *Babylina zeylanica*. On the other hand, the fungal pathogen *C. albicans* showed maximum susceptibility (7.1-mm inhibition zone) against methanol extracts of this species (Suresh et al., 2012). *B. zeylanica* showed the highest zone of inhibion (18 mm) against the fungal species *Aspergillus fumigtus* (Kanchana et al., 2014).

Buccinum undatum

Antibacterial activity: The acidic extracts of this species showed growth inhibition diameter of less than 7 mm against *Micrococcus luteus* and more than 10 mm against *Listonella anguillarum* (Defer et al., 2009).

Antiviral activity: This species showed 60.7% inhibition of viral activity at a protein concentration of 130 µg/mL (Defer et al., 2009).

Bufonaria crumena

Antiangiogenic activity: The methanolic extracts of *Bursa crumena* at the tested concentration of 200 µg exhibited fairly a higher degree of anti-angiogenic activity with an inhibitory percentage (60.48%) of the vascular endothelial growth factor (VEGF)-induced neovascularisation (Gupta et al., 2014). The significant antiangiogenic activity evinced by the extract of this species merits further investigation for ocular neovascular diseases.

Bufonaria rana

Antimicrobial activity: The whole body methanol extracts of this species showed 63% inhibition against 40 biofilm bacteria (Anon. https://www.

thefreelibrary.com/Potential+antimicrobial+activity+of+marine+moll uscs+ from+tuticorin%2c...-a01331081).

Bullacta exarata

Antioxidant activity: The compound, mannoglucan isolated from the foot muscle of this species exhibited positive antioxidant activity in scavenging superoxide radicals and reducing power (Liu et al., 2013).

Antioxidant and antitumor activities: The three polysaccharides (BEP1, BEP2, and BEP3) isolated from this species were found to possess antioxidant activities in a dose-dependent manner. The BEP3 exhibited stronger antioxidant activities than BEP1 and BEP2. Furthermore, BEP3 showed significant inhibitory effects on growth of Bcap37 breast cancer cells, SW1990 pancreatic cancer cells and HeLa cervical cancer cells, and the IC50 were 135.3, 147.5, and 172.6 µg/mL, respectively. The highest inhibition rates of BEP1 and BEP2 were approximately 10% against three cancer cells. The data obtained from in vitro models indicates that polysaccharides of this species could be explored as novel and potential natural antioxidants and cancer prevention agents for use in functional foods (Zhang et al., 2012).

Anticancer activity: The peptides isolated from these species namely BEPT II and BEPT II-1 significantly inhibited the proliferation of PC-3 cells in a time- and dose-dependent manner. BEPT II-1 for 24 h increased the percentage of the early stage of apoptotic cells from 11.22% to 22.09%. These data support that BEPT II-1 has anticancer properties and merits further investigation to understand the mechanisms of BEPT II-1-induced apoptosis in PC-3 cells (Ma et al., 2013).

Antitumor activity: The sulfated polysaccharide of this species exhibited highest inhibitory effects on growth of B-16 melanoma cells, and its IC50 was 31.1 l µg/mL (Zhang et al., 2013). Lin et al. (2012) reported that the peptide compound isolated from this species could inhibit the PC-3 cells growth and have antitumor activities on prostate cancer.

Antibacterial activity: Three isolated peptides of this species (coded as BEP-1, BEP-2 and BEP-3) were found to show activity against *Escherichia coli*, *S. aureus*, and *Bacillus subtilis* BEP-1 also showed activity against human pathogen strains (*Staphylococcus epidermidis*, *E. coli*, and Methecillin-resistant *S. aureus*) (Jian-yin et al., 2011).

Cerithidea obtusa

Antidiabetic activity: The methanol extracts containing of alkaloids, flavonoids and triterpenoids showed significant antidiabetic activity (IC50 = 36.40 mg/mL) (Cahyani et al., 2015).

Antioxidant activity: The methanol extract of this species showed potent antioxidant activity with a IC50 value of 58.19 ppm and vitamin C benchmark of 3.55 ppm (Purwaningsih, ISSN 0853-7291; e-ISSN: 2406-7598).

Charonia lampas

Enzymes (exoglycosidases): This species has been reported to produce the enzymes namely exoglycosidases which are essential tool for studies of the biological activity of sugar chains since they can specifically remove sugars from intact glycoproteins and glycolipids (Anon. http://www4. mpbio.com/ecom/docs/proddata.nsf/(webtds2)/32126).

Chicoreus ramosus

Antimicrobial activity: The acetone and chloroform extracts of both the tissues and eggs of this species were found to inhibit the growth of the tested pathogenic bacterial strains. The minimum inhibitory concentration of both the extracts ranged from 4 to 12 mg/mL (Ramasamy et al., 2013). Four different solvent extracts of tissue of this species were also tested for their activity against human and fish pathogens. The maximum zone of inhibition of about 12 mm was observed against *Proteus vulgaris* in the crude ethanol extract followed by 8 mm against *Salmonella paratyphi* at the concentration of 50 µL. Minimum inhibition zone of 2 mm was obtained by ethanol extract against *Streptococcus* mutants and by methanol extract against *Salmonella dysentriae* (Giftson et al., 2015). Benkendorff et al. (2015) studied on the antibacterial activity of the methanol extracts of whole body, digestive gland and egg mass of this species against human pathogenic bacteria (*S. aureus, E. coli,* and *P. aeruginosa*), aquatic pathogenic bacteria (*Vibrio anguillarum, Vibrio harveyi, Vibrio alginolyticus,* and *Enterococcus sericolicida*) and found the following results:

Whole body—MeOH—Inhibited 58% of the marine biofilm bacteria tested.

Digestive gland—MeOH, H_2O, DCM, Acetone—No activity against biofilm bacteria.

Egg mass—MeOH—Inhibited 100% of the marine biofilm bacteria tested.

Chicoreus sp.

Antiatherosclerotic activity: The methanolic crude extract of this unidentified species showed the highest activity at 25 µg/mL. Further, an active compound C35 has also been isolated to form this species (Sarizan, 2013).

Concholepas concholepas

Growth stimulating activity: A heparin binding factor with mitogenic growth stimulating activity in T3 fibroblasts has been isolated from this species. Mitogenic heparin binding can promote angiogenesis and increase the rate of dermal repair necessary for wound healing (Benkendorff et al., 2015).

Anticancer/antitumor activity: The hemocyanin subunits (CCHA and CCHB) of this species have been reported to possess immunotherapeutic effects and are of potential use in the treatment of bladder and prostate carcinoma (Benkendorff et al., 2015). Mice treated with the hemocyanin of this species showed a significant antitumor effect, with decreased tumor growth and incidence, prolonged survival and lack of toxic effects. Analysis of serum from treated mice showed an increased interferon-gamma and low interleukin-4, confirming that these hemocyanins induce a T helper type 1 cytokine profile. Therefore, the hemocyanin of the above species may be an alternative candidate for providing safe and effective immunotherapy for human superficial bladder cancer (Moltedo et al., 2006).

Conus spp.

Conotoxins produced by different species of *Conus* are widely used as pharmacological agents in ion channel research, and several have direct diagnostic and therapeutic potential. Since some conotoxins have high selectivity for pain targets, they are well suited for use as analgesics for chronic pain. They are providing a nonaddictive pain reliever 1000 times as powerful as, and possibly a replacement for, morphine. One conotoxin has been approved for use as an analgesic, several other conotoxins are in development, and still more are being explored for this indication (Layer and McIntosh, 2006). The edible species of *Conus* and their toxins are given below:

Conus gloriamaris: Delta-conotoxin GmVIA (Shon et al., 1994).

Conus miles: *O*-Superfamily conotoxins (Luo et al., 2007). These toxins may be of great use in drug development.

Conus planorbis: J-conotoxin, pl14a (Jin et al., 2015); toxins of P and T-Supefamiles (Pak, 2014).

Conus ventricosus: Conotoxin-Vn and Conotoxin-Vn2 (Spiezia et al., 2013).

Intensive research is needed to explore the possibilities of utilizing the above conotoxins for development of new drugs.

Crepidula fornicata

Antibacterial activity: The acidic extracts of *Crepidula fornicata* showed growth inhibition diameter of less than 7 mm against *M. luteus* and *L. anguillarum* (Defer et al., 2009).

Antiviral activity: This species showed 46.8% inhibition of viral activity at a protein concentration of 110 µg/mL (Defer et al., 2009).

Cypraea tigris

Antiatherosclerotic activity: This species has been reported to possess bioactive compounds as potential antiatherosclerotic substances (Sarizan, 2013). Further intensive research is, however, needed on this aspect.

Dicathais orbita

Anticancer activity: The indole-based compounds, 6-bromoisatin and tyrindoleninone, of this species are of interest because of their anticancer activity and ability to induce apoptosis in several cancer cell lines, both in vitro and in vivo. These novel bioactive compounds would exert selective cytotoxicity toward the reproductive cancerous cells while having minimal, or no effect on the reproductive primary cells (Edwards, 1967). The chloroform extracts from the egg masses and hypobranchial glands of this species have been reported to inhibit the proliferation of a range of lymphoma and adherent cell lines from solid reproductive and colon tumors (Benkendorff et al., 2015).

Antibacterial activity: The extracts of the egg mass of this species have been tested against human pathogenic bacteria (*S. aureus*, *E. coli*, and *P. aeruginosa*), aquatic pathogenic bacteria (*V. anguillarum*, *V. harveyi*, *V. alginolyticus*, and *E. sericolicida*) and the results are given below:

Extract or compound	Activity profile
CHCl3	Inhibits Gram +ve and Gram −ve human and marine pathogenic bacteria and *C. albicans* in the range of 0.1–1 mg/mL
Diethyl ether	Inhibits Gram +ve and Gram −ve human pathogens at 10 mg/mL EtOH Inhibits Gram +ve and Gram −ve human pathogens at 0.1 mg/mL
Tyriverdin	Inhibits human Gram +ve and Gram −ve pathogens at 0.0005 mg/mL, active against *C. albicans* and marine pathogens at 0.001 mg/mL
Tyrindoleninone	Inhibits human pathogens at 0.5–1 mg/mL, *C. albicans* and marine pathogens at 0.1 mg/mL
6-Bromoisatin	Inhibits Gram +ve and Gram −ve human pathogenic bacteria in the range of 0.1–1 mg/mL, but >1 mg/mL for *C. albicans* and marine pathogens

Source: Benkendorff et al. (2015).

Ficus ficus

Antifungal activity: The methanol extracts of this species were found to moderately inhibit the growth of the fungal species *A. fumigatus* with 4 mm zone of inhibition (Kanchana et al., 2014).

Hemolytic activity: The methanol extracts of this species when tested with the chicken red blood cells showed hemolytic activity (Kanchana et al., 2014).

Fissurella cumingi

Treating neoplastic diseases: The purified hemocyanins isolated from this species offer good scope for the treatment of neoplastic diseases (Anon. https://www.google.com/patents/US8436141).

Fissurella latimarginata

Treating neoplastic diseases: The purified hemocyanins isolated from this species offer good scope for the treatment of neoplastic diseases (https://www.google.com/patents/US8436141).

Antitumor activity: A novel hemocyanin (FLH) isolated from this species possesses increased antitumor activity. This FLH is a potential new marine adjuvant for immunization and possible cancer immunotherapy (Arancibia et al., 2014).

Fissurella maxima

Treating neoplastic diseases: The purified hemocyanins isolated from this species offer good scope for the treatment of neoplastic diseases (Anon. https://www.google.com/patents/US8436141).

Haliotis discus discus

Immune response against bacterial infection: Three novel proteins (designated as AbC1qDCs) isolated from this species are found involved in immune responses against invading bacterial pathogen (Bathige et al., 2016). Further, rAb-Antistasin (10 µM) isolated from this species was found to inhibit trypsin activity by 66% in a dose-dependent manner. Moreover, it exhibited low prolongation activity for coagulation with human blood (Nikapitiya et al., 2010). The abalone defensin isolated from this species was found to involve in the immune response reactions as a host defense against pathogenic bacteria such as *V. alginolyticus*, *Vibrio parahemolyticus*, and *Lysteria monocytogenes* (De Zoysa et al., 2010). A novel C-type lectin (designated CLHd) gene was isolated from this species. The recombinant CLHd specifically agglutinated *V. alginolyticus* at a concentration of 50 µg/mL in a calcium-dependent way. It is further suggested that CLHd is an important immune gene involved in the recognition and elimination of pathogens especially in abalones (Wang et al., 2008). The recombinant paramyosin isolated from this species has been reported for possible use as an alternative antigen of the natural counterpart for molecular studies and diagnosis of mollusk allergy (Suzuki et al., 2014).

Haliotis discus hannai

Taurine: This species is a good source of taurine which is isolated at the rate of 3.07 g from 1 kg abalone viscera. Taurine is a dietary supplement for epileptics, as well as for people who have uncontrollable facial twitches (Qian et al., 2014).

Antioxidant activity: Two sulfated polysaccharide conjugates (termed ACP I and ACP II) isolated from this species have been reported to possess antioxidant activity (Zhu et al., 2008). The enzymatic hydrolysates prepared from the viscera of this species have been reported to possess in vitro antioxidant activity (Zhou et al., 2012).

Antiinflammatory activity: The mucosubstance by-products (AM) of this species were found to significantly lower the nitric oxide (NO) production along the expressional suppression of inflammatory mediators such as cytokines TNF-α, IL-1β, and IL-6 and enzymes iNOS and COX-2. The AM was also shown to increase expression of antiinflammatory response mediator HO-1. The by-products of this species are therefore suggested to possess notable antiinflammatory potential which promotes the possibility of utilization as functional food ingredient (Rho et al., 2015).

Haliotis laevigata

Antiviral activity: The antiviral activity of this species assessed by Dang et al. (2015) is given below:

Antiviral extract or compound	Virus target(s)	Suggested mode of action
Lipophilic extract from the digestive gland	HSVs	Antiviral activity occurring postentry
Hemolymph plasma	HSV-1[a]	Prevention of viral attachment and entry into cells

[a]HIV-1, human immunodeficiency virus type 1.

The in vitro antiviral activity of this species was assessed against herpes simplex virus type 1 (HSV-1). The EC50 (effective concentration to inhibit HSV-1 plaque formation by ~50%) of the crude hemolymph (v/v) of this species was found to be 6.23% (Dang et al., 2011).

Haliotis rubra

Antiviral activity: The antiviral activity of this species is given below:

Antiviral extract or compound	Virus target(s)	Suggested mode of action
Lipophilic extract from the digestive gland	HSVs	Antiviral activity occurring postentry
Hemolymph plasma	HSV-1[a]	Prevention of viral attachment and entry into cells

[a]*HIV-1*, human immunodeficiency virus type 1.

Source: Dang et al. (2015).

Haliotis rufescens

Antiviral activity: The antiviral activity of this species is given below:

Antiviral extract or compound	Virus target(s)	Action
Aqueous extract	Polyomavirus, influenza A virus, and poliovirus	Unknown

Source: Dang et al. (2015).

Harpa conoidalis **Antimicrobial activity:** The human bacterial pathogen *S. paratyphi* was found to show maximum susceptibility (9.2-mm inhibition zone) against the ethanol extracts of Harpa major. On the other hand, the fungal pathogen *Aspergillus niger* showed maximum susceptibility (4.0 mm inhibition zone) against ethanol extracts of this species (Suresh et al., 2012).

Anticoagulant activity: The heparin sulphate of the body extract of this species showed significant anticoagulant activity which was estimated by using the activated partial thromboplastin time (APTT) and thrombin time (TT) clotting assay (Mohan et al., 2016).

Hexaplex trunculus

Antibacterial activity: The CHCl3 extract of the egg mass of this species inhibited *S. aureus* at a concentration of 1 mg/mL and *E. coli* at 10 mg/mL. 2,4,5-Tribromo-1*H*-imidazole inhibited the human pathogens at 0.1 mg/mL (Benkendorff et al., 2015).

Antiviral activity: The 6-bromoindirubin-3′-acetoxime (BIO-acetoxime), a synthetic derivative of a compound from this species, has been reported to suppress viral gene expression and protect oral epithelial cells from HSV-1 infection. The possible involvement of Glycogen synthase kinase 3 (GSK-3) is also suggested in HSV-1 infection (Hsu and Hung, 2013; Benkendorff et al., 2015).

Lambis chiagra

Antibacterial activity: The extracts of this species showed a growth inhibition of 13 mm against *S. paratyphi*. It also showed the highest activity against *Vibrio cholera* (Kanchana et al., 2014).

Lambis lambis

Cytotoxic and genotoxic activities: The *Lambis lambis* mucus (80% v/v) test solution had 65% cytotoxic activity. Further, with regard to the genotoxic effect, the mucus of this species (100% v/v) exhibited root growth of 0.367 cm. *Alkaloids*, terpenes, and proteins present in the mucus of this species account for the cytotoxic and genotoxic effects (See et al., 2016).

Antibacterial activity: The antibacterial activity of extracts of body tissue of *L. lambis* against eight bacterial pathogens namely, *B. subtilis, E. coli, Salmonella typhi, K. pneumoniae, Shigella flexneri, P. aeruginosa, S. aureus*, and *V. cholera* were carried out. Of all the fractions tested, Fraction 4 (methanolic extract) showed the maximum activity (Vimala and

Thilaga, 2012). According to Rohini et al. (2012), the water extract of this species was found to be active against *P. aeruginosa, P. vulgaris,* and *Klebsiella aerogenes.* On the other hand, the growth of *P. aeruginosa, Flavobacterium columnare,* and *S. aureus* was effectively controlled by the methanolic extracts of this species.

Littoraria angulifera

This species is used as a zootherapeutical product for the treatment of chesty cough and shortness of breath in traditional Brazilian medicine in the Northeast of Brazil (Wikipedia).

Littorina littorea

Antiviral activity: The antiviral activity of this species is given below:

Antiviral extract or compound	Virus target(s)	Suggested mode of action
Peptide extract from whole organism (littorein)	HSV-1[a]	Unknown

[a]*HIV-1*, human immunodeficiency virus type 1.

Source: Dang et al. (2015).

Littorina sitkana

Antitumor activity: β-D-Hydrolase enzyme complexes from the marine mollusk, *Littorina sitkana* have been used to obtain isoflavones from their conjugated forms. The β-D-glucanases of this species have been to exhibit antitumor activity (Kusaikin et al., https://dvfu.pure.elsevier.com/en/publications/deglycosylation-of-isoflavonoid-glycosides-from-maackia-amurensis). Further, the production of endo-1→3-β-D-glucanases by this species could also play an important role in the physiological and developmental processes of other animals. Further research however is needed on this aspect.

Margistrombus marginatus

Antibacterial activity: The methanol:water (1:1) extracts of this species exhibited antibacterial activity against more than 60% of marine biofilm bacteria (Anon. https://www.thefreelibrary.com/Potential+antimicrobial+activity+of+marine+molluscs+from+tuticorin%2c...-a0133108179).

Megathura crenulata

Anticancer activity: This species contains a hemocyanin (blood protein) called *Keyhole limpet hemocyanin* (KLH) which is a copper-containing respiratory protein, similar to hemoglobin in humans and is used as hapten carrier and immune stimulant (*vaccine carrier protein*). The major potential use of KLH is for bladder carcinoma by stimulating a specific immune response, but there are many other medical uses such as stress assessment, understanding inflammatory conditions, and treating drug addiction. It is also able to fight auto-immune disorders namely Alzheimer's disease and cancer (Kurokawa et al., 2002; McFadden et al., 2003; Anon. http://www. postranchkitchen.com/2014/03/keyhole-limpet-and-gumboot-chiton. html;Anon. http://www. united-academics.org/health-medicine/cancer-vaccines-from-mollusk/; Wikipedia).

Melo broderipii

Cytotoxic and genotoxic effects: The *Melo broderipii* mucus (80% v/v) test solution had 70% cytotoxic activity. Further, with regard to the genotoxic effect , the mucus of this species (100% v/v) exhibited root growth of 0.244 cm. *Alkaloids*, terpenes, and proteins present in the mucus of this species account for the cytotoxic and genotoxic effects (See et al., 2016).

Melo melo

Antibacterial and antifungal activities: The extracts from **Melo melo** is the potential source of antibacterial and antifungal bioactive compounds (Datta et al., 2015). In antibacterial activity, the maximum diameter of 24-mm zone of inhibition was recorded against *K. pneumoniae* strain of the mucus extract of this species, and minimum zone of inhibition of 11 mm was observed in *S. typhi* strain of body tissue extract. The antifungal activity of the extraction shows maximum activity against *Trichophyton mentagarophytes* (14 mm), and minimum activity was recorded in *Aspergillus flavus* (11 mm) (Kanagasabapathy et al., 2011).

Monetaria moneta

Antibacterial activity: The antibacterial activity of this species was studied in three different opportunistic human pathogens namely *Micrococcus* sp., *P. vulgaris*, and *Salmonella abory* in different concentrations (2, 3, 4, and 5% w/v) of shell powder extract of this species. Among these, *P. vulgaris* showed the maximum zone of inhibition (15 mm size) against 5% w/v concentration, followed by *Micrococcus* sp. (12 mm) and *S. abory* (10 mm) against the same concentration (Immanuel et al., 2012).

Wound healing: The wound healing effect of the shell powder of this species was found to be very effective in albino rats (Immanuel et al., 2012).

Murex tribulus

Antibacterial activity: The acetone extracts of the whole body tissues of this species inhibited 60% of the marine biofilm bacteria tested. The human pathogenic bacteria tested include *S. aureus*, *E. coli*, and *P. aeruginosa* (Benkendorff et al., 2015).

Nerita albicilla

Anticancer activity: Traditionally, this species had been used in Kei Island, Southern Maluku, and Indonesia to treat liver disease including cancer. All extracts of this species showed topoisomerase-I inhibitor activities relating to anticancer. Minimum inhibitory concentration (MIC) of methanol extract was 2.50 µg/mL. Chemical screening of the extracts showed that they contained steroidal and alkaloid compounds. This investigation revealed that *Nerita albicilla* contains active compounds that could be potential for nutraceutical or pharmaceutical development (Hardjito et al., 2012).

Antibacterial activity: The bactericidal activity of the different organs of this species in different species of bacteria is given below:

Acinetobacter baumannii	*Escherichia coli*	*Pseudomonas aeruginosa*
Gills	62.8	34.2	99.1
Gut	77.5	74.3	97.5
Gonad	28.17	84.8	90.0

Source: Kiran et al. (2014).

Patella rustica

Antimicrobial and antioxidant activities: The antimicrobial and antioxidant activities of crude peptide extracted from *Patella rustica* have been studied. The extracts were tested against eight strains of bacteria (*E. coli*, *S. aureus*, *B. subtilis*, *S. typhi*, *Enterococcus feacalis*, *K. pneumoniae*, *Streptococcus pneumonia*, and *P. aeruginosa*) and one strain of fungi (*C. albicans*) using agar well diffusion and broth dilution assays. *P. rustica* showed a markedly higher antifungal activity but with little antibacterial effect. The MIC of the extracts of this species was 13 mg/mL against all the strains of microorganisms tested except for *E. feacalis* (17 mg/mL), *K. pneumoniae* (17 mg/mL), and *C. albicans* (13 mg/mL). Antioxidant activity using the 2,2-diphenyl-1-picrylhydrazyl (DPPH) assay showed scavenging ability on the DPPH radical was 79.77% at 0.39 mg/mL for this species (Borquaye et al.,2015).

Patella vulgata

β-**Glucuronidase:** This enzyme produced by this species is reported to be more effective in hydrolyzing opioid-glucuronides than the *Helix pomatia*, bovine liver, and *E. coli* enzymes. Further, this enzyme is used as a reporter gene to monitor gene expression (Anon. http://www.sigmaaldrich.com/catalog/product/sigma/g2174?lang=en&region=IN).

Phalium glaucum

Anticoagulant activity: The bioactive compound glycosaminoglycans have been isolated from this species at 5.3 g/kg. This species showed anti-coagulant activity at 78 USP units/mg (Periasamy et al., 2013).

Planaxis sulcatus

Anticancer activity: The bioactive compounds dihydrosinularin and 11-epi-sinulariolide (which are the richest sources of marine cembra-noids) have been isolated from this species. These compounds have been reported to have marginal antineoplastic activity against P-388 lympho-cytic leukemia (Sanduja et al., 1986).

Cytotoxic activity: The bioactive compound "planaxool" (a novel Cembranoid) isolated from this species has shown significant cytotoxic activity (Alam et al., 1993).

Pleuroploca trapezium

Antioxidant activity: This species exhibited the highest antioxidant activity in DPPH and SOS assays with EC50 of 0.93 and 0.95 mg/mL, respectively (Gopeechund et al., Symposium.wiomsa.org/wp-content/uploads/2015/10/A.-GOPEECHUND.pdf). The meat of this species also has good antioxidant activity, justifying the need to popularize this meat as an important seafood. The extracts of this meat showed an antioxidant activity range of 10.3–50.3% at a concentration range of 810–4050 μg/mL (Anand et al., 2010).

Purpura bufo

Antibacterial activity: The MeOH extracts of the whole body of this species inhibited 25% of the marine biofilm bacteria tested. The human pathogenic bacteria tested were *S. aureus*, *E. coli*, and *P. aeruginosa* (Benkendorff et al., 2015).

Purpura persica

Antimicrobial activity: The whole body crude extracts of this species was tested against 10 bacterial pathogens namely *Aeromonas hydrophila, Bacillus cereus, E. coli, Pseudomonas aerogenosa, S. typhi, S. flexneri, V. cholera* 0139, *V. cholera classical, Vibrio cholerae* 01790, and *V. cholerae* EITOR and nine fungal pathogens namely *A. flavus, Aspergillus terreus, A. niger, Aspergillus fumigatus, Fusarium moniliforme, Trichoderma* sp., *Penicillium citrinum, Penicillium oxallicum,* and *Rhizopus* sp. A total of five different extracts namely Hexane: Chloroform (F1), Chloroform (F2), Benzene (F3), Benzene: Methanol (F4), and Methanol (F5) were prepared for this study. Among the tested bacterial pathogens *S. typhi, P. aerogenosa, S. flexneri,* and *B. cereus,* and fungal pathogens *A. fumigatus, A. terreus, F. moniliforme,* and *Trichoderma* sp. showed inhibition in growth by crude, F2, F3, and F5 fractions of *P. persica,* respectively (Santhi et al., 2016).

Analgesic, antipyretic, and antiinflammatory activities: The analgesic, antipyretic, and antiinflammatory effects of the 100% chloroform purified extracts of this species were studied with albino rats. At the concentration of 100 and 200 mg/kg, the pain was found to decrease in analgesic activity. With same dose, it reduced the body temperature in albino rats in antipyretic activity. The same dose in the antiinflammatory activity against carrageenan induced paw edema (Santhi, et al., 2011).

Rapana venosa

Antibacterial activity: The hemolymph which contains proline rich peptides of this species was found to inhibit the Gram +ve *S. aureus* and Gram −ve *K. pneumoniae* (Benkendorff et al., 2015). The hemolymph collected from the foot of this species showed prorich peptides which exhibited strong antimicrobial activities against tested microorganisms including Gram-positive and Gram-negative bacteria (Dolashka et al., 2011).

Antiviral activity: *Rapana venosa*—The hemolymph which contains hemocyanin of this species has been reported to inhibit the replication of Epstein–Barr virus at 1 µg/mL and Herpes simplex virus type 1 at 200 µg/mL (Benkendorff et al., 2015). The antiviral activity of this species as reported by Dang et al. (2015) is given below:

Antiviral extract or compound: Glycosylated functional unit of hemocyanin/RtH2

Virus target(s): Respiratory syncytial virus, HSV-1[a] and HSV-2, and EBV

Suggested mode of action: Prevention of virus attachment to cells by interaction with specific regions of HSV glycoproteins

[a]*HIV-1*, human immunodeficiency virus type 1.

Wound healing and antiinflammatory activity: Lipid extracts of this species were found to significantly improve the healing of induced skin burns in Wistar rats. These lipid extracts contain polyunsaturated fatty acids, Vitamin E, sterols and aromatic compounds. Amino acid extracts of this species were also found to accelerate skin wound healing by enhancing dermal and epidermal neoformation in Wistar rats. Healing occurred at least 10 days faster in rats treated with the above amino acid extracts compared to the untreated controls. The wounds treated with amino acid extracts from *R. venosa* also contained fewer inflammatory cells than the untreated control. Evidence for the antiinflammatory activity associated with the lipid extract was supported by normal blood cell counts in experimental rats treated with *R. venosa*'s extracts, compared to increasing quantities of leucocytes, lymphocytes, eosinophils and monocytes in the control rats (Benkendorff et al., 2015). The essential amino acids of this species have been reported to accelerate skin wounds healing via enhancement of dermal and epidermal neoformation in Wistar rats. The most abundant blood vessels, collagen fibers, basal, and stem cells were found only for treated animals with amino acids from the extracts of this species. The rich composition of amino acids of this mollusk merits consideration as therapeutic agents in the treatment of skin burns (Badiu et al., 2010).

Anticancer activity: The ethanol extracts of this species were found to control human leukemia HL-60 and human lung cancer A-549. Similarly, the purified hemocyanin of this species has been reported to control SiHa-cervical squamous cell carcinoma, CaOV-ovarian adenocarcinoma, MIA PaCa-pancreatic carcinoma, RD 64-rhabdomyosarcoma, EJ-urinary bladder carcinoma and Lep-nontumor human lung cell line and 647-V, T-24, and CAL_29 bladder cancer cells (Benkendorff et al., 2015).

Antioxidant activity: *R. venosa* appeared to be the best for a diet with relatively high protein and low lipid among the other examined molluscs (Celik et al., 2014). Consumption of the meat of this species has been reported to improve the lipid profiles and antioxidant capacities in serum of rats fed on atherogenic diet (Leontowicz et al., 2015).

Stramonita biserialis

Antibacterial activity: The whole body MeOH extracts of this species were found to inhibit 35% of the marine biofilm bacteria. The tested human pathogenic bacteria include *S. aureus, E. coli,* and *P. aeruginosa* (Benkendorff et al., 2015).

Tectus niloticus

Antiatherosclerotic activity: This species has been reported to possess bioactive compounds as potential antiatherosclerotic substances (Sarizan, 2013). Further, intensive research is, however, needed on this aspect.

Telescopium telescopium

Antimicrobial activity: Ammonium sulfate precipitated protein (SF-50) isolated from the spermatheca gland of this species showed antimicrobial effect on *E. coli.* This antimicrobial effect varied with the concentration of "SF-50" used, and the effect was found to be comparable to antibiotics like amikacin, contrimoxazole, and gentamycin in disc diffusion test. The "SF-50" was devoid of erythrocyte hemolysis property (Pakrashi et al., 2001).

Antiangiogenic activity: The methanolic extracts of *Telescopium telescopium* at the tested concentration of 200 µg showed fairly a higher degree of antiangiogenic activity with an inhibitiory percentage (64.63%) of the VEGF-induced neovascularisation. Further, in the experiment relating to in vivo antiangiogenic activity, the methanolic extracts of this species exhibited most noticeable inhibition (42.58%) of the corneal neovascularization in rats (Gupta et al., 2014). The significant antiangiogenic activity evinced by the extract of this species merits further investigation for ocular neovascular diseases.

Hypotensive effect: Intravenous administration of the tissue extract of this species produced a decrease in blood pressure (hypotensive effect) in anesthetized rats. The above extract has also been reported to produce potent esterase, cholinesterase, phospholipase, phosphatase, and protease activities (Samanta et al., 2008).

Tibia curta

Antimicrobial activity: The extract of this species possessed protein, amino acid, and carbohydrate. The diameter of inhibition zone (mm) as antibacterial activity recorded for the tested bacterial species is given below:

Escherichia coli	18.3
Pseudomonas aeruginosa	22.3
Proteus mirabilis	14.6
Klebsiella oxytoca	22.6
Serratia liquefaciens	22.3

The MIC of the above extract was found to be 10 mg/mL for all bacterial species tested except *P. mirabilis* which showed a value of 30 mg/mL (Degiam and Abas, 2010).

Turbinella pyrum

Antibacterial activity: The extracts of this species showed antibacterial activity against *S. paratyphi*, and value of growth inhibition was 8 mm (Kanchana et al., 2014).

Turbo bruneus

Antiinflammatory, antioxidant, analgesic, and antimicrobialactivities: The bioactive compound 6-(diphenylphosphoryl)-3,4-bis (diisopropylamino)-5-pyrrolidino pyridazine isolated from this species has found to possess these activities (Tamil Muthu and Selvaraj, 2015).

Antioxidant and antimicrobial properties: The bioactive compound 3,3,4,4-tetracyano-5,6-diphenyl-2-(cyclohexylimino)-2,3,4,5-tetrahydropyridine isolated from this species has found to possess these activities. Further, it is a neuroprotective agent used to prepare medicines for treating disease causing demyelination (Tamil Muthu and Selvaraj, 2015).

Antimicrobial and antifungal activities: The bioactive compound 1*H*-purin-6-amine,[(2-fluorophenyl)methyl]-(CAS) isolated from this species has found to possess these activities. Further, its derivatives are also known to possess antitubercular, antiinflammatory, antitumor, amoebic, antiparkinsonian, anthelmintic, antihypertensive, antihyperlipidemic, antiulcer, chemoprotective, and selective CCR3 receptor antagonist activity (Tamil Muthu and Selvaraj, 2015),

Antiandrogen, antiplatelet agent, antitubulin, and antimicrobial properties: The bioactive compound 3-(4-chlorophenyl)-4,6-dimethoxy-1-(prop-2'-enyl) indole-7 carbaldehyde isolated from this species has found to possess these properties (Tamil Muthu and Selvaraj, 2015).

The presence of various bioactive compounds justifies the potential use of meat of this species for various diseases. Further investigations on the purification and chemical elucidation of the above bioactive compounds pave the way for the development of new drugs in future.

Turbo setosus

Antioxidant activity: The crude extract of this species contains alkaloids, flavonoids, steroids, and triterpenoids (*n*-hexane, ethyl acetate, and methanol). The ethyl acetate of this extract has been reported to possess antioxidant activity with IC50 of 1578.43 ppm (Merdekawati, 2013).

Volegalea cochlidium

Antimicrobial activity: The ethanol extract of this species was found active against eight species of pathogenic bacteria. The inhibition zone ranged from 3 to 12 mm. The maximum inhibition zone was 12 mm against *Proteus mirabilis*. The methanol extract of this species showed maximum inhibition, that is, 11 mm against *E. coli*, and the minimum inhibition zone of 1 mm against *S. aureus*. Water extract of this species showed activities against all the four bacterial strains. The maximum inhibition zone was 4 mm against *Proteus mirabilis*, and the minimum inhibition zone was 1 mm against *K. pneumoniae* (Dhinakaran et al., 2011). Babar et al. (2012) reported that the extracts of this species exhibited good activity against *V. cholera*. Raj et al. (2014) made extensive studies on the antibacterial activities of this species. Different solvent extracts of this species were screened for their activity against *Vibrio parehaemolyticus* (J13300), *Aeromonas hydrophilla (*IDH1585), *S. typhi* (C6953), *S. paratyphi* A (C6915), *V. cholerae* (IDH5439), and *E. coli* (H10407). The results revealed that the ethyl acetate extract of the tissues inhibited the growth of the tested pathogenic bacterial strains. The minimum inhibitory of the tissue extract ranged from 05 to 20 mg/mL. Kanchana et al. (2014) reported that the extracts of this species were active against *S. paratyphi* with a growth inhibition of 13 mm. The antibacterial properties of the whole body extracts of this species were also studied by Sugesh et al. (2013) using 10 human pathogenic microorganisms such as *E. coli*, *K. oxytoca*, *Klebsiella pnuemoniae*, *Lactobacillus vulgaris*, *P. mirabilis*, *P. aeruginosa*, *S. typhi*, *S. paratyphi*, *S. aureus*, and *Vibrio paraheamolyticus*. Among the ethanol, methanol, and water extracts, the ethanolic extracts showed maximum antibacterial activities against *E. coli* (8 mm) and minimum activities against *Vibrio paraheamolyticus* (2 mm), methanolic extracts showed highest activity in *E. coli* (6 mm) and lowest activity against *S. paratyphi* (1 mm), and the extract of water showed antibacterial activities against *E. coli* and *S. paratyphi*.

KEYWORDS

- edible marine gastropods
- pharmaceutical compounds
- activity profile

CHAPTER 6

INDUSTRIAL USES OF EDIBLE MARINE GASTROPOD MOLLUSCS

CONTENTS

ABSTRACT

The uses of edible marine gastropods in the shellcraft industries; in the production of non-nacreous pearls such as queen conch pearls, Melo pearls, horse conch pearls, Kari pearls, and paua pearls; and the uses of their shells in the production of ornaments, lime, incense, fertilizer, and dyes are given in this chapter.

Several species of edible marine gastropods are of great industrial importance. While the meat of these species is a delicacy and is used in the preparation of many recipes in food industries, their shells are used in industries relating to the production of calcium carbonate for liming, incense, dyes, shellcraft products (such as souvenirs, curios, kitchenwares, home and office decors, personal ornaments, fashion accessories religious articles, etc.) Further, the gastropod shells including that of rare species have become the status symbol among the prominent people and fashion practioners in some parts of the world. These shells command high prices (e.g., *Cypraea annualus* and *Cypraea moneta*) and are known to have great value in the international market. A total of about 80 gastropod species are sold individually as decorative pieces or incorporated into utilitarian objects in shell trade (Jesily and Rooslin, 2015).

6.1 SHELL CRAFT INDUSTRIES

West Bengal in India is well known all over the world for its expertise in conch shell craft. Its artisans have skill in conch shell crafts. Not only are conch shell crafts beautiful and delicate, they are also considered to be extremely auspicious as per Hindu mythology.

6.1.1 SHELL CRAFT PRODUCTS

Several edible marine gastropod species are used in the manufacture of shell craft products, and the different types of such products are given below.

Kitchenwares: Spoons, fork, bowls, stirrer, and knives.

Home and office decors: Plates, flower vases, ashtrays, glass covers, chandelier, lamp shades, jewelry box, shell coins, pill boxes, pen holders, candle holder, curtains, table decors, and figurines.

Personal ornaments: Earrings, rings, combs, hair clips, bangles, buttons, chains, bracelets, hairpins, necklaces, cameos, tassels, bags, and footwear.

Other products: Inlays for guitars, fashion accessories, floor tiles, paint component, children's toys, etc.

6.1.2 COMMERCIALLY IMPORTANT SPECIES USED IN SHELL CRAFT INDUSTRIES

Babylonia spirata, Babylonia zeylanica, Bursa crumena, Cypraea tigris, Harpa major, Volegalea cochlidium, Lambis lambis, Lambis chiragra, Phalium glaucum, Tonna dolium, Tonna galea, and *Turbinella pyrum*—Large, colorful gastropods—used as a curios, souvenirs (mementos), and decorations.

Babylonia spirata, Babylonia zeylanica, Volegalea cochlidium, Lambis lambis, Lambis truncata, Chicoreus virgeneus, Phalium glaucum, and *Trochus radiatus*—"Rare" or specimen shells and collector's items.

Cassis cornuta—Kari pearl.

Cassis tuberosa, Voluta ebraea, Lobatus goliath (=*Eustrombus goliath*), *Monetaria moneta*—large shells as table lamps (Dias et al., 2011).

Cerithium spp., *Cerithedia* spp., *Phalium* sp., *Planaxis* sp., *Conus* spp., *Agaronia gibbosa*_(=*Oliva gibbosa*), *Strombus canarium, Umbonium vestiarium, Anadara* spp., and *Pecten* spp.—toys and dolls.

Charonia tritonis, Harpa major, Cassis cornuta, and *Syrinx aruanus* (large, polished shells)—decorative purposes.

Chicoreus ramosus (=*Murex ramosus*)—lamp shades.

Conus spp. (polished shells)—used as paper weights; often engraved with good wishes and greetings on the shells; small shells form pendants in garlands and key chains.

Cypraea tigris (beautiful, glossy shells)—for interior decoration on tables and shelves.

Cypraea annulus (sliced)—fashion accessories (Floren, 2003); place mats.

Cypraea annulus and *Nassarius* spp.—chandelier.

Cypraea moneta—used as a currency during ancient times; good tool of the astrologers and fortune tellers.

Cypraea tigris—used for interior decoration on tables and shelves; for making shell turtle toys.

Cypraea tigris and *Haliotis* spp.—Toys (snail boats, riders).

Cypraea tigris, Lambis lambis, Tonna spp., *Strombus* spp., *Conus* spp., and *Haliotis iris*—used as ornamental shells, curios, souvenirs, and collectors' items.

Cypraea spp. (cowries)—used by people in India for dice-playing.

Haliotis iris—non-nacreous *Haliotis iris* pearls or paua pearls.

Haliotis sorenseni—non-nacreous.

Haliotis spp.—bracelets.

Lambis lambis—Lamps and bathi stands.

Lobatus gigas (=*Strombus gigas*)—non-nacreous conch pearls.

Melo melo (=*Melo indica*) and *Turbo marmoratus* (large, beautiful shells)—mantel pieces and table decoratives.

Melo melo, Oliva spp. (black and brown colored), *Trochus* sp., *Haliotis asinine*, and *Strombus lunhaunus*—used in the manufacture of necklaces.

Melo volutes—non-nacreous Melo pearls.

Murex spp. (large shells)—lamp shades and ash trays.

Murex spp. and *Bursa* spp.—shell curtains.

Nassarius spp.—Nassa curtain.

Strombus canarium—toys and dolls.

Oliva spp.—used as raw material for pendants and rosettes for chains, garlands, and necklaces.

Pleuroploca gigantea—non-nacreous horse conch pearls.

Srtombus canarium—rings [they are made and are worn on fingers by some people in Tamil Nadu (India) and are made as chains for use by the people of Malabar and Karnataka (India) (Babu et al., 2011)].

Tectus niloticus (=*Trochus niloticus*): used in the manufacture of buttons and inlays—also as ornaments in the form of polished shell.

Tectus niloticus, Turbo marmoratus, and *Haliotis* spp.—for their nacre or mother-of-pearl.

Turbinella pyrum—baby milk feeders and blowing conches are made by boring an opening at the top of the spire of the chank.

Turbinella sp., *Tectus niloticus*, and *Chicoreus virgineus*—ash trays (mounted on wooden bases).

Shells as curios and souvenirs

Paper weights from *Cypraea* and trumpet shells

Polished shells

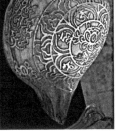

Engraved chank shells

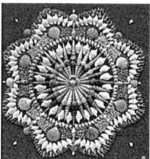

Place mats

Showcase material

Utilitarian objects

Key chains made of gastropod shells

6.2 NON-NACREOUS PEARL PRODUCTION

The non-nacreous pearls (organic gems) are normally formed when an irritant enters the mantle of molluscs as in the true pearl oysters. These molluscs may take several decades to produce these pearls of significant size.

6.2.1 QUEEN CONCH PEARLS

The non-nacreous conch pearls are the by-products of commercial queen conch (*Lobatus gigas*) farming. Good gem quality conch pearls are very rare, with a probability of occurrence of 1 in 1,000,000.

Queen conch pearl

The annual catch of queen conches is as so enormous that a considerable quantity of these pearls is produced annually (Anon. http://www.internet-stones.com/breakthrough-culturing-queen-conch-pearls-scientists-FAUs-harbor-branch-oceanographic-institute.html; Scarratt and Hann, http://www.ssef.ch/uploads/media/2004_Scarratt_Pearls_from_the_lion_s_paw_scallop_01.pdf).

6.2.2 MELO PEARLS

These Melo pearls are produced by the species *Melo melo* which are commonly found only in Southeast Asia, including China, Vietnam, Myanmar, Thailand, Malaysia, Indonesia, and the Philippines. As the shells of the *M. melo* are resembling coconuts, these pearls are also called "coconut pearls." Unlike the common pearls, the Melo pearls are not formed from layers of nacre. The chemical composition of this pearl is calcite and aragonite. Though these Melo pearls are non-nacreous, they are formed in the same way as more common pearls. It may take as long as several decades to grow a Melo pearl of significant size. The colors of these pearls range from tan to dark brown, with yellowish-orange or reddish-orange color. These pearls exhibit a porcelainlike luster that makes them very attractive. The surface of these pearls may also display a silky flame-like structure that makes a Melo pearl very valuable (Anon. Rare Melo Pearls—http://www.ajsgem.com/articles/rare-melo-pearls.html). These Melo pearls occur in a variety of shapes, ranging from irregular baroque to oval and egg shaped. Though spherical or round-shaped pearls are normally rare, spheres measuring 2–3 cm in diameter are not uncommon. The biggest pearl with a size of about 80 g has been reported from this species (Anon. http://www.wacht-troy.com/PTypeConchP.html).

Melo pearls

6.2.3   HORSE CONCH PEARLS

These non-nacreous pearls are produced by the edible gastropod species, *Peuroploca gigantea*. These pearls have soothing pink and orange colors.

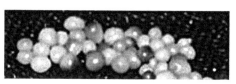

Horse conch pearls

6.2.4   KARI PEARLS

These pearls are produced by the species Cassis cornuta. A big and really great beautiful pearl with a weight of 22 g has been collected from the eastern island of Indonesia (Anon. http://www.karipearls.com/cassis-cornuta.html).

Kari pearl

6.2.5   HALIOTIS SORENSENI

This species has also been reported to produce non-nacreous abalone pearls especially in Californian seas. These pearls are also notable for their high luster and attractive colors (silvery-white, blue, green, or yellowish-green) (Joyce, http://www.gia.edu/gems-gemology/fall-2015-labnotes-large-natural-pearls-haliotis-abalone-species).

Pearl from *Haliotis sorenseni*

6.2.6 HALIOTIS IRIS PEARLS (PAUA PEARLS)

Among the different species of *Haliotis*, *Haliotis iris* is well known to produce natural, non-nacreous pearls which have a worldwide trade.

6.3 ORNAMENTS MADE OF SHELLS

Ring from the shell of
Turbo petholatus

Pendant from the operculum of
Turbo Petholatus

Ring from the operculum of
Turbo petholatus

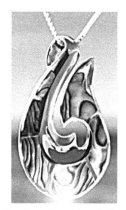

Pendant from the shell of
Haliotis iris

Chains from chank shells

6.4   CONSTRUCTION INDUSTRY

Gastropod shells along with gravel are used for building roads in some areas, and lime from these shells is a vital component in the production of concrete and plaster.

6.5   FERTILIZER MANUFACTURING

Owing to the presence of lime, the gastropod shells are used as a vital component of plant fertilize.

6.6   INCENSE INDUSTRIES

The operculum of *Pleuroploca trapezium* (popularly known as "Fish nail"), *Strombus lentiginosus*, *Strombus gigas*, *Lambis truncata*, *Rapana venosa*, and the nonedible *Chicoreus virgeneus* (=*Murex anguliferus*) has been reported to contain aromatic substances and smell when it is paced on smoldering charcoal. Hence, it is of great use as an incense material (Lev and Amar, 2008). Operculum powder is an important ingredient to Chinese, Japanese, and Middle East incense makers. Operculum is traditionally treated with vinegar, alcohol, and water to remove any fishy smell. The cleaned opercula are then ground to a powder and used as a scent fixative. In some countries, the operculum is rubbed with an alkali solution prepared from the plant bitter vetch to remove impurities, and it is then soaked in fermented berry juice of the Caper shrub or strong white wine, in order to enhance its fragrance. India is one of the major exporter countries of dried high-quality operculum. Operculum of species like *Babylonia spirata*, *Chicoreus virgeneus*, *Volegalea cochlidium* (*=Hemifusus cochlidium*), *Lambis lambis*, and *Chicoreus ramosus* is exported from India to other countries for industrial purposes (Sundaram and Deshmukh, http://eprints.cmfri.org.in/9617/1/17.pdf).

6.7   GASTROPOD SHELLS AS TOOLS

Drills, chisels, scrapers, sanders, etc. are made from various shells such as the Red helmet shell [*Cypraecassis rufa* (nonedible)] and are used in such trades as woodworking, farming, and tool making. Weapons such as spearheads and gouges were made from sharpened and shaped pieces of hard shells such as the Queen conch [*Strombus gigas* (Anon. http://www.manandmollusc.net/advanced_uses/advanced_uses-print.html)].

6.8   PRODUCTION OF DYES

Edible species of gastropods such as *Stramonita haemastoma*, *Nucella lappilus*, *Concholepas concholepas*, *Purpura persica*, *Galeodea echinophora*, *Plicopurpura pansa*, *Bolinus brandaris*, *Hexaplex trunculus*, *Drupa aperta*, *Dicathais orbita*, *Rapana bezoar*, and *Mitra* sp. are the sources of tyrian purple (royal purple, indigo, hyacinth purple, or imperial purple) dye which is secreted by the ink glands of these species. These dyes are used to beautify clothing and other items made from cloth with varying colors ranging from pink/red to violet/black. Many Mexican and South American natives still prefer these molluskan dyes for their garbs, since they produce more natural—looking and traditional hues (Tabugo et al., 2013; Verhecken, 1989; Anon. http://www.manandmollusc.net/ advanced_uses/advanced_uses-print.html).

6.9 CULINARY USES IN FOOD INDUSTRIES

Before cooking the edible gastropods, the chefs put them on fasting for 3 days with water and then feed them with flour and water for one week for further purification. After that, they are cooked for various recipes (Anon. http://www.snail-world.com/snails-as-food/).

6.9.1 PICKLE PREPARATION

Delicious pickles can be produced for commercial purposes from the edible portions, namely, foot, mantle, and columella muscle of *Chicorcus ramosus* (Patterson et al., 1995).

6.9.2 ABALONE WITH BUTTER AND GARLIC

Moderately cooked (meat will become very firm if overcooked) abalone (*Haliotis* spp.) topped with butter and garlic is much tastier and is a viable commercial product (Anon. Clay's Kitchen, http://www.panix.com/~clay/ cookbook/bin/table_of_contents.cgi?gastropods).

6.9.3 BAHAMIAN CONCH CHOWDER

It is a delicious preparation prepared from edible conchs and is served with wine and biscuit. The other preparations with conchs are Caribbean conch fritters with Island hot sauce, Crispy pan fried conch with citrus sauce, and Chinese conch soup recipe (Anon. Clay's Kitchen, http://www.panix. com/~clay/cookbook/bin/table_of_contents.cgi?gastropods).

6.9.4 PERIWINKLES IN GARLIC AND WINE

The slipper limpets (*Crepidula fornicata*) have a yellow disc of meat approx. 1 in. across which is high in protein and can be eaten raw or slightly cooked. The dish of this species in garlic and wine is increasingly popular in France (Anon. Cornwall Good Seafood Guide, http://www. eatsxm.com/sea-snails.html).

6.9.5   TURBAN SNAIL RECIPE

The sauce prepared form the black turban snail, *Tegula funebralis* is served over rice with a little bit of lime zest (Anon. http://www.spearboard.com/ showthread.php?t=159009).

6.10   MISCELLANEOUS USES

Shells are ground up and added to chicken feed (for stronger eggs).

For decades, the large *Strombus gigas* was used as ballast in ships returning from the Caribbean.

Strombus gigas was once consumed in such large quantities that their empty shells were used in the building of harbors and breakwaters.

South American Indians use shell lime to extract the narcotic (Cocaine) from cocoa leaves.

Some countries used powdered shells instead of calcium carbonate to make the liquid clay used in the production of ceramics. This adds a different effect to the finished product.

6.11   EDIBLE MARINE GASTROPODS IN RELIGION

Marine gastropod shells have played an important role in religion from prehistoric times onwards. Shells in some cultures even today are used as amulets, good luck charms, and as symbols for love, fertility, and life eternal. In the Hindu religious context, the very rare left-handed (sinistral) shells of *Turbinella pyrum* are known as Valampurich chanku in Tamil. These sinistral shells are highly valued in terms of its religious significance. Some examples of some these religious practices are given below:

Africa: Shells fetishes (fetish is an object which is treated with reverence and respect because it is either thought to have special powers or is where a god or spirit lives or is present in some special manner) were often used in worship. Ceremonial garbs are many times decorated with shells and were used in some religious ceremonies.

North America: North-American Indians also made fetishes of shells. The Canadian Ojibwa tribe maintained a Grand Medicine Society in which the sacred emblem was a shell.

India: The Hindus of India believe that their God Vishnu holds his staff crowned with a very rare and venerated sinistral (left-handed) *Turbinella pyrum* chank shell. These Hindus, when praying, often clasp a sacred chank or other venerated object in their hands, believing that it will help their petitions be heard. The Hindu priests also use it for blowing.

Turbinella pyrum (Sacred Chank) Shell Trumpet

Asia: Buddhists: The Chank or *Turbinella* also plays a significant role in the Buddhists ritual music and ceremonies (Anon. http://www.manand-mollusc.net/advanced_uses/advanced_uses-print.html).

KEYWORDS

- shell craft industries
- non-nacreous pearl production
- Melo pearls
- horse conch pearls
- Kari pearls

CHAPTER 7

AQUACULTURE OF EDIBLE MARINE GASTROPOD MOLLUSCS

CONTENTS

ABSTRACT

Farming practices of edible and commercially important gastropod species such as abalone and queen conch are dealt within this chapter.

Marine aquaculture, including shellfish culture, has the potential to supply an increasingly valuable contribution of high-quality protein-based foods for humans. Molluscan shellfish has traditionally been a major component of world aquaculture. For example, the molluscan shellfish production in 2012 accounted for about 60% of the world marine aquaculture production (Anon, 2014). Marine molluscs are cultured in numerous countries in both northern and southern hemispheres. However, culture activities have extensively developed in the Asian region, particularly in East and Southeast Asia. For the marine aquaculture production of molluscs, the contribution of bivalves is very significant (about 97%) and that of gastropods is very negligible (2.8%) (Anon, 2014). Although there are 28 species of marine gastropods have the potential for aquaculture, the major group which is commercially important is abalones. The list of edible marine gastropod species relating to commercial aquaculture and other potential cultivable species is given below:

Species under commercial aquaculture

Species	Cultivating countries
Chicoreus ramosus	India
Haliotis asinina	Philippines
Haliotis corrugata	California, Mexico
Haliotis discus	Korea
Haliotis diversicolor	Taiwan, Japan
Haliotis fulgens	California, Mexico
Haliotis gigantea	Japan
Haliotis midae	South Africa
Haliotis rubra	Australia
Haliotis rufescens	California, Mexico
Haliotis tuberculata	France
Littorina littorea	France, Ireland, Canada
Strombus gigas	Tropical West Atlantic
Tectus niloticus	Tropical West Pacific
Turbo cornutus	Korea, Japan

Source: Berthou et al. (http://archimer.ifremer.fr/doc/2001/publication-529.pdf).

Other potential cultivable species

Species	Countries
Babylonia areolata	Thailand, China
Buccinum undatum	France, Ireland, UK, Canada
Busycon canaliculatum	USA, Mexico
Busycon carica	USA, Mexico
Concholepas concholepas	Peru, Chile
Conomurex luhuanus	Papua N. Guinea, Philippines
Cymbium glans	Senegal
Cymbium pepo	Senegal
Haliotis iris	New Zealand
Haliotis ovina	Philippines, China
Lobatus galeatus	Mexico
Strombus gracilior	Mexico
Turbo marmoratus	Indo-West Pacific

Source: Berthou et al. (http://archimer.ifremer.fr/doc/2001/publication-529.pdf).

7.1 AQUACULTURE OF ABALONES, *HALIOTIS* SPP.

Global fish production continues to outpace world population growth, and aquaculture remains one of the fastest growing food producing sources. In 2012, global aquaculture production was 90.4 million tonnes. Although abalone contributes a relatively small proportion of the aquaculture production, it is one of the most highly prized seafood delicacies. The total volume of worldwide abalone fisheries has declined since the 1970s, but its farm production has increased significantly over the past few years (Cook, 2014).

Total global production of abalone from legal fisheries and from farms

Year	Fishery production	Farm production
2002	10,146	8700
2008	7869	30,760
2010	7424	65,344
2011	7424	85,344
2013	7486	103,464

Source: Cook (2014).

Farm production (mt) in various regions in 2010 and 2012–2013

Region	2010	2013–2014
China	56,511	90,694
Korea	6228	9300
South Africa	1015	1116
Chile	794	794
Australia	456	910
USA	250	250
Japan	200	200
Taiwan	171	171
New Zealand	80	90
Mexico	33	35
Europe	30	100
Thailand	10	15
Philippines	4	8

Source: Cook (2014).

Abalone farming was first developed in China and Japan where it was used for reseeding the wild-fishery. From here, it spread to various countries. In fact, abalone farms in various countries are still developing their own techniques to suit their production systems and market requirements. The Pacific abalone, *Haliotis discus hannai*, is the major species cultured in Japan, China, and Korea. Recent attempts to culture the pauas (chiefly *Haliotis iris*) in New Zealand as well as *Haliotis rubra* and *Haliotis laevigata* in Tasmania are showing encouraging results. The ormer, *Haliotis tuberculata*, is being produced in initial programs in Britain and France. Other experimental and incipient commercial attempts to culture native and introduced species are in progress in Canada, Mexico, and Chile. On the Pacific coast of the United States, several species of abalones such as chiefly the red, *Haliotis rufescens*, and the green, *Haliotis fulgens* have been subjects of experimental and commercial aquaculture for almost two decades. The red abalone is the largest member of the genus and, historically, the major fishery species. North American abalones are to be found only on the Pacific coast. There, seven species and two subspecies of haliotids occur. In the order of their historical value to California fisheries (prior to 1975) are the red, *H. rufescens*; pink, *Haliotis corrugata*; green,

H. fulgens; white, *Haliotis sorenseni*; and black, *Haliotis cracherodii*. The remaining species are small and of minor commercial interest.

Though the basic techniques for rearing abalone are similar the world over, specific details vary widely, often changing from one farm to the next. This diversity is a reflection of both the variety of conditions in which abalone are raised and the relative infancy of the industry. Many farms are still experimenting with methods and equipment, trying to find the perfect set-up to maximize production and minimize costs.

Commercial farming of abalones—an introduction

Land-based abalone farming

Most grow-out tanks in land-based abalone farming are raceways or circular tanks. These tanks are usually shallow and are made from strong black plastic. Floating baskets may be kept in outdoor tanks and each basket is provided with areas of solid surface (the white plastic) for the juvenile abalone to crawl upon. These grow-out tanks are also covered with greenhouse mesh to stop predators and provide shelter for the abalone. Sea water is pumped from the sea then through a sand filter before being pumped continuously through the tanks. The abalones grow quickly in the fast-flowing water provided by these flow-through systems. These abalones reach a size of about 10 cm in a period of 4–5 years. Most abalones are harvested at the end of their third year when they have reached a size of about 70 mm or 50 g. Most land-based abalone farms are 24 h operations involving continuous monitoring of the water systems and the stock. Any interruption to the water supply and water temperature can be disastrous with large-scale losses of abalone (Anon. http://www.mesa.edu.au/aqua-culture/aquaculture17.asp).

Sea-based abalone farming

Site selection: Optimum sites for abalone farming are coastal areas surrounded by islands, where seaweeds and algae, the natural food of abalones, are present abundantly. Farm sites should be away from fresh water inflows, have clear waters and a slow current. Stony and rocky sea beds are ideal for abalone culture.

Preparation of the culture site: The ecological condition of the selected sites is important for farming the abalones to the marketable size. If the selected site has the suitable physical and chemical conditions but lacks rocks and stones, then various types of structures including stones should be made available to meet the habitat requirements of abalones which normally cling to rocks and stones. It is also recommended to grow their food items namely seaweeds such as *Undaria, Laminaria,* and so on.

Cage method of releasing abalones into the sea: Releasing methods are of great importance in the survival rate of the growing abalones. The best method is as follows:

About 500–800 juvenile abalones are placed on a board, and a plastic setting board measuring 50 × 50 cm is then placed over it. The board is wrapped with net and placed in a tank with flowing water for 1–2 days. The growing abalones prefer to cling the plastic board. The plastic board still wrapped in the net is placed into a net cage and sunk to the bottom of the sea. One net cage with a mesh size of 4 cm may contain 3–5 boards. This method prevents high mortality that usually occurs during initial release.

Survival rate of abalone: The survival rate of abalones released in the sea depends largely on the size of juvenile abalones at the time of release. The survival rate reported in the Korean technique is as follows:

Size (mm)	Survival rate (%)
20	<10
25	25–30
30	30–60
40	70–80

When juvenile abalones are released into a new farm, the survival rate increases, and in wild habitat they suffer from predation while searching for proper places to hide (Anon. http://www.fao.org/3/contents/129a1845-3d53-5d6a-b431-b6518739a964/ab731e00.htm).

Growth of abalones under culture conditions: The growth rate of abalones during culture depends largely on the water temperature and availability of feed. It is reported that the abalones grow fast at 20°C, but when the temperature is below or above 20°C, the growth is reduced. The growth of abalones is also related to feed quality. They normally grow well in waters with abundant species of seaweeds like *Laminaria* and *Undaria*. Abalones feeding on seaweeds species usually have 3.5 times body weight than abalones cultured in sites with other flora. It is also reported that an abalone needs 10 g wet *Undaria* daily to increase its body weight by 1 g. After release into the sea, an abalone takes 2–3 years to reach the marketable size.

Farm maintenance: Maintenance activities in abalone culture site should be carried out periodically. The amount of feed should be controlled according to the bottom condition. Necessary measures should also be taken to maintain good water quality. It is always necessary to take suitable measures for controlling pests and predators (Anon. http://www.fao.org/3/contents/129a1845-3d53-5d6a-b431-b6518739a964/ab731e00.htm).

7.1.1   FARMING OF DISK ABALONE HALIODISCUS DISCUS IN KOREA

Grow-out facilities

Land-based raceways and circular tanks: These tanks are generally shallow and are made of black plastic. They are often covered with greenhouse mesh to stop predation and provide shelter for the abalone. Seawater is pumped from the sea and through a sand filter before being pumped continuously through the tanks.

   Barrel and cage culture: Both these techniques are similar in that they are divided into a number of sections in which the abalone are placed inside and allowed to attach. The barrels and cages are placed in the open sea and secured to the sea-floor by long-lines and anchors or attached to rafts. The barrels and cages are open. That is, they are covered with mesh, allowing water to circulate while serving to keep the abalone inside and protected from potential predators. The main source of food for the abalone is kelp and such seaweeds are kept inside the holding facilities (Ezzo Abalone, https://www.lib.noaa.gov/retiredsites/korea/main_species/abalone.htm).

7.1.2   FARMING OF ABALONES HALIOTIS SPP. IN AUSTRALIA

Species cultivated: Greenlip abalone, *H. laevigata*; brownlip abalone, *Haliotis conicopora*; blacklip abalone, *H. rubra*; Roe's abalone, *Haliotis* *roei*; donkey ear abalone, *Haliotis asinine*; and staircase abalone or /ridged abalone, *Haliotis scalaris* (Freeman, 2001).

Grow-out facilities

Land-based grow-out systems: These systems include large, deep concrete tanks, specialized tanks) and outdoor ponds. The tanks are very shallow with high flow rate. Water through this system flows as a unit which ensures that bacteria and wastes can be easily flushed from the system. Abalone production in the above tanks is estimated at 1000 kg/tank/year.

Sea-based grow-out method: In this method, inexpensive old juice concentrate barrels or polyvinyl chloride (PVC) manufactured tubes are used. These facilities, however, can only hold a small number of abalones. Alternatively, small to medium size cages are used in a range of depths. They can be attached to long lines and rafts, or placed on the sea-floor. These cages can hold more abalone and can provide better water circulation than the aforesaid barrels. Large sea cages can be placed in areas with a large supply of drift seaweed. The advantages of sea-based systems over land-based facilities include lower capital costs, better water exchange, more stable water temperature and feed supplementation from algal growth within the culture unit.

Species	Size at stocking
Haliotis laevigata	7 mm
Haliotis rubra	3–7 mm
Haliotis asinina	15 mm

Growth rate: The growth rate of abalone in both land and sea-based culture systems is highly variable depending on the quality and quantity of food provided. However, cultured individuals with a constant food supply showed the greatest growth during the warmer months. Growth rates are also affected by several factors including, density, type and amount of feed, water flow, water quality, and handling techniques. It has been estimated that Donkey-ear abalone can grow in maximum shell length to 70 mm in 1 year. Roe's abalone grows rapidly in the first year and reaches up to 40 mm in shell length (size of maturity) but slow-down in the following years. This reduction in growth is largely due to energy expenditure during gonad development. Greenlip and blacklip abalones reach market/harvest size (50–80 mm) in a period of 2–3 years. The growth rates of these species may improve in grow-out facilities when artificial diets are used and culture facilities are more refined (Freeman, 2001).

7.1.3 FARMING OF HALIOTIS SPP. IN CALIFORNIA

Status of abalone culture in California: All California species have now been cultured experimentally, but the red abalone, *H. rufescens* remains as the principal species for mariculture. The green abalone *H. fulgens* has shown special potential for mariculture in systems utilizing thermal effluent. Pink abalone *H. corrugata* and white abalone *H. sorenseni* have been cultured on a small scale. The black abalone *H. cracherodii*, a shallow-water species of lower commercial grade, is broadly tolerant of temperature in adult stages and has been reared from laboratory-spawned eggs. The flat abalone *Haliotis walallensis* and threaded abalone *Haliotis assimilis* have also been cultured in research projects in Southern California. The pinto abalone, *Haliotis kamtschatkana*, is receiving attention as a mariculture subject in Washington and British Columbia (Leighton, 1989).

Culture technology

In American culture technology, abalones are reared either in specialized tanks on shore or in containments held in protected ocean waters until a marketable size of 7–10 cm. [In Japan, however, juvenile abalones are usually released to the natural sea environment at a size of 2–5 cm for a continued growth over a period of 2–4 years until a marketable size (7–10 cm) is reached.]

Raceway culture: In this farm practice, juvenile abalones are reared in a series of concrete troughs through which seawater cascades from upper to lower troughs. Vigorous aeration is supplied intermittently and seawater flow rates are maintained from 0 to 200 L/min, depending on the pumping schedule. The seaweed, *Macrocystis pyrifera*, is the major food provided, although many species of red, green, and brown algae are supplied supplementally. Juvenile red abalones of about 1 cm size are transferred to containment structures held in the channel. Large polyethylene drums (200 L capacity) with plastic or stainless steel mesh capped ends are secured in a horizontal position to braces in racks. All these structures are immersed to a depth of a few meters beneath a pier. Each drum is stocked with several thousand juvenile abalones initially. Thinned with growth, the young abalones are reared to a marketable size of 4–5 cm (Leighton, 1989).

7.1.4 AQUACULTURE OF Haliotis discus hannai IN CHINA

Two species of abalones, namely, the Pacific abalone, *Haliotis discus hannai* and the small abalone *Haliotis diversicolor supertexta*, have been heavily fished in China, and their natural populations have become depleted. *Haliotis discus hannai* is the predominantly farmed abalone species in coastal regions. In 2013, the farmed yield of this species has been reported to exceeded 110,000 metric tons, and it accounted for more than 95% of total production. Recently, the industry relating to this species has developed rapidly. In northern regions, the abalone seeds of greater than 2 cm size are overwintered at high density in hatcheries. After overwintering, 1-year-old animals (with an average size of 4.0–4.5 cm) are transferred to sea-based farms and harvested after a further 12–24 months, when they have reached a market size of 10–14 individuals/kg (Wu and Zhang, 2016).

Northern regions

Sea-based culture systems: In northern regions, there are three sea-based abalone grow-out systems, namely, the intertidal pond culture system in Qingdao, the long-line hanging net cage system in Rongcheng, and the release or bottom culture method in Changhai, Dalian.

Intertidal pond culture system: In intertidal pond grow-out system, suitable sites and stocking density are crucial. Clear water quality in the culture site is essential for the occurrence of seaweed, food of growing abalones. In this culture system, an initial stocking density of 20–25 individuals/m^2 (average size, 4–4.5 cm) is followed. There is no culling or sorting during the grow-out phase of this culture system.

Long-line hanging net cage system: In long-line hanging net cage grow-out, regular control of biofouling by oysters (mainly *Crassostrea gigas*) and sea squirts (mainly *Ciona intestinalis*) is required. Net replacement and density sorting are also conducted four times until harvesting.

Bottom culture: In the release or bottom culture method, seeds with a size of 4 cm are used. The survival rate of *Haliotis discus hannai* (from 1-year-old animals to market size) in this culture method has been reported to be 50–70% (Wu and Zhang, 2016).

Southern regions of China

While large Pacific abalone-farming corporations are present in the northern regions, small farms are more popular in the southern regions of China. In southern grow-out systems, 6–7-month-old animals (1.5 cm size) or 1-year-old seeds (2.2 cm) are transferred to sea-based multitier culture systems. The former sized animals grow over two summers to reach a market size of 24 individuals/kg after 18–24 month of culture. On the other hand, 1-year-old seeds will reach the market size after 9- to 21-month culture. Annual production in southern regions is largely affected by summer mortality. The summer survival rate has been reported to be 50–70%. A new submerged cage culture system has also been developed in Fujian, southern China. In this system, the cage consists of five vertical slots oriented perpendicular to the flow of water. These slots are separated by six vertically connected plastic plates at the bottom of the cage, for abalone attachment and shelter. The advantages of this cage culture system include enhanced growth performance and improved rearing conditions, such as increased water flow velocity and higher concentrations of dissolved oxygen (Wu and Zhang, 2016).

A novel submerged cage culture system for abalone rearing was attempted in Fujian, South China. The growth of abalone in the above system was found to be density dependent. In terms of wet weight biomass, it was 1.48–3.01 times higher in the cage system compared with the traditional system. Abalone survival was also found to be more than 87.5% in these culture systems. Advantages of this newly established cage culture system include improved water flow velocity and dissolved oxygen (Wu and Zhang, 2013).

7.1.5 EXPERIMENTAL AQUACULTURE OF Haliotis asinina (FERMIN AND BUEN, 2001)

Source of experimental animals: Prior to the grow-out experiment, hatchery-bred abalone juveniles of 10-mm shell length were reared for 90 days in perforated plastic baskets suspended in flow-through tanks. They were fed with the macroalga *Gracilariopsis bailinae*, and they grew to 30 mm after a period of 3 months.

Stocking and culture system: Juvenile abalones of 32 mm size (7.5 g) were stocked ($n = 50$ individuals) in 12 cages (38 × 38 × 28 cm) double-lined with black nylon mesh (3 mm × 5 mm). Each cage was provided with either one, two, or three pieces of 30-cm-long PVC gutter as shelters. All the above cages were suspended in a 6-ton indoor tank which was continuously receiving sand-filtered sea water. Water depth was maintained at 90 cm. Aeration was provided through perforated PVC pipes laid at the tank bottom. The predetermined quantity of food algae was added to each cage at weekly intervals. At the end of each week, uneaten seaweed was collected, drained, and weighed (Fermin and Buen, 2001).

Results: The mean shell lengths, wet body weights, and mean daily growth rates of abalone reared in mesh cages did not differ significantly during the first 13 weeks of culture. Abalone reared in the cages with 0.66 m² were found to grew significantly faster (132 m/day, 188 mg/day). In this experiment, *Haliotis asinina* attained a final size of 58–60 mm within 9 months (sea-cage trials showed that *H. asinina* with an initial size range of 35–40 mm and stocked at 43 m⁻² attained a harvest size of ca. 60 mm over 180 days) (Fermin and Buen, 2001).

7.1.6 EXPERIMENTAL FARMING OF Haliotis asinina IN SUSPENDED PLASTIC CAGES (MINH ET AL., 2010)

Plastic cages (30 × 40 × 30 cm) were suspended at a depth of 0.8 m in a 15-m³ indoor tank that continuously received sand-filtered sea water. The water depth in the tank was maintained at 1 m and aeration was provided through the tank bottom. Stocking densities in these cages were 40, 60, 80, and 100 pcs/cage, with an initial size range of 4–5, 7–8, and 10–11 mm. During the experiment, predetermined amounts of seaweed (*Gracilaria verrucosa*) were added as food to the abalones once every 2–3 days. The juvenile abalones were reared in constant darkness with light only being turned on during feeding. Water quality measurements were taken weekly, and the mortality rate of the abalone was observed in the morning.

First experiment: This involved growing out for 6 months with four density classes of abalone at 40, 60, 80, and 100 pcs/cage. The initial size of the abalone per trial was 7–8 mm, and the initial mean shell lengths were 7.3, 7.2, 7.4, and 7.8 mm, respectively, for the four density classes.

Second experiment: It involved abalone that were reared out for 6 months, with three different initial mean shell length classes of 4–5, 7–8, and 10–11 mm. The initial number of abalone per trial was 60 pcs/cage, and the initial mean shell lengths were 3.9, 7.2, and 10.8 mm, respectively.

Growth and survival rate with different stocking densities of abalones: Abalones growing at low stocking density (40–60 pcs/cage) in the plastic cages showed significantly greater growth than at the higher stocking density (80–100 pcs/cage). The highest survival rate was at a stocking density of 60 pcs/cage, followed by 40, 80, and 100 pcs/cage. The average daily growth for 180 days of shell length and weight of the group initially 4–5 mm with density 60 pcs/cage was 0.04 mm/day and 2.21 g/day, respectively. For the 7–8-mm size group, the values were 0.02 mm/day and 0.55 g/day, respectively. The 10–11-mm group showed the values of 0.01 mm/day and 0.33 g/day, respectively (Fermin and Buen, 2001).

7.1.7 EXPERIMENTAL GROW-OUT CULTURE OF ABALONE *Haliotis tuberculata coccinea*

Experiments were conducted in abalone cages installed in a commercial open-sea cages fish farm. Growing abalones were fed with the red algae *Gracilaria cornea* and the green one *Ulva rigida*. Survival rates were found to be very high (94–98%) regardless the density used. A high linear growth of shell and weight was recorded. However, a reduction of 17–19% in weight gain was found by doubling the initial stocking density. However, the high growth performance (70–94 µm/day; 250–372% weight gain) and survival attained, even at high densities, suggested the suitability of the offshore mariculture system as well as the biofilter produced macroalgae for this grow-out culture of *H. tuberculata coccinea* which reached a commercial size in 18–22 months (Mgaya, 1995; Viera et al., 2016).

7.1.8   EXPERIMENTAL FARMING OF Haliotis midae

While *Haliotis midae* could reach a maximum size of about 20 cm shell length at an age of over 30 years in the wild, farm produced animals were found to reach an average size of 100 mm in about 5 years. Growth rates of 0.08–4.5% body weight/day for abalone of 10–17 mm shell length have been found with the formulated diets (Sales and Britz, 2001).

7.2   GROW-OUT CULTURE OF QUEEN CONCH, *Lobatus gigas*

Larger juveniles (7 cm size) can be grown up to 18 cm in the coastal waters with a predator-exclusion fence or in cages. The fence extends from the bottom to the high-tide line. A variety of fencing or cage material may be used to exclude predators such as brachyuran crabs, hermit crabs, spiny lobster, octopus, fish, sharks, rays, turtle, and so on. Large, circular cages (25 m diameter, 500 m²) with mesh openings (5–20 cm in diameter) will help to exclude predators. The stocking density for juvenile conch may be 1–2/m² or about 10,000 per ha. Cages can normally be stocked at a higher density (10 conch/m²) than open sea environment. However, the natural food supply may need to be supplemented with a prepared diet in the event of increasing the stocking density. Stock must also be rotated in the tidal environment to avoid overgrazing. The best grow-out sites are nursery grounds. These sites have a sandy bottom with suitable stocking density of seagrass and high algal productivity. Further, the culture sites may be 2–4 m deep and have strong tidal currents to flush the area and bring in new feed materials. Grow-out size of 7–18 cm will take about 1.5–2 years at a growth rate of 0.15–0.20 mm per day. Survival in the tidal culture environment may be 60–90%, depending on the nature of management practices (Davis, http://www2.ca.uky.edu/wkrec/QueenConch.pdf).

7.3   PEARL CULTURE IN QUEEN CONCH, *Lobatus gigas*

Among the saltwater porcelaneous pearls, those of *Lobatus gigas* have been reported to be the best known. Unlike an actual oyster pearl, the conch pearl is nonnacreous and is produced by the conch in its digestive tract. Also known as "pink pearl," "conch pearl," or "Queen conch pearl," these pearls are produced in a variety of colors (deep pink, orange, white, yellow, and brown). The conch pearls which are worthy of use in jewelry tend to be symmetrical in shape. Most conch pearls have an elongated, oval, or baroque shape, and near-round specimens are very rare. Although conch pearls can be found over 10 carats, larger sizes (above 2 carats) of conch pearls are uncommon, with the average size being under 3 mm in diameter. Average carat weights are from 0.2 to 0.3 carats. Due to the high value of conch pearls, drilling and/or gluing should be avoided in the mounting, as this will devalue the pearl. Additionally, prolonged exposure to sunlight (ultraviolet light) can have a dulling and fading effect and should be avoided (http://www.wacht-troy.com/PTypeConchP.html).

Beaded and nonbeaded pearl culture: Although queen conch aquaculture methods are well established, attempts to develop pearl culture techniques for this species were unsuccessful until conventional seeding techniques for pearl oysters and freshwater mussels were modified.

Cultured pearls from *Lobatus gigas*

Hatchery-produced queen conch were seeded with either one nucleus and mantle tissue (i.e., beaded) or mantle tissue only (i.e., nonbeaded). Nonbeaded cultured pearls were sampled after 6 weeks, 12 weeks, and 6 months of culture, and beaded cultured pearls were sampled after 6 weeks, 6 months, and 1 year of culture. Mean growth rates of nonbeaded cultured pearls remained within a narrow range (0.41–0.58 mg/d), whereas rates of beaded cultured pearl growth increased with time from 0.45 mg/d after 6 weeks to 1.52 mg/d after 1 year of culture. With a mean shell deposit rate of 0.5 mg/d, it will take more than a year to produce queen conch nonbeaded cultured pearls with a mean weight of 1 ct (200 mg). Beaded cultured pearls larger than 7 mm in diameter can be produced in the queen conch in 1 year of culture if beaded with 5.1 mm (1.7 bu) nuclei. Furthermore, postsurgery survival of the queen conch was 100%, and no sacrifice of the animal is required to produce a conch cultured pearl (Acosta-Salmón and Davis, https://accreditedgemologists.org/pastevents/2011abstracts/SalmonDavis.php).

Nonbeaded pearl culture: The colors of the nonbeaded conch pearls cultured from a new farm based in Honduras ranged from white and pinkish white to yellowish orange. Their shapes varied from oval to baroque, and weights ranged from 0.21 ct (2.7–3.4 × 4.1 mm) to 3.13 ct (6.5–7.6 × 9.3 mm).

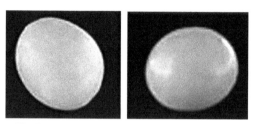

Nonbead-cultured pearls from *Lobatus gigas*

The luster of the above cultured was porcelaneous, and the surface was very smooth and homogenous (Segura and Fritsch, https://www.gia.edu/gems-gemology/summer-2015-gemnews-nonbead-cultured-pearls-strombus-gigas).

7.4 AQUACULTURE OF OTHER EDIBLE GASTROPOD MOLLUSCS

7.4.1 AQUACULTURE OF Tectus niloticus (WIKIPEDIA)

Experimental farming of *Tectus niloticus* has been conducted in Vanuatu, Australia, Indonesia, and Vietnam.

Cage-based culture systems: These systems have been found to be effective for *Tectus niloticus* as they allow control of food and predators, both of which contribute greatly to the growth and survival of juveniles. A stocking size of 35 mm (basal shell width) is recommended for this culture system. The culture period for marketable size is about 6 months under optimum conditions.

Cage design: Cages with 8 × 8 mm plastic mesh and an aluminum frame are suitable for the culture of *Tectus niloticus*. Suitable locations for reef cages are sand and rubble bottom reef bases exposed to less wave action with water depths of 0–2.5 m during spring-tides. Floating cages may need frequent cleaning to remove fouling species and allow high water flow.

Growth: The growth of the cultured individuals depends largely on the density of juveniles in the cages. At an initial density of 100 trochus/m², the growth in terms of basal diameter was 10–20 mm and at 50 trochus/m², the basal diameter was 25–40 mm. For more growth, a density of less than 10 trochus/m² is recommended. Seaweed and algae serve as the main food for the juvenile trochids. These algae grow on the dead corals and rocks which may occur naturally at the cultured sites or collected from adjacent areas and placed at the bottom of the cages.

7.4.2   AQUACULTURE OF Babylonia areolata

7.4.2.1   INDOOR CULTURE TANKS

Hatchery-reared juveniles (mean initial shell length—12.8 mm) were cultured intensively to marketable size in three 3.0 × 2.5 × 0.7 m indoor canvas rectangular tanks. The duplicate treatments of flow-through and semiclosed recirculating sea-water systems were compared at an initial stocking density of 300 individuals/m² (2250 juveniles per tank). These animals were fed ad libitum with fresh carangid fish *Selaroides leptolepis* once daily. During 240 culture days, average growth rates in shell length and body weight were 3.86 mm/month and 1.47 g/month for the flow-through system and 3.21 mm/month and 1.10 g/month for those in the semiclosed recirculating system. Survival in the flow-through system (95.77%) was significantly higher than that in the semiclosed recirculating system (79.28%). Feed conversion ratios were found to be 1.7 and 2.0 for flow-through and semiclosed recirculating systems, respectively.

7.4.2.2   MONOCULTURE OF Babylonia areolata IN INDOOR TANKS

Hatchery-reared juvenile spotted babylon *Babylonia areolata* (mean initial shell length—12.8 mm) were cultured intensively to marketable size in three 3.0 × 2.5 × 0.7 m indoor canvas rectangular tanks with flow-through and semiclosed recirculating sea-water facilities. Initial stocking density was 300 individuals/m^2 (2250 juveniles per tank). The animals were fed ad libitum with fresh carangid fish *S. leptolepis* once daily. During the 8-month culture period, the average growth rates in shell length and body weight were found to be 3.86 mm/month and 1.47 g/month for the flow-through system and 3.21 mm/month and 1.10 g/month for those in the semiclosed recirculating system. Survival in the flow-through system (95.77%) was significantly higher than that in the semiclosed recirculating system (79.28%). Feed conversion ratios were found to be 1.68 and 1.96 for flow-through and semiclosed recirculating systems, respectively (Chaitanawisuti and Kritsanapuntu, 1999).

7.4.2.3 MONOCULTURE OF Babylonia areolata IN FLOW-THROUGH CANVAS PONDS AND EARTHEN PONDS

Large-scale farming operation of *B. areolata* was attempted in flow-through canvas ponds and earthen ponds. The average growth rates in body weight were 0.9–1.1 g/month and 0.8–1 g/month for the canvass pond and earthen pond trials, respectively. At the end of the experiment, final body weights of the animals ranged from 5.6 to 6.6 and 5.2 to 6.2 g for the canvas pond and earthen pond trials, respectively. Total yields per production cycle were found to be 1.9 and 1.8 t for the canvas pond and earthen pond trials, respectively. This study indicated that grow-out of juvenile *B. areolata* in earthen ponds was better than those in flow-through canvas ponds (Kritsanapuntu and Santaweesuk, 2011).

7.4.2.4 MONOCULTURE OF Babylonia areolata AND POLYCULTURE WITH Lates calcarifer

Monoculture of *B. areolata* and polyculture of *B. areolata* with sea bass (*Lates calcarifer*) were undertaken in earthen ponds. Each pond was stocked with juveniles of 0.3 g initial weight at a density of 200 snails/m^2. The average growth rates of *B. areolata* were found to be 0.67 and 0.51 g/month in the monoculture and polyculture, respectively. At the end of the experiment, total yield of spotted babylon held in the monoculture and polyculture ponds was 10,520 and 10,450 kg/ha, respectively. The present study indicated the technically feasible and economically attractive for monoculture and polyculture of *B. areolata* to marketable sizes in earthen ponds (Kritsanapuntu et al., 2006).

7.4.2.5 MONOCULTURE OF Babylonia areolata AND POLYCULTURE WITH Lates calcarifer OR Chanos chanos

The growth, survival, and water quality for monoculture of spotted babylon *B. areolata* were compared with the polyculture trials made with sea bass (*L. calcarifer*) or milkfish (*Chanos chanos*). The mean body weight gain of the above snails held in the monoculture was 5.4 g, and 4.1 and 4.3 g for those held in the polyculture with sea bass or milkfish, respectively. Food conversion ratios were found to be 2.7, 3 and 2.7 for snails held in the monoculture and polyculture with sea bass and milkfish, respectively. The final survival rates were found to be 84.9, 74.3, and 81.2%, respectively. The present study showed the feasibility for monoculture and polyculture of *B. areolata* to marketable sizes in earthen ponds (Kritsanapuntu et al., 2008).

7.4.2.6 POLYCULTURE OF Babylonia areolata AND MILKFISH, Chanos chanos

Polyculture field experiments were conducted in a total farm area of 0.8 ha which comprised of 0.3 ha grow-out earthen ponds. Each pond was stocked with spotted babylon juveniles of 0.3-g initial body weight at a density of 200 snails/m² with milkfish juveniles. They were harvested at a 7-month period and the yield of spotted babylon and milkfish was 9875 and 6875 kg/ha, respectively (Kritsanapuntu et al., 2006).

7.4.2.7 MIXED OUTDOOR CULTURE OF HORSESHOE CRAB WITH SPOTTED BABYLON

An outdoor mixed culture of juvenile horseshoe crabs (*Tachypleus tridentatus*) and juvenile spotted babylon (*B. areolata*) was performed for a period of 150 days.

Management of culture ponds: Two cement ponds (2365 and 2150 m²) were selected for the mixed culture of juvenile horseshoe crabs (*T. tridentatus*) and juvenile spotted babylon (*B. areolata*). Each pond was 1.5 m in height, and the seawater was maintained at 60–100 cm in depth. Before experimentation, the moss, macroalgae, and mud within the ponds were removed, and the ponds were exposed to air for 10 days. The bottom of the pond was then covered with clean coarse sand grains ranging from 0.3 to 0.5 mm in diameter, to a thickness of 5 cm. After the ponds were filled with sea water, aqueous crude extract of tea seed (*Camellia* sp.) oil cake and bleach were added in water to kill unwanted fishes, shrimps, and crabs. After the ponds were treated with the said cleansing procedures, juveniles of both species, *T. tridentatus* and *B. areolata*, were introduced to the ponds. Seawater temperature, salinity, and pH were measured every day during culture. A proportion of 40–50% of seawater in each pond was changed using high tide flows twice every month; a total of eight changes were performed throughout the culture period of 150 days. When the water pH was lower than 8.0, calcium oxide was added to maintain the pH. During the mixed culture, water temperature ranged from a low of 14°C to a high of 23.5°C. The salinity was varying from 32 to 33 ppt.

At the final harvest time, the horseshoe crab juveniles exhibited survival of 32.5%. Spotted babylon had a survival rate of 87.3%. The resultant survival and abundance levels were greater than those previously reported. These profound results were attributed to (1) abundant benthic algae grown in outdoor ponds under sunlight; (2) the commensal lifestyle of juvenile horseshoe crab and spotted babylon, which likely prefer different food particle sizes; and (3) the behavior of spotted babylon which is burrowing and moving in the sediment substrata thereby creating an environment that favors horseshoe crab growth (Chen et al., 2016)].

Water quality management and the growth of Babylonia areolata, in earthen ponds

Influence of water quality on the growth of *B. areolata* was studied in earthen ponds at three water-exchange regimes of 7-, 15-, and 30-day intervals over a 5-month culture period. The higher body weight gains and shell length increments were observed in animals maintained at water exchange of 7- and 15-day intervals when compared with those maintained at water exchange of 30-day intervals. At the end of the experiment, average body weight gains were found to be 4.2, 3.7, and 2.8 g for animals maintained in water-exchange treatments of 7-, 15-, and 30-day intervals, respectively, and 16.6, 15.0, and 13.8 mm for those of shell length increments, respectively. The average final survivals were found to be 83.6, 80.9, and 74.2% for animals maintained in water-exchange treatments of 7-, 15-, and 30-day intervals, respectively (Kritsanapuntu et al., 2009).

KEYWORDS

- **abalone farming**
- **culture conditions**
- **farm maintenance**
- **barrel and cage culture**
- **grow-out facilities**

CHAPTER 8

DISEASES AND PARASITES OF EDIBLE MARINE GASTROPOD MOLLUSCS

CONTENTS

ABSTRACT

The disease-causing agents such as microbes, parasitic protozoans, helminths, copepods, and crabs infecting the different species of edible marine gastropod molluscs, and the resulting social and economic significance are given in this chapter.

In molluscs, infectious diseases associated with various kinds of parasites such as bacteria, virus, fungi, protozoans, trematodes, turbellarians, nematodes, and copepods may devastate wild populations, fisheries, and aquaculture industries. There are indications that as the oceans are changing, marine diseases are becoming more frequent or severe, and these changes might be associated with climate change and human activities. The diseases and parasites infecting the different species of edible marine gastropod molluscs are given below.

8.1 BACTERIA

(i) Pathogen: *Xenohaliotis californiensis* (abalone rickettsia)

Host(s): *Haliotis cracherodii, Haliotis rufescens, Haliotis corrugata, Haliotis fulgens, Haliotis sorenseni, Haliotis rubra, Haliotis tuberculata, Haliotis discus hannai, Haliotis wallalensis*, and *Haliotis diversicolor supertexta*.

Disease: Withering syndrome of abalone.

Disease signs at the farm, tank, or pond level: Reduced feeding; inability of individuals to right themselves when upside down; weakness and lethargy (clinging to horizontal surfaces rather than vertical or inverted); and inability to adhere to the substrate.

Pathological signs: Wasting of body mass; retraction of mantle; atrophy of the foot muscle; decreased response to tactile stimuli; diminished reproductive output; mottling of digestive gland (dark brown with small foci of tan-colored tissue) (Anon. http://www.agriculture.gov.au/SiteCollectionDocuments/animal-plant/aquatic/field-guide/4th-edition/molluscs/infection-xenohaliotis-californiensis.pdf).

Social and economic significance: The susceptibility and mortality vary with species. For example, *H. sorenseni* and *H. cracherodii* are highly susceptible to this infection, and it is known to cause disease with nearly 100% mortality in these species. On the other hand, in *H. rufescens*, only up to 35% mortality has been observed. The magnitude of mortality, however, is not well documented in *H. corrugata* and *H. fulgens*. Clearly, host factors are involved in this differential suscepti-bility to disease, but the exact nature of these factors remain unknown (Anon. http://www.cabi.org/isc/datasheet/90287). A reduction in the abalone population caused by this fatal disease may have significant ecological effects.

(ii) Pathogen: *Oceaniserpentilla haliotis* Schlosser, Lipski, Schmal-fuss, Kugler & Beckmann, 2008 has been isolated from the hemolymph serum of this species (Schlösser et al., 2008).

Host(s): *H. rubra*

Disease: Not reported

Clinical signs: No information

Social and economic significance: No information

(iii) Pathogen: *Vibrio carchariae*

Disease: French abalone disease

Host(s): *H. tuberculata*

Clinical signs: Not known

Social and economic significance: Since 1997, mass mortality of the abalone *H. tuberculata* has occurred in the natural environment along the French coast. The outbreak of disease started on the south coast of Brit-tany near Concarneau in 1997, then spread to the north of Brittany (in 1998) and the west coast of Normandy (in 1999). About 60–80% of the abalone died. In 1999, mortality also affected a land-based abalone farm in Normandy during the summer (Nicolas et al., 2002).

(iv) Pathogen: *Vibrio fluvialis*-II.

Disease: Pustule disease

Host(s): *Haliotis discus hannai*

Clinical sings: The pustule disease is transmitted through lesions in the foot. It affects different growth stages of the abalone.

Social and economic significance: A serious pustule disease has been reported to spread among several abalone hatcheries in the Dalian area (China) with 50–60% mortality (Li et al., 1998).

(v) Pathogen: *Vibrio harveyi*-related bacteria

Disease: Vibriosis

Host(s): *H. diversicolor*

Clinical signs: Various factors including presence of flagella and production of extracellular enzymes, such as lipase, phospholipase, and hemolysin, could be responsible for pathogenesis.

Social and economic significance: Mass mortality of farmed small abalones of this species occurred in Fujian, China, from 2009 to 2011 (Jiang et al., 2013).

(vi) Pathogens: *Vibrio alginolyticus*, *V. harveyi*, and *Vibrio parahaemolyticus*

Disease: Vibriosis

Host(s): *Haliotis* spp.

Clinical signs: The main clinical signs observed in the affected juvenile and adult individuals were abscessing or ulceration in the mantle, white spots on the foot, general whitening, as well as loss of ability to adhere to surfaces.

Social and economic significance: Mass mortalities of the above abalones have been reported (Romalde and Barja, 2010).

(vii) Pathogens: *V. fluvialis* and *V. harveyi*

Disease: Blister disease of cultured abalone (Anon. http://www.cabi.org/isc/datasheet/81169)

Host(s): *Haliotis asinina*, *Haliotis discus hannai*, and *Haliotis midae*

Clinical signs: No report

Social and economic significance: Not known

(viii) Pathogen: *V. parahaemolyticus*

Disease: Postlarval abalone disease

Host(s): *H. diversicolor supertexta*

Clinical signs: Extracellular products of the above bacterial species may be involved in the pathogenesis of the disease

Social and economic significance: Outbreaks of mass mortality of postlarvae of the above species has been recorded. This has led to the closure of many abalone farms (Cheng et al., 2008).

(ix) Pathogen: *Vibrio superstes*—It is an alginolytic, non-motile and facultatativley anaerobic species isolated from the gut of the host species (Hayashi et al., 2003).

Disease: Vibriosis

Host(s): *Haliotis laevigata* and *H. rubra*

Clinical signs: No report

Social and economic significance: Not known

(x) Pathogens: *Vibrio comitans, Vibrio rarus* and *Vibrio inusitatus.* All these species have been isolated from the gut of the host species (Sawabe et al., 2007).

Disease: Vibriosis

Host(s): *Haliotis discus discus, Haliotis gigantea, Haliotis madaka*, and *H. rufescens*

Clinical signs: No information

Social and economic significance: No report

(xi) Pathogens: *V. harveyi* and *Vibrio splendidus*

Disease: *Septicaemic vibriosis* and pustule disease (blister disease) (due to *V. harveyi*)

Host(s): *H. rubra*

Clinical signs for *V. harveyi*: Formation of pustules that are analogous to abscesses in vertebrates; abscesses in the foot muscle; protein precipitation and bacterial concentration in the left kidney; and kidneys showed marked eosinophilia from protein precipitation and basophilia from bacterial aggregates (Hanlinger et al., 2005).

Clinical signs for *V. splendidus*: Perigut interstitial tissue was found affected despite an intact gut epithelium; and bacteria were often trapped within protein deposits, which were common in the gills as well as left kidney, and were occasionally seen in other locations (Hanlinger et al., 2005).

Social and economic significance: Outbreaks of the above disease typically occurred as peaks of mortality in summer, with few gross or clinical signs of infection, following a sharp increase in water temperature, or other stresses related to water quality. Secondary surface infection with mixed *Vibrio*-like bacteria was seen following shell damage by spionid polychaetes (Hanlinger et al., 2005).

(xii)　Pathogen: *Flavobacterium*-like bacteria

Disease: *Flavobacterium* syndrome 1 and *Flavobacterium* syndrome 2

Host(s): *H. laevigata*

Clinical signs of *Flavobacterium* syndrome 1: Loss of pigmentation and contact avoidance; lesions progressed to pale flaking epithelium; and shallow epithelial erosions of the foot and epipodium (Handlinger et al., 2005).

Clinical signs of *Flavobacterium* syndrome 2: Erosions progressed rapidly within one to several days to deep necrotic lesions (Handlinger et al., 2005).

Social and economic significance: No report

(xiii)　Pathogens: *Nuadamonas halioticida* and *Nuadamonas abalonii* (Romalde and Barja, 2010)

Disease: No report

Host(s): *H. tuberculata*

Clinical signs: No report

Social and economic significance: No report

(xiv)　Pathogen: *Francisella* sp.

Disease: Not known

Host(s): *H. gigantea*

Clinical signs: In histological observations, bacteria-like spherical particles were found in affected animals, suggesting a bacterial infection. Focal lesions were also found in the basal part of the mantle.

Social and economic significance: Mass mortality of the above species of abalone has been reported from a private abalone farm in Shimane Prefecture, Japan. The cumulative mortality rate reached about 84% (Kamaishi et al., 2010).

(xv)　Pathogen: *Formosa haliotis* (Tanaka et al., 2015) (This species has been isolated from the gut of the abalone *H. gigantea* obtained in Japan).

Disease: No report

Host(s): *H. gigantea*

Clinical signs: Not known

Social and economic significance: No report

(xvi)   Pathogen: Rickettsiales-like prokaryotes (found in the epithelia of the digestive gland of all moribund abalone)

Disease: Withering syndrome

Host(s): *H. rufescens*

Clinical signs: No report

Social and economic significance: 84% of apparently healthy abalone were found infected although no clinical signs were observed (Caceres-Martinez and Tinoco-Orta, 2001).

(xvii)   Pathogen: *Shimia haliotis* sp. nov. (This novel Gram-stain-negative, motile, rod-shaped bacterium, was isolated from the intestinal tract of an abalone, *Haliotis discus hannai*, which was collected from the northern coast of Jeju in Korea.) (Hyun et al., 2013).

Disease: No information

Host: *Haliotis discus hannai*

Clinical signs: No information

Social and economic significance: Not reported

(xviii)   Pathogen(s): *V. alginolyticus*, *V. fluvialis*, *Vibrio vulnificus*, *Vibrio parahemolyticus*, and *Vibrio cholerae*

Disease: Vibriosis

Host(s): *Babylonia areolata*

Clinical signs: Gross lesions appeared in the infected animals. Further, the proboscis was protruded, flaccid, red, and swollen (Koeypudsa et al., 2008).

Social and economic significance: No report

(xix)   Pathogen: *V. parahaemolyticus* (Wikipedia)

Disease: Vibriosis

Host(s): *Nerita albicilla*

Clinical signs: Not reported

Social and economic significance: Not known

8.2 VIRUS

(i) Pathogen: Abalone herpes virus (AbHV)

Disease: Abalone viral ganglioneuritis

Host(s): *H. rubra*, *H. diversicolor*, *H. laevigata*, *H. rubra*, and *H. rubra conicopora*).

Clinical signs: Inflammation and necrosis of neural tissue; swollen and protruding mouth parts; reduced activity of the pedal muscle; edges of the foot curled inwards, leading to exposure of clean, shiny shell; hard foot; excessive mucous production; and abnormal spawning and bloating.

Other disease signs: Clean (empty) shells on substrate due to predation of moribund and dead abalone; and inability to adhere to the substrate.

Social and economic significance: Rapid and high cumulative mortality of up to 90% has occurred due to the above viral disease (Corbeil et al., 2016; Anon., 2012).

(ii) Pathogen: Virus-like particles

Disease: Not known

Host(s): Postlarvae of *H. diversicolor supertexta*

Clinical signs: Not reported

Social and economic significance: Mass mortalities have occurred in the cultured postlarvae of this species during 10–30 days of postfertilization. These mortalities, which are characterized by a white appearance and falling of the diatom films on which they grow, have occurred since 2002 on the south coast of China, including Taiwan (Zhou et al., 2013).

8.3 FUNGI

(i) Pathogen type: Fungal infection

Disease: Shell disease

Host(s): *H. tuberculata*

Clinical signs: Conchiolin deposits which are found only under the shell spire result from a localized fungal infection of the shell. This conchiolin deposit may increase rapidly in a shell length of over 85 mm and reached over 90% in the 110–120-mm size class. The positive correlation of the number of successive infections found in the shells with the level of infestation of the shell by borers suggests that boring polychaetes and sponges may be vectors of the disease, or that the parasite infestation may increase the susceptibility of the animal to this infection.

Social and economic significance: This infection causes mortality in abalone (Huchette et al., 2006).

(ii)   Pathogen: Fungal hyphae within lesions and the shell matrix of affected host species.

Disease: Shell lesions

Host(s): *Haliotis iris*, *Haliotis australis*, and *Haliotis virginea virginea*.

Clinical signs: Blisters of conchiolin and occasionally nacreous material appear on the inside of the abalone shell near the apex (Grindley et al., 1998).

Social and economic significance: Not reported

(iii)   Pathogen type/agent: Unidentified fungus (member of the Class Deuteromycotina)

Disease: Shell lesions

Host(s): *H. iris* and *H. australis*

Clinical signs: Shell irregularities characterized by deposition of conchiolin are seen on the inner shell surface of the infected abalone species (Friedman et al., 2013).

Social and economic significance: Not reported

(iv)   Pathogen type/agent: *Haliphthorous milfordensis* (A fungus provisionally identified as a member of the Class Deuteromycotina)

Disease: Fungal disease

Host(s): *Haliotis sieboldii*, *H. iris*, *H. australis*, and *H. virginea virginea*

Clinical signs: (1) Infection caused tubercle-like swellings (up to 5 mm in length) on the mantle and foot of *H. sieboldii* held in captivity. (2) Blisters of conchiolin and occasionally nacreous material were found on the inner shell surface near the shell apex. Lesions in some severely affected abalone were found extending into the foot-muscle attachment site. In some cases, the external surface of the shell apex was seen crumbling and caving-in.

Social and economic significance: The presence of infection in cultured abalone is of concern because the chronic progressive nature of the mycosis may become problematic if the abalones are reared to a large size (Anon. http://www.dfo-mpo.gc.ca/science/aah-saa/diseases-maladies/fungusab-eng.html).

8.4 PROTOZOANS

(i)   Pathogen type/agent: *Perkinsus olseni*

Disease: Perkinsosis (*Perkinsus* sp. infections)

Host(s): *H. rubra, H. laevigata, Haliotis scalaris*, and *Haliotis cyclobates*

Clinical signs: Reduced growth rate occurs in the above species prior to the onset of mortality (Reece and Dungan, http://www.afs-fhs.org/perch/resources/14069252415.2.1perkinsus2014.pdf).

In *H. rubra* and *H. laevigata*, this parasite may produce pustules in the foot and mantle; lesions appear as pale brown circles in processed abalones; 30–40% mortality in *H. laevigata* (Anon. http://library.enaca.org/Health/FieldGuide/pdf/Infection%20with%20Perkinsus%20olseni.pdf).

Social and economic significance: Infections have been reported in moribund southern Australia abalones. *P. olseni* is the only species known to cause this disease in the Asia-Pacific region and is responsible for perkinsosis in abalones.

(ii)   Pathogen type/agent: *Haplosporidium babyloniae*

Disease: Haplosporidium disease

Host(s): *B. areolata*

Clinical signs: The organs of *B. areolata*, including the proboscis, foot muscle, intestine, digestive gland, stomach, liver, and gill, were found badly damaged and a series of serious lesions occurred (Peng et al., 2011).

Social and economic significance: No information

(iii)   Pathogen type/agent: A novel haplosporidian parasite

Disease: Haplosporidium disease

Host(s): *H. iris*

Clinical signs: Heavily infected animals exhibited behavioral abnormalities including lethargy and loss of righting reflex. Some heavily infected individuals exhibited edema and pale lesions in the foot and mantle (Diggles et al., 2002).

Social and economic significance: The presence of the haplosporidian in the affected facility was found associated with mortalities of slow growing animals during the summer months.

(iv) Pathogen: *Margolisiella haliotis* (Apicomplexa) (González et al., 2014).

Disease: No information

Host(s): *H. rufescens*

Clinical signs: No report

Social and economic significance: The renal coccidian *Margolisiella haliotis*, was prevalent in 72% of the moribund and 10% of apparently healthy abalone (Caceres-Martinez and Tinoco-Orta, 2001).

(v) Pathogens: Ciliate protozoans

Disease: No report

Host(s): *H. rufescens*

Clinical signs: No information

Social and economic significance: Ciliate protozoans were observed in the lumina of the esophageal pouches of both healthy (78% prevalence) and moribund animals (57% prevalence). Other ciliates were observed on the bronchi and within the branchial cavity of healthy abalone (17%) and moribund abalone (88% prevalence) (Caceres-Martinez and Tinoco-Orta, 2001).

(vi) Pathogen: Apicomplexa parasite (Eimeriidae)

Disease: No report

Host(s): *Lobatus gigas*

Clinical signs: This parasite may be affecting the gametogenesis activity in queen conch. It disperses through the feces of the host and may spread to other benthic detritus feeders also (Anon. http://www.cio.noaa.gov/services_programs/prplans/pdfs/ID236_Queen_Conch_Final_Status_Report.pdf).

Social and economic significance: The infection of the above parasite has been reported to occur year round causing greater damage for queen conch in culture systems.

8.5   SPONGES

Pathogen: *Cliona celata*

Disease: Not reported

Host(s): *H. tuberculata*

Clinical signs: No information

Social and economic significance: Only shells greater than 50 mm were found infested. The mean incidence of parasite increased with shell length and reached 80% in old individuals. Very severe infestations weakened the shell. This parasite initially settled on the protoconch and progressively extended from this point. The mechanism for the dissolution of calcium carbonate by this parasite involved enzymatic activity from etching cells (Clavier, 1989).

8.6   TREMATODES

(i)   Pathogen type/agents: *Himasthla elongata, Renicola roscovita, Cercaria emasculans, Paramonostomum chabaudi, Cryptocotyle lingua, Microphallus similis, Microphallus pygmaeus, Podocotyle atomon,* and *Cercaria littorinae* (Thieltges et al., 2009).

Disease: Not reported

Host(s): *Littorina littorea*

Clinical signs: No information

Social and economic significance: No information

(ii)   Pathogen type/agents: *M. pygmaeus, H. elongata, C. lingua, Cercaria lebouri, R. roscovita,* and *P. atomon* (Thieltges et al., 2009).

Disease: No information

Host(s): *L. littorea*

Clinical signs: Not reported

Social and economic significance: No information

(iii)   Pathogen type/agent: *C. lingua* (Pechenik et al., 2012)

Disease: No report

Host(s): *L. littorea*

Clinical signs: Not reported

Social and economic significance: No information

(iv)   Pathogen type/agents: *R. roscovita, C. lingua, M. pygmaeus,* and *Himasthla* sp.

Disease: Not reported

Host(s): *L. littorea*

Clinical signs: More studies are needed before any conclusions regarding stunting of the host. Further, no effect or enhanced growth rate due to parasites can be reached. It seems that growth rate is primarily affected on available food and time available for feeding, and not on parasites (Costa et al., 2016).

Social and economic significance: No information

(v)   Pathogen type: *C. emasculans*

Disease: No report

Host(s): *L. littorea*

Clinical signs: No information

Social and economic significance: Mortality rate of up to 94% per annum has been observed for the first two months, followed by up to 60% per annum for the rest of the first year. Older individuals above 15 months old seem to have a mortality of 23% only per annum. *C. emasculans* is known to be fatal to the snail, but this does not account for the observed mortality (Costa et al., 2016).

(vi)   Pathogen type/agents: *Zoogonoides viviparous, Cercaria buccini, Stephanostomum baccatum,* and sporocysts of *Renicola* sp.

Disease: No information

Host(s): *Buccinum undatum*

Clinical signs: No information

Social and economic significance: There was no significant difference in growth between stations, sexes or parasitized and uninfected *Buccinum*. There was no pronounced host tissue response to infections with larval digeneans in this species although layers of flattened amoebocytes occasionally formed compact capsules around a few, possibly senescent or dead, sporocysts. Further, there was no difference between reference or dump site *Buccinum* in the pathology associated with larval digenean infections or in the host response to infection (Siddall et al., 1993).

(vii) Pathogen type/agents: Rediae and ophtalmotrichocercous cercariae probably belonging to *Opechona* sp. were identified in the gonad and digestive gland of the host.

Disease: No information

Host(s): *Buccinanops cochlidium*

Clinical signs: Rediae affected male and female snails equally, but prevalence increases along with host size. The parasite caused the complete castration of the host. Parasitized adult snails also showed a reduction of penis size in comparison with healthy males.

Social and economic significance: Overall prevalence of infection was found to be 15.5%. This prevalence rate varied seasonally, rising during the warm months after the host oviposition period. Cercariae were expelled at the same time as the hatching of snail embryos (during the higher water temperature period) (Averbuj et al., 2010).

(viii) Pathogen type/agent: *Opechona* sp., a digenean trematode

Disease: No report

Host(s): *B. cochlidium* (first intermediate host)

Clinical signs: Not reported

Social and economic significance: The parasite sets up shop within the snail's gonads where it starts cloning itself, eventually castrating the snail through physical destruction of the gonad tissue (Anon. http://dailyparasite.blogspot.in/2011/02/opechona-sp.html).

(ix) Pathogen type/Agent: Cercarial, redial, metacercarial, and adult stages of a digenean trematode (family, Hemiuridae) were found in the digestive glands as well as on the gill filaments. these hemiurids have a single-host life-cycle (Peters and Basson, file:///E:/My%20 Downloads/518-1282-1-SM.pdf).

Disease: Not reported

Host(s): *Haliotis spadicea*

Clinical signs: Not reported

Social and economic significance: No information

(x)   Pathogen type/agent: *Proctoeces lintoni* (Family: Fellodistomidae)

Disease: No information

Host(s): *Fissurella crassa* and *Fissurella maxima* (Archaeogastropoda)

Clinical signs: The infected gonads of this host species show altered structure and the gametogenic processes is aborted. There is no evidence of hemocytic response, but leucocite infiltration is evident at least in male infected gonads. An increased content of polysaccarides is evident in infected gonads. This parasite caused parasitic castration in *F. crassa* (Oliva, 1992).

Social and economic significance: These parasites live unencysted in the gonads, and the main mechanical damage is originated by the action of a well-developed acetabulum. Chemical actions of parasitic secretions may also be involved. Studies relating to marine protected areas in central Chile, the parasitized limpets showed a greater shell length, muscular foot biomass, and gonadosomatic index compared to nonparasitized limpets of the same age. Further, the life cycle of *P. lintoni* and its trophic links have been found strengthened. The increased growth rate could reduce the time required to reach the minimum catch size and increase the reproductive and muscular output of the host population. Thus, parasitism should be considered in the conservation and management of economically important mollusk hosts (Aldana et al., 2014).

(xi)   Pathogen type/agent: *Maritrema lintoni* (Family: Microphallidae) (Anon. host-parasite.myspecies.info/taxonomy/term/48793)

Disease: Not reported

Host(s): *Fissurella picta and Fissurella nigra*

Clinical signs: No report

Social and economic significance: No information

(xii)   Pathogen: *Opecoeloides columbellae* (Jousson and Bartoli, 2000).

Disease: No information

Host(s): *Columbella rustica*

Clinical signs: No information

Social and economic significance: It is also a digestive parasite of some fish species.

8.7   TURBELLARIANS

(i)   Pathogen type/Agent: *Stylochoplana pusilla* (Polyclad)

Disease: No information

Host(s): *Monodonta labio* and *Omphalius rusticus*

Clinical signs: The mantle, which provides habitat space for this polyclad, was proportionally the longest in *M. labio*.

Social and economic significance: The prevalence of the above polyclad in the mantle cavity was 82.6% in *M. labio*) and 15.4% in *O. rusticus*). In an evolutionary sense, the relationship between this polyclad and its host may have developed as a mechanism for the flatworm to avoid predation by living in the eulittoral and to avoid desiccation by establishing commensal relationships with certain snails (Fujiwara et al., 2014).

(ii)   Pathogen type/agent: *Bivesiculoplana lamothei* (Family: Leptoplanida) (host-parasite.myspecies.info/taxonomy/term/42480).

Disease: No information

Host(s): *Fissurella gemmata*

Clinical signs: Not reported

Social and economic significance: No information

8.8   NEMATODES

(i)   Pathogen type/agent: *Echinocephalus pseudouncinatus* n. sp. (larval forms)

Disease: No information

Host(s): *H. corrugata*

Clinical signs: These parasites encyst in the ventral portion of the foot of the host producing a blister-like effect on the outside of this structure.

Further, they weaken the muscle and decrease the efficacy of this structure as a hold-fast organ (Millemann, 1951).

Social and economic significance: Not reported

(ii)   Pathogen: *Thynnascaris* sp. (larvae)

Disease: No information

Host(s): *Cantharus cancellarius* and *Thais haemasloma*

Clinical signs: No information

Social and economic significance: If it is a homeotherrn, then the larvae of these molluscs may be potentially infective to humans (Cheng, http:// spo.nmfs.noaa.gov/mfr4010/mfr401019.pdf).

(iii)   Pathogen: *Thynnascaris adunca*

Disease: No information

Host(s): *Tritia reticulata*

Clinical signs: No information

Social and economic significance: If it is a homeotherrn, then the larvae of these molluscs may be potentially infective to humans (Cheng, http:// spo.nmfs.noaa.gov/mfr4010/mfr401019.pdf).

8.9   POLYCHAETES

(i)   Pathogen: *Polydora* spp. (Spionid)

Disease: No information

Host(s): *H. tuberculata*

Clinical signs: No information

Social and economic significance: Shells smaller than 5 cm size were not found infected. On the other hand, the infestation rates increased rapidly until 8 cm, and almost all specimens greater than this size were infested with a mean number of *Polydora* tubes of about 40. Several species of *Polydora* are able to damage molluskan shells by their boring activities. The species of *Polydora* involved and their frequency of occurrence were: *Polydora hoplura*, 77%, *Polydora ciliata*, 19%; *Polydora caeca*, 2%, *Polydora armata*, 1% and *Polydora flava*, 1% (Clavier, 1989).

(ii)   Pathogens: *Polydorid P. hoplura, Pseudopolydora dayii, Dipolydora capensis, Dipolydora caeca, Dipolydora normalis, Dipolydora keulderae, Dipolydora giardi, Boccardia proboscidea, Boccardia pseudonatrix,* and *Boccardia polybranchia.*

Disease: No information

Host(s): *H. midae*

Clinical signs: Not recorded

Social and economic significance: Among the different species of parasites, *B. proboscidea* and *D. capensis* were dominating the farm sites. On the other hand, *D. capensis* was recorded at all wild sites and was usually present on more than 85% of the samples (Boonzaaier et al., 2014).

(iii)   Pathogens: *Boccardia knoxi* and *P. hoplura* (spionids)

Disease: Not reported

Host(s): *Haliotis* spp.

Clinical signs: Shell blister coverage was estimated at 30%, and mortality rates of 50% or greater of total populations were recorded. *B. knoxi* mudworm prevalence was 36.5 per abalone. The percentage of total abalone weight comprising the soft tissues was significantly reduced from 71.1 to 64.5 and 49.6% at two affected farms. Follow up trials from 1998 showed a significant inverse relationship between percentage of shell affected by blisters and both length and whole weight of abalone (Lleonart et al., 2003).

Social and economic significance: Severe spionid infestations of farmed abalone (*Haliotis* spp.) were reported from several sea-based culture facilities.

(iv)   Pathogen: *P. ciliata* (Thieltges et al., 2006)

Disease: Not reported

Host(s): *Crepidula fornicata*

Clinical signs: No information

Social and economic significance: Not reported

(v)   Pathogen: *P. ciliata*

Disease: Not reported

Host(s): *L. littorea*

Clinical signs: This worm has been reported to excavate burrows in the shell of the common periwinkle when the snail is matured and is above 1 cm long.

Social and economic significance: The infection by this parasite does not seem to alter the growth of the host and proportions of the snail shell (Buschbaum et al., 2007).

(vi)   Pathogen: *Terebrasabella heterouncinata* (Sabellid)

Disease: No information

Host(s): *H. midae*

Clinical signs: Not reported

Social and economic significance: The above pathogen has been reported to impede the growth of the host by causing irritation beneath the mantle in abalone (Sales and Britz, 2001).

(vii)   Pathogen: *T. heterouncinata* (sabellid)

Disease: No information

Host(s): *H. rufescens*

Clinical signs/impact: This parasite grossly deforms the abalone's shell and causes temporary or permanent cessation of growth. Respiratory holes may fuse shut, thus interfering with gas exchange, urinary release, and gamete dissemination. The burrows of this polychaete are oriented perpendicular to the growing edge of the shell, and that densities within a shell may be quite high. The worm lives inside apertures, such as shell holes in abalones, or main apertures of other gastropods, at the growing edge of the shell. When the worm initially positions itself as a larva at the aperture edge of a potential host, it covers itself in a mucous sheath. The host recognizes the worm as nonself and secretes nacreous shell material around it, in the same way that pearls or blister pearls are formed. The host actually creates the tube of the parasite. Settlement of the worm at the shell aperture stops prismatic growth of the shell and only nacre is secreted.

Social and economic significance: *T. heterouncinata* has become a serious pest in some abalone *H. rufescens* mariculture facilities in southern California are affected (Caceres-Martinez and Tinoco-Orta, 2001; Anon. http://www.asnailsodyssey.com/LEARNABOUT/TUBEWORM/tubePara.php).

(viii) Pathogen: *T. heterouncinata* (sabellid)

Disease: No information

Host(s): *Tegula funebralis* (=*Chlorostoma funebralis*)

Clinical signs: This parasitic worm may cause growth deformities or a complete cessation of linear growth.

Social and economic significance: *T. funebralis* was found infested with the sabellid *T. heterouncinata* at Cayucos, California near the site of an abalone culture facility (Anon. http://www.asnailsodyssey.com/LEARN-ABOUT/TUBEWORM/tubePara.php).

8.10 COPEPODS

(i) Pathogen type/agent: *Panaietis haliotis* (Wikipedia)

Disease: Not reported

Host(s): *H. gigantea*

Clinical signs: No information

Social and economic significance: Not reported

(ii) Pathogen type/agent: *Neanthessius renicolis* (ectoparasite) (Cyclopoida) (Anon. http://indiabiodiversity.org/species/show/262615).

Disease: Not reported

Host(s): *Pleuroploca trapezium*

Clinical signs: No information

Social and economic significance: Not reported

(iii) Pathogen type/agents: *Panaietis incamerata, Panaietis doraconis* n. sp., and *Panaietis satsuma* n. sp (Copepoda: Cyclopoida: Anthessiidae) (live in the mantle cavity of the host)

Disease: Not reported

Host(s): Tegulid top shells, *Tegula* spp. (Uyeno, 2016)

Clinical signs: No information

Social and economic significance: Not reported

(iv) Pathogen type/agent: *P. incamerata*

Disease: Not reported

Host(s): *Tectus niloticus* (=*Trochus niloticus*) [Archaeogastropoda, Trochidae]; *Batillus cornutus* [Archaeogastropoda, Tubinidae] (Izawa, 1976).

Clinical signs: In *B. cornutus*, the parasite was collected from the buccal cavity and in *T. niloticus*, it was from the buccal cavity and the esophagus. Owing to the site of this parasite the digestive system of these species may be damaged.

Social and economic significance: Not reported

(v) Pathogen type/agent: *Anthessius isamusi*

Disease: Ectoparasite

Host(s): *Turbo marmoratus*

Clinical signs: Not reported

Social and economic significance: Not reported

KEYWORDS

- bacteria
- virus
- fungi
- protozoans
- sponges

REFERENCES

Abdullah, A.; Nurjanah, N.; Hidayat, T.; Gifari, A. Characterize Fatty Acid of *Babylonia spirata, Meretrix meretrix, Pholas dactylus. Int. J. Chem. Biomol. Sci.* **2016**, *2*, 38–42.

Acosta-Salmón, H.; Davis, M. *Cultivation of the Conch Pearl: A Comparison to Natural & What the Future Holds.* https://accreditedgemologists.org/pastevents/2011abstracts/SalmonDavis.php.

Alam, M.; Martin, G. E.; Zektzer, A. S.; Weinheimer, A. J.; Sanduja, R.; Ghuman, M. A. Planaxool: A Novel Cytotoxic Cembranoid from the Mollusk *Planaxis sulcatus. J. Nat. Prod.*; **1993**, *56*, 774–779.

Aldana, M.; Pulgar, J. M.; Orellana, N.; Ojeda, F. P.; García-Huidobro, M. R. Increased Parasitism of Limpets by a Trematode Metacercaria in Fisheries Management Areas of Central Chile: Effects on Host Growth and Reproduction. *EcoHealth* **2014**, *11*, 215–226.

Anand, T. P.; Chellaram, C.; Kumaran, S.; Shanthini, C. F. Biochemical Composition and Antioxidant Activity of *Pleuroploca trapezium* meat. *J. Chem. Pharm. Res.* **2010**, *2*, 526–535.

Anand, T. P.; Chellaram, C.; Chandrika, M.; Rajamalar, C. G.; Parveen, A. N.; Shanthini, F. Nutritional Studies on Marine Mollusk *Pleuroploca trapezium* (Gastropoda: Fasciolariidae) from Tuticorin Coastal Waters. *J. Chem. Pharm. Res.* **2013**, *5*, 16–21.

Anon. *Aquatic Animal Diseases Significant to Australia: Identification Field Guide.* 4th Edition. 4. Diseases of Molluscs, Australian Government Department of Agriculture, Fisheries and Forestry, 2012, p.128–133.

Anon. Daily Diet Guide, 2006.

Anon. Dietfacts.com.

Anon. Fishflies, http://www.fishfiles.com.au/knowing/species/molluscs/abalones/Pages/Brownlip-Abalone.aspx.

Anon. http://dailyparasite.blogspot.in/2011/02/opechona-sp.html.

Anon. http://library.enaca.org/Health/FieldGuide/pdf/Infection%20with%20Perkinsus%20olseni.pdf.

Anon. http://museum.wa.gov.au/research/collections/aquatic-zoology/baler-shells.

Anon. http://www.cabi.org/isc/datasheet/81169.

Anon. http://www.cio.noaa.gov/services_programs/prplans/pdfs/ID236_Queen_Conch_Final_Status_Report.pdf.

Anon. http://www.dfo-mpo.gc.ca/science/aah-saa/diseases-maladies/fungusab-eng.html.

Anon. http://www.internetstones.com/breakthrough-culturing-queen-conch-pearls-scientists-FAUs-harbor-branch-oceanographic-institute.html.

Anon. http://www.lbaaf.co.nz/nz-aquaculture-species/modelled-species/paua/.

Anon. http://www.manandmollusc.net/advanced_uses/advanced_uses-print.html.

Anon. http://www.manandmollusc.net/advanced_uses/advanced_uses-print.html.

Anon. http://www.manandmollusc.net/advanced_uses/advanced_uses-print.html.

Anon. http://www.sigmaaldrich.com/catalog/product/sigma/g2174?lang=en&region=IN.

Anon. http://www.snail-world.com/snails-as-food/.

Anon. http://www.spearboard.com/showthread.php?t=159009.

Anon. http://www.wacht-troy.com/PTypeConchP.html.

Anon. http://www4.mpbio.com/ecom/docs/proddata.nsf/(webtds2)/32126.

Anon. https://researcharchive.lincoln.ac.nz/bitstream/handle/10182/2863/Shi_MApplSci.pdf?sequence=5.

Anon. https://researcharchive.lincoln.ac.nz/bitstream/handle/10182/2863/Shi_MApplSci.pdf?sequence=5&isAllowed=y.

Anon. https://www.thefreelibrary.com/Potential+antimicrobial+activity+of+marine+molluscs+from+tuticorin%2c...-a0133108179.

Anon. https://www.thefreelibrary.com/Potential+antimicrobial+activity+of+marine+molluscs+from+tuticorin%2c...-a01331081.

Anon. Melo Melo Pearls- http://www.internetstones.com/melo-melo-pearls.html.

Anon. Rare Melo Pearls- http://www.ajsgem.com/articles/rare-melo-pearls.html.

Anon. *The State of World Fisheries and Aquaculture Opportunities and Challenges*. FAO, 2014, p. 243.

Anon. USDA National Nutrient Database.

Anon. https://researcharchive.lincoln.ac.nz/bitstream/handle/10182/2863/Shi_MApplSci.pdf?sequence=5&isAllowed=y.

Anon. Clay's Kitchen, http://www.panix.com/~clay/cookbook/bin/table_of_contents.cgi?gastropods.

Anon. Cornwall Good Seafood Guide, http://www.eatsxm.com/sea-snails.html.

Anon. host-parasite.myspecies.info/taxonomy/term/48793.

Anon. http://indiabiodiversity.org/species/show/262615.

Anon. http://nopr.niscair.res.in/handle/123456789/28650.

Anon. http://www.agriculture.gov.au/SiteCollectionDocuments/animal-plant/aquatic/field-guide/4th-edition/molluscs/infection-xenohaliotis-californiensis.pdf.

Anon. http://www.asnailsodyssey.com/LEARNABOUT/TUBEWORM/tubePara.php.

Anon. http://www.cabi.org/isc/datasheet/90287.

Anon. http://www.fao.org/3/contents/129a1845-3d53-5d6a-b431-b6518739a964/ab731e00.htm.

Anon. http://www.fao.org/3/contents/129a1845-3d53-5d6a-b431-b6518739a964/ab731e00.htm.

Anon. http://www.fao.org/docrep/t8365e/t8365e05.htm.

Anon. http://www.karipearls.com/cassis-cornuta.html.

Anon. http://www.mesa.edu.au/aquaculture/aquaculture17.asp.

Anon. http://www.postranchkitchen.com/2014/03/keyhole-limpet-and-gumboot-chiton.html.

Anon. http://www.united-academics.org/health-medicine/cancer-vaccines-from-mollusk/.

Anon. https://au.nutrihand.com/Nutrihand/pctools/showFoodFacts1.do;jsessionid=3727F0B3FBF3EAF76514434892C9992B?nodeID=&customerUserID=&mealTypeID=&mealPlanID=&styleID=&inFramed=&foodLogMiscID=&courseID=&spicinessID=&source=&recipeID=&foodID=1120441&recordType=.

Anon. https://sta.uwi.edu/fst/lifesciences/documents/Littoraria_angulifera.pdf.

Anon. https://www.google.com/patents/US8436141.

Anon. https://www.tradeindia.com/fp3221803/Babylonia-Zeylanica-Medium-Shells.html.

Arancibia, S.; Espinoza, C.; Salazar, F.; Campo, M. D.; Tampe, R.; Zhong, T.; De Ioannes, P.; Moltedo, B.; Ferreira, J.; Lavelle, E.C.; Manubens, A.; De Ioannes, A.E.; Becker, M.

I. A Novel Immunomodulatory Hemocyanin from the Limpet *Fissurella latimarginata* Promotes Potent Anti-Tumor Activity in Melanoma. *PLoS One.* **2014**, *9*, e87240.

Asano, M.; Itoh, M. Salivary Poison of a Marine Gastropod, *Neptunea arthritica bernardi,* and the Seasonal Variation of its Toxicity. *Biochem. Pharmacol. Compd. Derived from Mar. Organ.* **1960**, *90*, 674–688.

Atalah, J.; Newcombe, E. M.; Hopkins, G. A.; Forrest, B. M. Potential Biocontrol Agents for Biofouling on Artificial Structures. *Biofouling.* **2014**, *30*, 999–1010.

Averbuj, A.; Cremonte, F. Parasitic Castration of *Buccinanops cochlidium* (Gastropoda: Nassariidae) Caused by a Lepocreadiid Digenean in San José Gulf, Argentina. *J. Helminthol.* **2010**, *84*, 381–389.

Aziz, U. S. *Nutritional Value of Babylonia areolata.* Undergraduate project report, Faculty of Agro – Based Industry. 2016, http://umkeprints.umk.edu.my/id/eprint/5667.

Babar, A. G.; Pande, A.; Kulkarni, B. G. Bioactive Potential of Some Intertidal Molluscs Collected from *Mumbai coast,* West Coast of India. *Asian Pac. J. Trop. Biomed.* 2012, *2* (Supplement), S1060–S1063.

Babu, A.; Kesavan, K.; Annadurai, D.; Rajagopal, S. *Bursa spinosa*—A Mesogastropod Fit for Human Consumption. *Adv. J. Food Sci. Technol.* **2010**, *2*, 79–83.

Babu, A.; Venkatesan, V.; Rajagopal, S. Contribution to the Knowledge of Ornamental Molluscs of Parangipettai, Southeast Coast of India. *Adv. Appl. Sci. Res.* **2011**, *2*, 290–296.

Babu, A.; Venkatesan, V.; Rajagopal, S. Fatty Acid and Amino Acid Compositions Of The Gastropods, *Tonna dolium* (Linnaeus, 1758) and *Phalium glaucum* (Linnaeus, 1758) from the Gulf of Mannar, Southeast Coast of India. *Ann. Food Sci. Technol.* **2011**, *12*, 159–163.

Badiu, D. L.; Luque, R.; Dumitrescu, E.; Craciun, A.; Dinca, D. Amino Acids from *Mytilus galloprovincialis* (L.) and *Rapana venosa* Molluscs Accelerate Skin Wounds Healing via Enhancement of Dermal and Epidermal Neoformation. *Protein J.* **2010**, *29*, 81–92.

Bannister, W. H.; Bannister, J. V.; Micallef, H. Bile Pigment in the Shell of *Monodonta turbinata* (Mollusca: Gastropoda). *Comp. Biochem. Physiol.* **1968**, *27*, 451–454.

Barraza, J. E. Food Poisoning due to Consumption of the Marine Gastropod *Plicopurpura columellaris* in El Salvador. *Toxicon.* **2009**, *54*, 895–896.

Bathige, S. D. N. K.; Umasuthan, N.; Jayasinghe, J. D. H. E.; Godahewa, G. I.; Park, H. C.; Lee, J. Three Novel C1q Domain Containing Proteins from the Disk Abalone *Haliotis discus discus*: Genomic Organization and Analysis of the Transcriptional Changes in Response to Bacterial Pathogens. *Fish Shellfish Immunol.* **2016**, *56*, 181–187.

Benkendorff, K.; Rudd, D.; Nongmaithem, B. D.; Liu, L.; Young, F.; Edwards, V.; Avila, C.; Abbott, C. A. Are the Traditional Medical Uses of Muricidae Molluscs Substantiated by Their Pharmacological Properties and Bioactive Compounds? *Mar. Drugs.* **2015**, *13*, 5237–5275.

Berthou, P.; Poutiers, J. M.; Goulletquer, P.; Dao, J. C. Shelled Molluscs. *Encyclopedia of Life Support Systems* (EOLSS). http://archimer.ifremer.fr/doc/2001/publication-529.pdf, 24p.

Boonzaaier, M. K.; Neethling, S.; Moutond, A.; Simon, C. A. Polydorid Polychaetes (Spionidae) on Farmed and Wild Abalone (*Haliotis midae*) in South Africa: An Epidemiological Survey. *Afr. J. Mar. Sci.* **2014**, *36*, 369–376.

Borquaye, L. S.; Darko, G.; Ocansey, E.; Ankomah, E. Antimicrobial and Antioxidant Properties of the Crude Peptide Extracts of *Galatea paradoxa* and *Patella rustica*. *SpringerPlus.* **2015**, *4*, 6.

Buschbaum, C.; Buschbaum, G.; Schrey, I.; Thieltges, D. W. Shell-Boring Polychaetes Affect Gastropod Shell Strength and Crab Predation. *Mar. Ecol. Prog. Ser.* **2007**, *329*, 123–130.

Caceres-Martinez, J.; Tinoco-Orta, G. D. Symbionts Of Cultured Red Abalone *Haliotis rufescens* from Baja California, Mexico. *J. Shellfish Res.* **2001**, *20*, 875–881.

Cahyani, R. T.; Purwaningsih, S.; Azrifitria. Antidiabetic Potential and Secondary Metabolites Screening of Mangrove Gastropod *Cerithidea obtusa*. *J. Coastal Life Med.* **2015**, *3*, 356–360.

Castro, I. B.; de Meirelles, C. A. O.; Matthews-Cascon, H.; Rocha-Barreira, C. A.; Penchaszadeh, P.; Bigatti, G. Imposex in Endemic Volutid from Northeast Brazil (Mollusca: Gastropoda). Braz. Arch. Biol. Technol. *2008*, 51, On-line version ISSN 1678-4324.

Celik, M. Y.; Culha, S. T.; Culha, M.; Yildiz, H.; Acarli, S.; Celik, I.; Celik, P. Comparative Study on Biochemical Composition of Some Edible Marine Molluscs at Canakkale Coasts, Turkey. *Indian J. Geo-Mar. Sci.* **2014**, *43*, 601–606.

Chaitanawisuti, N.; Kritsanapuntu, S.; Santaweesuk, W. Comparisons between two Production–Scale Methods for the Intensive Culture of Juveniles Spotted Babylon, *Babylonia areolata*, to Marketable Sizes. *Int. J. Fish. Aquacult.* **2011**, *3*, 79–88.

Chakravarty, M. S.; Dogiparti, A.; Sudha, B. S.; Ganesh, P. R. C. Biochemical Composition of Three Potamidid Snails-*Telescopium telescopium*, *Cerithidea cingulate* and *C. obtusa* of Tekkali Creek (Bhavanapadu Mangroves), Andhra Pradesh, India. *Adv. Appl. Sci. Res.* **2015**, *6*, 50–53.

Chen, C.; Chen, R.; Chen, P.; Liu, H.; Hsieh, H. Intermediate Culture of Juvenile Horseshoe Crab (*Tachypleus tridentatus*) Mixed with Juvenile Spotted Babylon (*Babylonia areolata*) for Restocking Horseshoe Crab Populations. *Bioflux.* **2016**, *9*, 623–633.

Cheng, L.; Huang, J.; Shi, C. Y.; Thompson, K. D.; Mackey, B., Cai, J. P. *Vibrio parahaemolyticus* Associated With Mass Mortality of Postlarval Abalone, *Haliotis diversicolor supertexta* (L.), in Sanya, China. *J. World Aquacult. Soc.* **2008**, *39*, 746–757.

Cheng, T. C. Larval Nematodes Parasitic in Shellfish. MFR Paper 1345, http://spo.nmfs. noaa.gov/mfr4010/mfr401019.pdf.

Clavier, J. Infestation of *Haliotis tuberculata* shells by *Cliona celata* and *Polydora species*. In: *Abalone of the world. Biology, Fisheries and Culture* 1989, p. 16–20.

Cook, P. A. The Worldwide Abalone Industry. *Modern Econ.* **2014**, *5*, 1181–1186.

Corbeil, S.; Williams, L. M.; McColl, K. A.; Crane, M. S. Australian Abalone (*Haliotis laevigata, H. rubra* and *H. conicopora*) are Susceptible to Infection by Multiple Abalone Herpesvirus Genotypes. *Dis. Aquat .Organ.* **2016**, *119*, 101–106.

Costa, G.; Soares, S.; Carvalho, F.; Bela, J. Digenean Parasites of the Marine Gastropods *Littorina littorea* and *Gibbula umbilicalis* in the Northern Portuguese Atlantic Coast, with a Review of Digeneans Infecting the Two Gastropod Genera. *J. Coastal Life Med.* **2016**, *4*, 345–352.

Cumplido, M.; Averbuj, A.; Bigatti, G. Reproductive Seasonality and Oviposition Induction in *Trophon geversianus* (Gastropoda: Muricidae) from Golfo Nuevo, Argentina. *J. Shellfish Res.* **2010**, *29*, 423–428.

Czeczuga, B. Carotenoid Contents in *Diodora graeca* (L.) (Gastropoda: Fissurellidae from the Mediterranean (Monaco). *Comp. Biochem. Physiol., B: Comp. Biochem.* **1980**, 439–441.

Dagorn, F.; Buzin, F.; Couzinet-Mossion, A.; Decottignies, P.; Viau, M.; Rabesaotra, V.; Barnathan, G.; Wielgosz-Collin, G. Multiple Beneficial Lipids Including Lecithin Detected in the Edible Invasive Mollusk *Crepidula fornicata* from the French Northeastern Atlantic Coast. *Mar. Drugs.* **2014**, *12*, 6254–6268.

Dang, V. T.; Benkendorff, K.; Speck, P. In Vitro Antiviral Activity against Herpes Simplex Virus in the Abalone *Haliotis laevigata. J. Gen. Virol.* **2011**, *92*, 627–637.

Dang, V. T.; Benkendorff, K.; Green, T.; Speck, P. Marine Snails and Slugs: A Great Place To Look for Antiviral Drugs. *J. Virol.* **2015**, *89*, 8114–8118.

D'Armas, H.; Yáñez, D.; Reyes, D.; Salazar, G. Fatty Acids Composition of the Marine Snails *Phyllonotus pomum* and *Chicoreus brevifrons* (Muricidae)]. *Rev. Biol. Trop.* **2010**, *58*, 645–654.

Datta, D.; Talapatra, S. N.; Swarnakar, S. Bioactive Compounds from Marine Invertebrates for Potential Medicines—An Overview. *Int. Lett. Nat. Sci.* **2015**, *34*, 42–61.

Davies I. C.; Jamabo, N. A Proximate Composition of Edible Parts of Shellfishes from Okpoka Creeks in Rivers State, Nigeria. *Int. J. Life Sci. Res.* **2016**, *4*, 247–252.

Davis, M. Species Profile Queen Conch, *Strombus gigas*. SRAC Publication 2005, 7203, http://www2.ca.uky.edu/wkrec/QueenConch.pdf.

De Zoysa, M.; Whangb, I.; Lee, Y.; Lee, S.; Lee, J.; Lee, J. Defensin from Disk Abalone *Haliotis discus discus*: Molecular Cloning, Sequence Characterization and Immune Response against Bacterial Infection. *Fish Shellfish Immunol.* **2010**, *28*, 261–266.

Defer, D.; Bourgougnon, N.; Fleury, Y. Screening for Antibacterial and Antiviral Activities in Three Bivalve and Two Gastropod Marine Molluscs. *Aquaculture.* **2009**, *293*, 1–7.

Degiam, Z. D.; Abas, A. T. Antimicrobial Activity of Some Crude Marine Mollusca Extracts against Some Human Pathogenic Bacteria. *Thi-Qar Med. J.* **2010**, *4*, 142–147.

Dhinakaran, A.; Sekar, V.; Sethubathi, G. V.; Suriya, J. Antipathogenic Activity of Marine Gastropoda (*Hemifusus pugilinus*) from Pazhayar, South East Coast of India. *Int. J. Environ. Sci.* **2011**, *2*, 536–542.

Dias, T. L. P.; Neto, N. A. L.; Alves, R. R. N. Molluscs in the Marine Curio and Souvenir Trade in NE Brazil: Species Composition and Implications for their Conservation and Management. *Biodivers. Conserv.* **2011**, *20*, 2393–2405.

Diggles, B. K.; Nichol, J.; Hine, P. M.; Wakefield, S.; Cochennec-Laureau, N.; Roberts, R. D.; Friedman, C. S. Pathology of Cultured Paua *Haliotis iris* Infected with a Novel Haplosporidian Parasite, with Some Observations on the Course of Disease. *Dis. Aquat. Organ.* **2002**, *50*, 219–231.

Dolashka, P.; Moshtanska, V.; Borisova, V.; Dolashki, A.; Stevanovic, S.; Dimanov, T.; Voelter, W. Antimicrobial Proline-Rich Peptides from the Hemolymph of Marine Snail *Rapana venosa. Peptides.* **2011**, *32*, 1477–1483.

Domínguez-Ojeda, D.; Patrón-Soberano, O. A.; Nieto-Navarro, J. T.; Robledo-Marenco, M. L.; Velázquez-Fernández, J. B. Imposex in *Plicopurpura pansa* (Neogastropoda: Thaididae) in Nayarit and Sinaloa, Mexico. Rev. Mex. Biodiversidad. **2013**, *86*, 531–534.

Dong, F. M. The nutritional value of shellfish. https://wsg.washington.edu/aquaculture/pdfs/Nutritional-Value-of-Shellfish.pdf.

Durazo, E.; Viana, M. T. Fatty Acid Profile of Cultured Green Abalone (*Haliotis fulgens*) Exposed to Lipid Restriction and Long-Term Starvation. *Cienc. Mar.* **2013**, *39*, 363–370.

Edwards, V.; Vicki, E. The Effects of Bioactive Compounds from the Marine Mollusc *Dicathais orbita* on Human Reproductive Cells and Human Reproductive Cancer Cells [Manuscript], **1967**.

El Universal, Science and Health, 2016.

Ezzo Abalone, https://www.lib.noaa.gov/retiredsites/korea/main_species/abalone.htm.

Fermin, A. C.; Buen, S. M. Grow-Out Culture of Tropical Abalone, *Haliotis asinina* (Linnaeus) in Suspended Mesh Cages with Different Shelter Surface Areas. *Aquacult. Int.* **2001**, *9*, 499–508.

Floren, A. S. *The Philippine shell industry withspecial focus on Mactan, Ceba.* 2003, http://www.oneocean.org/download/db_files/philippine_shell_industry.pdf.

Flores-Garza, R.; García-Ibáñez, S.; Flores-Rodríguez, P.; Torreblanca-Ramírez, C.; Galeana-Rebolledo, L.; Valdés-González, A.; Suástegui-Zárate, A.; Violante-González, J. Commercially Important Marine Mollusks for Human Consumption in Acapulco, México. *Nat. Res.* **2012**, *3*, 11–17.

Francesconi, K. A.; Edmonds, J. S.; Hatcher, B. G. Examination of the Arsenic Constituents of the Herbivorous Marine Gastropod *Tectus pyramis*: Isolation of Tetramethylarsonium Ion. *Comp. Biochem. Physiol., C Comp. Pharm.* **1998**, *90*, DOI: 10.1016/0742-8413(88)90004-7.

Freeman, K. A. Aquaculture and Related Biological Attributes of Abalone Species in Australia—A Review. *Fish. Res. Rep. West. Aust.* **2001**, *128*, 1–48.

Freije, A. M.; Awadh, M. N. Fatty Acid Compositions of *Turbo coronatus* Gmelin 1791. http://dx.doi.org/10.1108/00070701011080195.

Friedman, G. C.; Grindley, R.; Keogh, J. A. Isolation of a Fungus from Shell Lesions of New Zealand Abalone, *Haliotis iris* Martyn and *H. australis. Molluscan Res.* **2013**, *18*, 313–324.

Fryda, J. *Fossil Invertebrates: Gastropods, Reference Module in Earth Systems and Environmental Sciences*, 2013. DOI: 10.1016/B978-0-12-409548-9.02806-2.

Fujiwara, Y.; Urabe, J.; Takeda, S. Host Preference of a Symbiotic Flatworm in Relation to the Ecology of Littoral Snails. *Mar. Biol.* **2014**, *161*, 1873–1882.

Gharsallah, I. H.; Vasconcelos, P.; Zamouri-Langar, N.; Missaoui, H. Reproductive Cycle and Biochemical Composition of *Hexaplex trunculus* (Gastropoda: Muricidae) from Bizerte Lagoon, Northern Tunisia. *Aquat. Biol.* **2010**, *10*, 155–166.

Giftson, H.; Mani, A. E.; Jayasanta, I.; Kailasam, S.; Patterson, J. Antibacterial, Biochemical Composition and FTIR Analysis of *Chicoreus ramosus* from Kanyakumari Coast. *Ijppr. Hum.* **2015**, *4*, 171–181.

González, R.; Lohrmann, K. B.; Pizarro, J.; Brokordt, K. Differential Susceptibility to the Withering Syndrome Agent and Renal Coccidia in Juvenile *Haliotis rufescens, Haliotis discus hannai* and the interspecific hybrid. *Invertebr. Pathol.* **2014**, *116*, 13–17.

Gopeechund, A.; Bhagooli, R.; Sharadha, V.; Bhujun, N.; Bahoun, T. *Antioxidant Activities of Edible Marine Molluscs from a Tropical Indian Ocean Island.* Symposium.wiomsa. org/wp-content/uploads/2015/10/A.-Gopeechund.pdf.

Gorain, B.; Chakraborty, S.; Pal, M. M.; Sarkar, R.; Samanta, S. K.; Karmakar, S.; Sen, T. Arylamine *N*-Acetyl Transferase (NAT) in the Blue Secretion of Telescopium

Telescopium: Xenobiotic Metabolizing Enzyme as a Biomarker for Detection of Environmental Pollution. *SpringerPlus.* **2014**, *3*, 666. doi:10.1186/2193-1801-3-666.

Govindarajalu, J.; Muthusamy, A.; Gurusamy, C.; Mani, K.; Arumugam, K. Comparative Studies on Biochemical Analysis of Some Economically Important Marine Gastropods along Gulf of Mannar Region, Southeast Coast of India. *J. Coastal Life Med.* **2016**, *4*, 444–447.

Grindley, R. M.; Keogh, J. A.; Friedman, C. S. Shell Lesions in New Zealand *Haliotis* spp. (Mollusca, Gastropoda). *J. Shellfish Res.* **1998**, *17*, 805–811.

Guerra-García, J. M.; Corzo, J.; Espinosa, F.; García-Gómez, J. C. Assessing Habitat Use of The Endangered Marine Mollusc *Patella ferruginea* (Gastropoda, Patellidae) in Northern Africa: Preliminary Results and Implications for Conservation. *Biol. Conserv.* **2004**, *116*, 319–326.

Gupta, P.; Arumugam, M.; Azad, R. V.; Saxenab, R.; Ghoseb, S.; Biswasd, N. R.; Velpandiana, T. Screening of Antiangiogenic Potential of Twenty Two Marine Invertebrate Extracts of Phylum Mollusca from South East Coast of India. *Asian Pac. J. Trop. Biomed.* **2014**, *4*, (Supplement) S129–S138.

Hanlinger, J.; Carson, J.; Donachie, L.; Gabor, L.; Taylor, D. Bacterial Infection in Tasmanian Farmed Abalone: Causes, Pathology, Farm Factors and Control Options. In P. Walker, R. Lester and M. G. Bondad-Reantaso (Eds). *Dis. Asian Aquacult.* **2005**, *5*, 289–299.

Hardjito, L., Royani, D. S.; Santoso, J. Nutritional Composition and Topoisomerase Inhibitor Activity of Ethnomedicinal Marine Mollusk *Nerita albicilla. J. Food Sci. Eng.* **2012**, *2*, 550–556.

Hayashi, A.; Matsubara, T.; Matsuura, F. Biochemical Studies on the Lipids of *Turbo cornutus.* I. The Conjugated Lipids of Viscera. (Part 1). *Fisheries Research Board of Canada Translation Series* 1969, 1328, http://www.dfo-mpo.gc.ca/Library/139275.pdf.

Hayashi, K.; Moriwaki, J.; Sawabe, T.; Thompson, F. L.; Swings, J.; Gudkovs, N.; Christen, R.; Ezura, Y. *Vibrio superstes* sp. nov.; Isolated from the Gut of Australian Abalones *Haliotis laevigata* and *Haliotis rubra. Int. J. Syst. Evol. Microbiol.* **2003**, *53*, 1813–1817.

Hexaplex trunculus—Wikipedia

Hsu, M.; Hung, S. Antiherpetic Potential of 6-Bromoindirubin-3′-Acetoxime (BIO-Acetoxime) in Human Oral Epithelial Cells. *Arch. Virol.* **2013**, *158*, 1287–1296.

http://www.fisheat.it/murex-bolinus-brandaris.

http://www.marbef.org/wiki/TBT_and_Imposex.

http://www.ospar.org/site/assets/files/34390/tbt_example_sheet.pdf.

https://researcharchive.lincoln.ac.nz/bitstream/handle/10182/2863/Shi_MApplSci.pdf;jsessionid=8CEB19C444DA715A3BE8BE8640F37503?sequence=5.

Hua, N. T.; Ako, H. Reproductive Biology and Effect of Arachidonic Acid Level in Broodstock Diet on Final Maturation of the Hawaiian Limpet *Cellana sandwicensis. J. Aquac. Res. Dev.* **2014**, *5*(5), 1000256.

Huchette, S.; Paillard, C.; Clavier, J.; Day, R. Shell Disease: Abnormal Conchiolin Deposit in the Abalone *Haliotis tuberculata. Dis. Aquat. Organ.* **2006**, *68*, 267–271.

Hyun, D.; Kim, M.; Shin, N.; Kim, J. Y.; Kim, P. S.; Whon, T. W.; Yun, J.; Bae, J. *Shimia haliotis* sp. nov.; A Bacterium Isolated from the Gut of an Abalone, *Haliotis discus hannai. Int. J .Syst. Evol. Microbiol.* **2013**, *63*, 4248–4253.

Immanuel, G.; Thaddaeus, B. J.; Usha, M.; Ramasubburayan, R.; Prakash, S.; Palavesam, A. Antipyretic, Wound Healing and Antimicrobial Activity of Processed Shell of the Marine Mollusc *Cypraea moneta*. *Asian Pac. J. Trop. Biomed.* **2012**, S1643–S1646.

Izawa, K. Two New Parasitic Copepods (Cyclopoida: Myicolidae) from Japanese Gastropod Molluscs. *Publ. Seto Mar. Biol. Lab.* **1976**, *23*, 213–227.

Jang, M.; Jang, J. R.; Park, H.; Yoo, H. Overall Composition, and Levels of Fatty Acids, Amino Acids, and Nucleotide-type Compounds in Wild Abalone *Haliotis gigantea* and Cultured Abalone *Haliotis discus hannai*. *Korean J. Food Preserv.* **2010**, *17*, 533–540.

Jayalakshmi, K. Biochemical Composition and Nutritional Value of Marine Gastropod *Babylonia zeylanica* from Puducherry, South East Coast of India. *Indo—Asian J. Multidiscip. Res.* **2016**, *2*, 478–483.

Jesily, S.; Rooslin, R. A Survey of Gastropods from Tuticorin Coast. *Golden Res. Thoughts* **2015**, *5*, 1–10.

Jiang, Q.; Shi, L.; Ke, C.; You, W.; Zhao, J. Identification and Characterization of *Vibrio harveyi* Associated with Diseased Abalone *Haliotis diversicolor*. *Dis. Aquat. Organ.* **2013**, *103*, 133–139.

Jian-yin, M.; Xiao-mian, G.; Jun-feng, H.; Xian, W. New Antimicrobial Peptides Purified Directly from *Bullacta exarata*. *Afr. J. Pharm. Pharmacol.* **2011**, *5*, 1508–1512.

Jiménez-Arce, G. Chemical and Nutritional Composition in the Marine Snail *Strombus gracilior* (Mesogastropoda: Strombidae) of Various Sizes and Sexes in Playa Panamá, Costa Rica. *Rev. Biol. Trop.* **1993**, *41*, 345–349.

Jin, A. H.; Vetter, I.; Himaya, S. W.; Alewood, P. F.; Lewis, R. J.; Dutertre, S. Transcriptome and Proteome of *Conus planorbis* Identify the Nicotinic Receptors as Primary Target for the Defensive Venom. *Proteomics.* **2015**, *15*, 4030–4040.

João, C.; Armindo, R.; de Frias, M. A. M. The reproductive cycle of *Patella candei gomesii* Drouët, 1858 (Mollusca: Patellogastropoda), an Azorean Endemic Subspecies. *Costa Manuel.* **2005**, doi:10.1080/07924259.2005.9652180.

Joyce, W. Y. http://www.gia.edu/gems-gemology/fall-2015-labnotes-large-natural-pearls-haliotis-abalone-species.

Jolivet, A.; Chauvaud, L.; Thébault, J.; Robson, A. A.; Dumas, P.; Amos, G.; Lorrain, A. Circadian Behaviour of *Tectus (Trochus) niloticus* in the Southwest Pacific Inferred from Accelerometry. *Mov. Ecol.* **2015**, *3*, 26. doi: 10.1186/s40462-015-0054-5. eCollection 2015.

Jousson, O.; Bartoli, P. The Life Cycle of *Opecoeloides columbellae* (Pagenstecher, 1863) n. comb. (Digenea, opecoelidae): Evidence from Molecules and Morphology. **2000**, *30*, 747–760.

Kamaishi, T.; Miwa, S.; Goto, E.; Matsuyama, T.; Oseko, N. Mass Mortality of Giant Abalone *Haliotis gigantea* Caused by a *Francisella* sp. Bacterium. *Dis. Aquat. Organ.* **2010**, *89*, 145–154.

Kamarazaman, N. S.; Misnan, R.; Yadzir, Z. H. M. Identification of Major Allergenic Spots of *Cerithidea obtusa* (Obtuse horn shell) by Two Dimensional Electrophoresis (2-de) and Immunoblotting. *Int. J. Sci. Environ. Technol.* **2016**, *5*, 222–228.

Kanagasabapathy, S.; Samuthirapandian, R.; Kumaresan, M. Preliminary Studies for a New Antibiotic from the Marine Mollusk *Melo melo* (Lightfoot, 1786). *Asian Pac. J. Trop. Med.* **2011**, *4*, 310–314.

Kanchana, S.; Vennila, R.; Kumar, K. R.; Arumugam, M.; Balasubramanian, T. Antagonistic and Cyto-Toxicity Activity of Mollusc Methanol Extracts. *J. Biol. Sci.* **2014**, *14*, 60–66.

Kiran, N.; Siddiqui, G.; Khan, A. N.; Ibrar, K.; Tushar, P. Extraction and Screening of Bioactive Compounds with Antimicrobial Properties from Selected Species of Mollusk and Crustacean. *J. Clin. Cell Immunol.* **2014**, *5*, 189. doi: 10.4172/2155-9899.1000189.

Koeypudsa, W.; Kitkamthorn, M.; Chaitanawisuti, N.; Kritsanapuntu, A.; Tantawanich, T.; Tangtrongpiros, J. Natural Infection on Farmed Spotted Babylon (*Babylonia areolata* Link 1807) Proceedings, In *The 15th Congress of FAVA FAVA—OIE Joint Symposium on Emerging Diseases,* 2008, p. 27–30.

Kritsanapuntu, S.; Chaitanawisuti, N.; Santhaweesuk, W.; Natsukari, Y. Pilot Study on Polyculture of Juveniles Spotted Babylon, *Babylonia areolata*, with Milkfish, *Chanos chanos*, to Marketable Sizes using Large-Scale Operation of Earthen Ponds in Thailand. *Aquacult. Res.* **2006**, *37*, 618–624.

Kritsanapuntu, S.; Chaitanawisuti, N.; Natsukari, Y. Growth and Water Quality for Growing-Out of Juvenile Spotted Babylon, *Babylonia areolata*, at Different Water-Exchange Regimes in a Large-Scale Operation of Earthen Ponds. *Aquacult. Int.* **2009**, *17*, 77–84.

Kritsanapuntu, S.; Chaitanawisuti, N.; Santhaweesuk, W.; Natsukari, Y. Growth, Production And Economic Evaluation of Earthen Ponds for Monoculture and Polyculture Of Juveniles Spotted Babylon (*Babylonia areolata*) to Marketable Sizes using Large-Scale Operation. *J. Shellfish Res.* **2006**, *25*, 913–918.

Kritsanapuntu, S.; Chaitanawisuti, N.; Santhaweesuk, W.; Natsukari, Y. Growth Performances for Monoculture and Polyculture of Hatchery-Reared Juvenile Spotted Babylon, *Babylonia areolata* Link, 1807, in Large-Scale Earthen Ponds. *Aquacult. Res.* **2008**, *39*, 1556–1561.

Krzynowek, J.; Murphy, J. Proximate Composition, Energy, Fatty Acid, Sodium, and Cholesterol Content of Finfish, Shellfish, and their Products. *NOAA Technical Report* 1987, 55, p. 53.

Kurokawa, T. M.; Wuhrer, M.; Lochnit, G.; Geyer, H.; Markl, J.; Geyer, R. Hemocyanin from the Keyhole Limpet *Megathura crenulata* (KLH) Carries a Novel Type of N-Glycans with Gal(beta1-6). *Eur. J. Biochem.* **2002**, *269*, 5459–5473.

Kusaikin, M. I.; Zakharenko, A. M.; Ermakova, S. P.; Veselova, M. V.; Grigoruk, E. V.; Fedoreev, S. A.; Zvyagintseva, T. N. Deglycosylation of Isoflavonoid Glycosides from *Maackia amurensis* Cell Culture by β-d-Glucosidase from *Littorina sitkana* Hepatopancrease. https://dvfu.pure.elsevier.com/en/publications/deglycosylation-of-isoflavonoid-glycosides-from-maackia-amurensis.

Lah, R. A.; Smith, J.; Savins, D.; Dowell, A.; Bucher, D.; Benkendorff, K. Investigation of Nutritional Properties of Three Species of Marine Turban Snails for Human Consumption. *Food Sci. Nutr.* **2016**, DOI: 10.1002/fsn3.360.

Layer, R. T.; McIntosh, J. M. Conotoxins: Therapeutic Potential and Application. **2006**, *4*, 119–142.

Lee, J. H. Gonadal Development and Reproductive Cycle of the Top Shell, *Omphalius rusticus* (gastropoda: Trochidae). *J. Korean J. Biol. Sci.* **2001**, *5*, 37–44.

Leighton, D. L. Abalone (Genus *Haliotis*) Mariculture on the North American Pacific Coast. *Fish. Bull.* **1989**, *87*, 689–702.

Leontowicz, M.; Leontowicz, H.; Namiesnik, J.; Apak, R.; Barasch, D.; Nemirovski, A.; Moncheva, S.; Goshev, I.; Trakhtenberg, S.; Gorinstein, S. *Rapana venosa* Consumption Improves the Lipid Profiles and Antioxidant Capacities In Serum of Rats Fed on Atherogenic Diet. *Nutr. Res.* **2015**, *35*, 592–602.

Lev, E.; Amar, Z..*Practical Materia Medica of the Medieval Eastern Mediterranean According to the Cairo Genizah.* 2008, Science, 619 p.

Li, T.; Ding, M.; Zhang, J.; Xiang, J.; Liu, R. Studies on the Pustule Disease of Abalone (*Haliotis discus hannai* Ino) on the Dalian Coast. *J. Shellfish Res.* **1998**, *17*, 707–711.

Limaverdea, A. M.; Wagenera, A. L. R.; Fernandezb, M. A.; Scofielda, A. L.; Coutinhoc, R. *Stramonita haemastoma* as a Bioindicator for Organotin Contamination in Coastal Environments. *Mar. Environ. Res.* **2007**, *64*, 384–398.

Lin, H.; Li, Z.; Chen, B.; Yi, Z.; Wang, Z.; Hans, J. M. Anti-Tumor Activity of the Peptide from *Bullacta exarata. J. Biomed.* **2012**, *2*, 19–22.

Littoraria angulifera—Wikipedia

Liu, D.; Liao, N.; Ye, X.; Hu, Y.; Wu, D.; Guo, X.; Zhong, J.; Wu, J.; Chen, S. Isolation and Structural Characterization of a Novel Antioxidant Mannoglucan from a Marine Bubble Snail, *Bullacta exarata* (Philippi). *Mar. Drugs.* **2013**, *11*, 4464–4477.

Liu, L.; Wang, S. Population Dynamics and Mantle Autotomy of the Figsnail *Ficus ficus* (Gastropoda: Mesogastropoda: Ficidae). *Zoolog. Stud.* **1999**, *38*, 1–6.

Lleonart, M.; Handlinger, J.; Powell, M. Spionid Mudworm Infestation of Farmed Abalone (*Haliotis* spp.). *Aquaculture.* **2003**, *221*, 85–96.

López, D. A.; González, M. L.; Pérez, M. C. Feeding and Growth In The Keyhole Limpet, *Fissurella picta* (Gmelin, 1791). *J. Shellfish Res.* **2003**, *22*, 165–169.

Lowry, J. K.; Stoddart, H. E. New Genus of Eusiridae (Crustacea, Amphipoda), associated with the Abalone *Haliotis rubra* Leach, in South-Eastern Australia: *Zoosystema. Paris [Zoosystema].* **1998**, *20*, 307–314.

Luo, S.; Zhangsun, D.; Feng, J.; Wu, Y.; Zhu, X.; Hu, Y. Diversity of the O-Superfamily Conotoxins from *Conus miles. J. Pep. Sci.* **2007**, *13*, 44–53.

Ma, J.; Huang, F.; Lin, H.; Wang, X. Isolation and Purification of a Peptide from *Bullacta exarata* and Its Impaction of Apoptosis on Prostate Cancer Cell. *Mar. Drugs.* **2013**, *11*, 266–273.

Mahmoud, M. A. M.; Mohammed, T. A. A.; Yassien, M. H. Spawning Frequency, Larval Development and Growth of Muricid Gastropod *Chicoreus ramosus* (Linnaeus, 1758) in the Laboratory at Hurghada, Northern Red Sea, Egypt. *Egypt. J. Aquat. Res.* **2013**, *39*, 125–131.

Margret, M. S.; Jansi, M. J. *Thais bufo* (Lamarck), A Neogastropod-Fit for Human Consumption. *Int. J. Res. Biotechnol. Biochem.* **2013**, *3*, 27–30.

Martin, G. E.; Sanduja, R.; Alam, M. Isolation of Isopreropodine from the Marine Mollusk *Nerita albicilla*: Establishment of the Structure via Two Dimensional NMR Techniques. *J. Nat. Prod.* **1986**, *49*, 406–411.

Massilia, G. R.; Schininà, M. E.; Ascenzi, P.; Polticelli, F. Contryphan-Vn: A Novel Peptide from the Venom of the *Mediterranean snail Conus ventricosus. Biochem. Biophys. Res. Commun.* **2001**, *288*, 908–913.

McFadden, D. W.; Riggs, D. R.; Jackson, B. J.; Vona-Davis, L. Keyhole Limpet Hemocyanin, A Novel Immune Stimulant with Promising Anticancer Activity in Barrett's Esophageal Adenocarcinoma. *Am. J. Surg.* **2003**, *186*, 552–555.

Megathura crenulata—Wikipedia

Merdekawati, D. *Nutrition Content and Antioxidant Compound of the Rough Turban Snails (Turbo setosus Gmelin 1791) (* Supervised by Tati Nurhayati and Agoes M. Jacoeb) 2013, http://repository.ipb.ac.id/bitstream/handle/123456789/66859/2013dme.pdf?sequence=1&isAllowed=y

Mgaya, Y. D. Synopsis of biological data on the European abalone (ormer) *Haliotis tuberculata* Linnaeus, 1758 (Gastropoda: Haliotidae), *FAO Fisheries Synopsis* 1995,156, http://www.fao.org/docrep/017/v7050e/v7050e.pdf

Millemann, R. E. *Echinocephalus pseudouncinatus* n. sp.; a Nematode Parasite of the Abalone. *J. Parasitol.* **1951**, *37*, 435–439.

Miller, A. C.; Boxt, M. A. A Nutritional Study of Seven Molluscan Species Recovered from CA-LAN 2630, Long Beach, California. http://www.pcas.org/documents/Boxtand-Millerweb.pdf.

Minh, N. D.; Petpiroon, S.; Jarayabhand, P.; Meksumpun, S.; Tunkijjanukij, S. Growth and Survival of Abalone, *Haliotis asinina* Linnaeus 1758, Reared in Suspended Plastic Cages. *Kasetsart J. (Nat. Sci.).* **2010**, *44*, 621–630.

Misra, S.; Choudhury, S. M. A.; Ghosh, A. Lipids and Fatty Acids of the Gastropod Mollusc *Cerethidea obtusa. Food Chem.* **1986**, *22*, 251–258.

Mohan, K.; Abirami, P.; Kanchana, S.; Arumugam, M. Isolation and Characterization of *Molluscan glycosaminoglycans* from Pazhayar, South-East Coast of India. *Int. J. Fish. Aquat. Stud.* **2016**, *4*, 100–105.

Moltedo, B.; Faunes, F.; Haussmann, D.; De Ioannes, P.; De Ioannes, A. E. Immuno-therapeutic Effect of *Concholepas* Hemocyanin in the Murine Bladder Cancer Model: Evidence for Conserved Antitumor Properties among Hemocyanins. *Puente J. Becker Ml. J. Urol.* **2006**, *176*, 2690–2695.

Moraes, R. B. C.; Mayr, L. M. *Strombus pugilis* (Mollusc: Gastropode) as a Potential Indicator of Co-60 in a Marine Ecosystem. In *Radionuclides in the Study of Marine Processes*; 1991, pp 385–385.

Moreno, C. A.; Asencio, G.; Duarte, W. E.; Marin, V. Settlement of the Muricid *Concholepas concholepas* and Its Relationship with El Nino and Coastal Upwellings In Southern Chile. *Mar. Ecol. Prog. Ser.* **1998**, *167*, 171–175.

Nelson, T.; Oxenford, H. A. The Whelk (*Cittarium pica*) Fishery of Saint Lucia: Description and Contribution to the Fisheries Sector. In *Proceedings of the 65th Gulf and Caribbean Fisheries Institute*; 2012, p.61–68.

Nicolaidou, A.; Nott, J. A. The Role of the Marine Gastropod *Cerithium vulgatum* in the Biogeochemical Cycling of Metals. *Biogeochem. Cycling Sediment Ecol.* **1999**, *59*, 137–146.

Nicolas, J. L.; Basuyaux, O.; Mazurie, J.; Thebault, A. *Vibrio carchariae*, A Pathogen of the Abalone *Haliotis tuberculata. Dis. Aquat. Org.* **2002**, *50*, 35–43.

Nikapitiya, C.; De Zoysa, M.; Oh, C.; Lee, Y.; Ekanayake, P. M.; Whang, I.; Choi, C. Y.; Lee, J. S.; Lee, J. Disk Abalone (*Haliotis discus discus*) Expresses a Novel Antistasin-Like Serine Protease Inhibitor: Molecular Cloning and Immune Response against Bacterial Infection. *Fish Shellfish Immunol.* **2010**, *28*, 661–667.

Noble, W. J. Aspects of Life History and Ecology of Dicathais Orbita Gmelin, 1781 Related to Potential Aquaculture for Bioactive Compound Recovery. Ph.D. Thesis, Flinders University, 2014, 156p.

Noguchi, T.; Onuki, K.; Arakawa, O. Tetrodotoxin Poisoning Due to Pufferfish and Gastropods, and Their Intoxication Mechanism. *ISRN Toxicology* **2011**, Article ID 276939, 10 p.

Nurjanah Nurjanah, N.; Asadatun Abdullah, A.; Taufik Hidayat, T.; Yulianti, I. Characteristics of Minerals and Vitamin B12 by Tiger Snails, Shellfish Snow. *Meretrix Meretrix. Agric. Biol. Sci. J.* **2015**, *1*, 186–189.

Oliva, M. E. Parasitic Castration in *Fissurella crassa* (Archaeogastropoda) Due to an Adult Digenea, Proctoeces lintoni (Fellodistomidae). *Mem. Inst. Oswaldo Cruz* **1992**, *87*, http://dx.doi.org/10.1590/S0074-02761992000100007.

Ozyegin, L. S.; Ristoscu, S. F. C.; Kiyici, I. A.; Mihailescu, I. N.; Meydanoglu, O.; Oktar, F. N. Sea Snail: An Alternative Source for Nano-Bioceramic Production. *Key Eng. Mater.* **2012**, *493–494*, 781-786.

Pak, A. *Isolation and Characterization of Conotoxins from the Venom of Conus Planorbis and Conus Ferrugineus*. Florida Atlantic University, 2014, 83p.

Pakrashi, A.; Roy, P.; Datta, V. Antimicrobial Effect of Protein(s) Isolated from a Marine Mollusc *Telescopium telescopium. Indian J. Physiol. Pharmacol.* **2001**, *45*, 249–252.

Palpandi, C.; Vairamani, S.; Shanmugam, A. Proximate Composition and Fatty Acid Profile of Different Tissues of the Marine Neogastropod *Cymbium melo* (Solander, 1786). *Indian J. Fish.;* **2010**, *57*, 35–39.

Panaietis haliotis - Wikipedia

Panda, D.; Jawahar, P.; Venkataramani, V. K. Growth and Mortality Parameters of *Turbinella pyrum* (Linnaeus, 1758) Exploited Off Thoothukudi, South-East Coast of India. *Indian J. Fish.* **2011**, *58*, 29–33.

Patterson, J.; Ramesh, M. X.; Ayyakkannu, K. Processing Meat of *Chicoreus ramosus* into Pickle. *Phuket Mar. Bioi. Cent.* **1995**, Spec. Publn. *15*, 17–19.

Pechenik, J. A.; Fried, B.; Bolstridge, J. The Marine Gastropods *Crepidula plana* and *Crepidula convexa* Do Not Serve as First Intermediate Hosts for Larval Trematode Development. *Comp. Parasitol.* **2012**, *79*, 5–8.

Penchaszadeh, P. E.; Rocha-Barreira, C.; Matthews-Cascon, H.; Bigatti, G. Description of Egg Capsules of *Voluta ebraea* Linnaeus, 1758 (Gastropoda: Neogastropoda), *Comunicaciones de la Sociedad Malacológica del Uruguay* **2010**, *9*, 237–244.

Peng, J.; Xian-Ping, G. E.; Ming, L.; Zhou, Q.; Xu, P.; Xie, J. Study on Haplosporidium Disease of *Babylonia areolata. Acta Hydrobiol. Sin.* **2011**, *35*, 803–807.

Pérez, M. C.; González, M. L.; López, D. A. Breeding Cycle and Early Development of the Keyhole Limpet *Fissurella nigra* Lesson, 1831. *J. Shellfish Res.* **2007**, *26*, 315–318.

Perez-Estrada, C. J.; Civera-Cerecedo, R.; Hernandez-Llamas, A.; Serviere-Zaragoza, E. Compositionof Juvenile Green Abalone, *Haliotis fulgens* Fed Rehydrated Macroalgae. *Aquacul. Nutr.* **2011**, *17*, 62–69.

Periyasamy, N.; Srinivasan, M.; Devanathan, K.; Balakrishnan, S. Nutritional Value of Gastropod *Babylonia spirata* (Linnaeus, 1758) from Thazhanguda, Southeast Coast of India. *Asian Pac. J. Trop. Biomed.* **2011**, S249–S252.

Periyasamy, N.; Murugan, S.; Bharadhirajan, P. Anticoagulant Activity of Marine Gastropods *Babylonia spirata* Lin, 1758 and *Phalium glaucum* Lin, 1758 Collected from Cuddalore, Southeast Cost of India. *Int. J. Pharm. Pharm. Sci.* **2013**, *5*, 117–121.

Periyasamy, N.; Srinivasan, M.; Balakrishnan, S. Antimicrobial Activities of the Tissue Extracts of *Babylonia spirata* Linnaeus, 1758 (Mollusca: Gastropoda) from Thazhanguda, Southeast Coast of India, *Asian Pac. J. Trop. Biomed.* **2012**, *2*, 36–40.

Pesentseva, M. S.; Sova, V. V.; Sil'chenko, A. S.; Kicha, A. A.; Sil'chenko, A. S.; Haertle, T.; Zvyagintseva, T. N. A New Arylsulfatase from the Marine Mollusk *Turbo chrysostomus*. *Chem. Nat. Compd.* **2012**, *48*, 853–859.

Peters, H.; Basson, L. Notes on the Life-Cycle of the Digenean (Trematoda: Hemiuridae) Occurring in *Haliotis spadicea* (Donovan, 1808). file:///E:/My%20Downloads/518-1282-1-SM.pdf.

Primost, M. A.; Bigatti, G.; Márquez, F. Shell Shape as Indicator of Pollution in Marine Gastropods Affected by Imposex. *Mar. Freshw. Res.* **2016**, *67*, 1948–1954.

Prusina, I.; Peharda, M.; Ezgeta-Balić, D.; Puljas, S.; Glamuzina, B.; Golubić, S. Life-History Trait of the Mediterranean Keystone Species *Patella rustica*: Growth and Microbial Bioerosion. *Medit. Mar. Sci.* **2015**, *16*, 393–401.

Prusina, I.; Ezgeta-Balić, D.; Ljubimir, S.; Dobroslavić, T. On the Reproduction of the Mediterranean Keystone Limpet *Patella rustica*: Histological Overview. *J. Mar. Biol. Assoc. United Kingdom* **2014**, *94*, 1651–1660.

Purwaningsih, S. ISSN 0853-7291; e-ISSN: 2406-7598.

Purwaningsih, Antioxidant Activity and Chemical Composition of Red Matah Keong (*Cerithidea obtusa*) (Antioxidant Activity and Nutrient Composition of Matah Red Snail (Cerithidea obtusa). *Indonesian J. Mar. Sci.* (p- ISSN 0853-7291; e-ISSN: 2406-7598).

Qian, Z.; FuLai, Z.; Ling, W.; LingJing, Z.; Jie, C. M. Extraction and Determination of Taurine from Viscera of Abalone (*Haliotis discus hannai*).. *J. Food Saf. Qual.* **2014**, *5*, 70–76.

Ragi, A. S., Leena, P. P.; Nair, S. M. Study of Lipids and Amino Acid Composition of Marine Gastropod, *Tibia curta* Collected from the Southwest Coast of India. *World J. Pharm. Pharm. Sci.* **2016**, *5*, 1058–1076.

Raj, et al.; Antibacterial Attributes of Marine Gastropod *Hemifusus pugilinus* (Born, 1778) against Human Pathogenic Bacteria. *Int. J. Mar. Sci.* **2014**, *4*, 1–4.

Rakshit, S.; Bhattacharyya, D. K.; Misra, K. K. Distribution of Major Lipids and Fatty Acids of the Estuarine Gastropod Mollusc, *Telescopium telescopium*. *Folia Biol.* **1997**, *45*, 83–87.

Ramasamy, M. S.; Murugan, A. Imposex in Muricid Gastropod *Thais biserialis* (Mollusca: Neogastropoda: Nuricidae from Turicorin Harbour, Southeast Coast of Inida. *Indian J. Mar. Sci.* **2002**, *31*, 243–245.

Ramasamy, P.; Thampi, D. P. K.; Chelladurai, G.; Gautham, N.; Mohanraj, S.; Mohanraj, J. Screening of Antibacterial Drugs from Marine Gastropod *Chicoreus ramosus* (Linnaeus, 1758). *J. Coastal Life Med.* **2013**, *1*, 181–185.

Rambli, M.; Badii, F.; Yousr, M.; Howell, N. K. Nutritional and Rheological Properties of Underutilised Shellfish (Mollusc), Limpet (*Patella vulgata*). https://www.researchgate.net/.

Reece, K.; Dungan, C. *Perkinsus* sp. Infections of Marine Molluscs. http://www.afs-fhs.org/perch/resources/14069252415.2.1perkinsus2014.pdf.

Rho, H.; Kim, H.; Kim, J.; Karadeniz, F.; Ahn, B.; Nam, K.; Seo, Y.; Kong, C.. Anti-Inflammatory Effect of By-Products from *Haliotis discus hannai* in RAW 264.7 Cells. *J. Chem.* 2015, Article ID 526439, 7p.

Rohini, B.; Priya, C. S.; Lavanya, A.; Kalpana, K.; Karthika, V. Potential of Water and Methanol Extracts of *Lambis lambis* Against Fish and Human Pathogens. *Biol. Rhythm Res.* **2012**, *43*, 205–213.

Romalde, J. L.; Barja, J. L. Bacteria in Molluscs: Good and Bad Guys *Current Research, Technology and Education Topics in Applied Microbiology and Microbial Technology* (A. Mendez Vilas, Ed), 2010, p. 136–147.

Romero, M. S.; Gallardo, C. S.; Bellolio, G. Egg Laying and Embryonic-Larval Development in the Snail *Thais (Stramonita) chocolata* (Duclos, 1832) with Observations on its Evolutionary Relationships within the Muricidae. *Mar. Biol.* **2004**, *145*, 681–692.

Rossato, M.; Castro, I. B.; Pinho, G. I. I. Imposex in *Stramonita haemastoma*: A Preliminary Comparison Between Waterborne and Dietborne Exposure. *Ecotoxicol. Environ. Contam.* **2014**, *9*, 87–92.

Saglam, H.; Duzgunes, E.; Ogut, H. Reproductive Ecology of the Invasive Whelk *Rapana venosa* Valenciennes, 1846, in the Southeastern Black Sea (Gastropoda: Muricidae). *ICES J. Mar. Sci. Adv. Access* 2009, 3p.

Sales, J.; Britz, P. J. Research on Abalone (*Haliotis midae* L.) Cultivation in South Africa. *Aquacult. Res.* **2001**, *32*, 863–874.

Samanta, S. K.; Adhikari, D.; Karmakar, S.;Dutta, A.; Roy, A.; Manisenthil, K. T.; Roy, D.; Vedasiromoni, J. R.; Sen, T. Pharmacological and Biochemical Studies on *Telescopium telescopium* – A Marine Mollusk from the Mangrove Regions. *Oriental Pharm. Exp. Med.* **2008**, *8*, 386–394.

Sanduja, R.; Sanduja, S. K.; Weinheimer, A. J.; Alam, M.; Martin, G. E. Isolation of the Cembranolide Diterpenes Dihydrosinularin and 11-epi-sinulariolide from the Marine Mollusk *Planaxis sulcatus*. *J. Nat. Prod.* **1986**, *49*, 718–719.

Santhanam, R. *Nutritional Marine Life*. CRC Press: Taylor & Francis, 2015, 278p.

Santhi, V.; V. Sivakumar, S.; Jayalakshmi, R. D.; Thilaga, M.; Mukilarasi. Isolating Bioactive Compound from Marine Prosobranch *Purpura persica* from Tuticorin Coast. *Int. J. Environ. Protect. Policy* **2016**, *4*, 64–76.

Santhi, V.; Sivakumar, V.; Thangathirupathi, A.; Thilaga, R. D. Analgesic, Anti-Pyretic and Anti-Inflammatory Activities of Chloroform Extract of Prosobranch Mollusc *Purpura persica*. *Int. J. Pharm. Biol. Sci.* **2011**, *5*, 9.

Sarizan, N.M.B. Identificaiton of Anti-Atherosclerotic Compounds from Marine Molluscs of Bidong Archipelago. M.Sc. Thesis, Universiti, Malaysia, 2013.

Sato, M.; Jensen, G. C. Shell Selection by the Hermit Crab, *Pagurus hartae* (McLaughlin & Jensen, 1996) (Decapoda, Anomura). http://people.oregonstate.edu/~satomei/MeiSato/Publications_files/SatoJensen2005.pdf.

Sawabe, T.; Fujimura, Y.; Niwa, K.; Aono, H. *Vibrio comitans* sp. nov.; *Vibrio rarus* sp. nov. and *Vibrio inusitatus* sp. nov.; from the Gut of the Abalones *Haliotis discus discus, H. gigantea, H. madaka* and *H. rufescens* . *Int. J. Syst. Evol. Microbiol.* **2007**, *57*, 916–922.

Scarratt, K.;Hann, H. A. Pearls from the Lion's Paw Scallop. http://www.ssef.ch/uploads/media/2004_Scarratt_Pearls_from_the_lion_s_paw_scallop_01.pdf.

Schlösser, A.; Lipski, A.; Schmalfu, J.; Kugler, F.; Beckmann, G. *Oceaniserpentilla haliotis* gen. nov., sp. nov., a Marine Bacterium Isolated from Haemolymph Serum of Blacklip Abalone. *Int. J. Syst. Evol. Microbiol.* **2008**, *58*, 2122–2125.

See, G. L. L.; Deliman, Y. C.; Arce, J. F. V.; Ilano, A. Cytotoxic and Genotoxic Studies on the Mucus of Indian Volute *Melo broderipii* (Gmelin 1758) and Spider Conch Lambis Lambis (Linn 1758). *J. Pharmacogn. Nat. Prod.* **2016**, *2*, 120. doi:10.4172/2472-0992.1000120.

Segura, O.; Fritsch, E. Nonbead-Cultured Pearls from *Strombus gigas*. https://www.gia.edu/gems-gemology/summer-2015-gemnews-nonbead-cultured-pearls-strombus-gigas.

Selvi, K. G.; Jeevanandham, P. Protein, Carbohydrate and Lipid Analysis of *Ficus ficoides* (Lamarck, 1822) from Vanjiure, Southeast Coast of India. *Int. J. Curr. Microbiol. Appl. Sci.* **2016**, *5*, 284–292.

Serra, G.; Chelazzi, G.; Castilla, J. C. Temporal and Spatial Activity of the Key-Hole Limpet *Fissurella crassa* (Mollusca: Gastropoda) in the Eastern Pacific. *J. Mar. Biol. Assoc. United Kingdom* **2001**, *81*, 485–490.

Shepherd, S. A.; Huchette, S. Studies on Southern Australian Abalone (genus *Haliotis*) XVIII. Ring Formation in *H. scalaris*. *J. Molluscan Res.* **1997**, *18*, 247–252.

Shi, https://researcharchive.lincoln.ac.nz/bitstream/handle/10182/2863/Shi_MApplSci. pdf;jsessionid=8CEB19C444DA715A3BE8BE8640F37503?sequence=5.

Shon, K. J.; Hasson, A.; Spira, M. E.; Cruz, L. J.; Gray, W. R.; Olivera, B. M. Delta-Conotoxin GmVIA, a Novel Peptide from the Venom of *Conus gloriamaris*. *Biochemistry* **1994**, *33*, 11420–11425.

Siddall, R.; Pike, A. W.; McVicar, A. H. Parasites of *Buccinum undatum* (Mollusca: Prosobranchia) as Biological Indicators of Sewage-Sludge Dispersal. *J. Mar. Biol. Assoc. United Kingdom* **1993**, *73*, 931–948.

Spiezia, M. C.; Chiarabelli, C.; Schininà, M. E.; Polticelli, F. : *Bioactive peptides from the venom of the Mediterranean cone snail Conus ventricosus.* Nova Science Publishers Inc. 2013, pp. 135–148.

Sri Kantha, S. Carotenoids of Edible Molluscs: A Review. *J. Food Biochem.* **1989**, *13*, 429–442.

Sri Kumaran, N.; Bragadeeswaran, S.; Thangaraj, S. .Screening for Antimicrobial Activities of Marine Molluscs *Thais tissoti* (Petit, 1852) and *Babylonia spirata* (Linnaeus, 1758) against Human, Fish Andbiofilm Pathogenic Microorganisms. *Afr. J. Microbiol. Res.* **2011**, *5*, 4155–4161.

Sugesh, S.; Mayavu, P.; Ezhilarasan, P.; Sivashankar, P.; Arivuselvan, N. Screening of Antibacterial Activities of Marine Gastropod *Hemifusus pugilinus*. *Curr. Res. J. Biol. Sci.* **2013**, *5*, 49–52.

Sundaram, S.; Deshmukh, V. D. *Gastropod Operculum – An Unique Trade*. http://eprints. cmfri.org.in/9617/1/17.pdf.

Suresh, M.; Arularasan, S.; Sri Kumar, N. Screening on Antimicrobial Activity of Marine Gastropods *Babylonia zeylanica* (Bruguière, 1789) and Harpa Conoidalis (Lamarck, 1822) from Mudasalodai, Southeast Coast of India. *Int. J. Pharm. Pharm. Sci.* **2012**, *4*, 552–556.

Suzuki, M.; Shimizu, K.; Kobayashi, Y.; Ishizaki, S.; Shiomi, K. Paramyosin from the Disc Abalone *Haliotis discus discus.. J. Food Biochem.* **2014**, *38*, 444–451.

Tabugo, S. R. M.; Pattuinan, J. O.; Sespene, N. J. J.; Jamasal, A. J. Some Economically Important Bivalves and Gastropods found in the Island of Hadji Panglima Tahil, in the Province of Sulu, Philippines. *Int. Res. J. Biol. Sci.* **2013**, *2*, 30–36.

Tamil Muthu, P.; Selvaraj, D. Analysis of Bioactive Constituents from the Flesh of *Turbo brunneus* (Roding, 1798). *Int. J. Fish. Aquat. Stud.* **2015**, *3*, 257–259.

Tanaka, R.; Cleenwerck, I.; Mizutani, Y.; Iehata, S.; Shibata, T.; Miyake, H.; Mori, T.; Tamaru, Y.; Ueda, M.; Bossier, P.; Vandamme, P. *Formosa haliotis* sp. nov., a Brown-Alga-Degrading Bacterium Isolated from the Gut of the Abalone *Haliotis gigantea*. *Int. J. Syst. Evol. Microbiol.* **2015**, *65*, 4388–4393.

Tanikawa, E.; Yamashita, J.; Chemical Studies on the Meat of Abalone (*Haliotis discus hannai*). *Bull. Fac. Fish.; Hokkaido Univ.* **1961**, *12*, 210–238.

Teng, F.; Tanioka, Y.; Hamaguchi, N.; Bito, T.; Takenaka, S.; Yabuta, Y.; Watanabe, F. Determination and Characterization of Vitamin B12 Compounds in Edible Sea Snails, Ivory Shell *Babylonia japonica* and Turban Shell Turdo *Batillus cornutus. Fish. Sci.* **2015**, *81*, 1105–1111.

Terlizzi, A.; . Delos, A. L.; Garaventa, F.; Faimali, M.; Geraci, S. Limited Effectiveness of Marine Protected Areas: Imposex in *Hexaplex trunculus* (Gastropoda, Muricidae) Populations from Italian Marine Reserves. *Mar. Pollut. Bull.* **2004**, *48*, 164–192.

Teshima, S.; Kanazawa, A.; Hyodo, S.; Ando, T. Sterols of the Triton, *Charonia tritonis. Comp Biochem Physiol Part B: Compar. Biochem.* **1979**, *64*, 225–228.

Thieltges, D. W.; Krakau, M.; Andresen, H.; Fottner, S.; Reise, K. Macroparasite Community in Molluscs of a Tidal Basin in the Wadden Sea. *Helgoland Mar. Res.* **2006**, *60*, 307.

Thieltges, D. W.; Ferguson, M. A. D.; Jones, C. S.; Noble, L. R.; Poulin, R. Biogeographical Patterns of Marine Larval Trematode Parasites in Two Intermediate Snail Hosts in Europe. *J. Biogeogr.* **2009**, *36*, 1493–1501.

Turner, A. D.; Tarnovius, S.; Goya, A. B. Paralytic Shellfish Toxins in the Marine Gastropods *Zidona dufresnei* and *Adelomelon beckii* from Argentina: Toxicity and Toxin Profiles. *J. Shellfish Res.* **2014**, *33*, 519–530.

Udayantha, H. M.; Godahewa, G. I.; Bathige, S. D.; Wickramaarachchi, W. D.; Umasuthan, N.; De Zoysa, M.; Jeong, H. B.; Lim, B. S.; Lee, J. A Molluscan Calreticulin Ortholog from *Haliotis discus discus*: Molecular Characterization and Transcriptional Evidence for Its Role in Host Immunity. *Biochem. Biophys. Res. Commun.* **2016**, *474*, 43–50.

Udotong, J. I. R.; Ukot, C. A. Microbiological and Nutritional Quality of *Cymbium glans* from qua Iboe river Estuary, Nigeria. *Int. J. Curr. Res.* http://www.journalcra.com/article/microbiological-and-nutritional-quality-cymbium-glans-qua-iboe-river-estuary-nigeria.

Uyeno, D. Copepods (Cyclopoida) Associated with Top Shells (Vestigastropoda: Trochoidea: Tegulidae) from Coastal Waters in southern Japan, with Descriptions of Three New Species. *Zootaxa* **2016**, doi: 10.11646/zootaxa.4200.1.4.

Velayutham, S.;, Sivaprakasam, R. M.; Williams, S. F.; Samuthirapandian, R. Potential Activity of In Vitro Antioxidant on Methanolic Extract of *Babylonia zeylanica* (Bruguiere, 1789) from Mudasalodai, Southeast Coast of India. *Int. J. Pharm. Pharm. Sci. Res.* **2014**, *4*, 60–64.

Venkatesan, V. *Marine Ornamental Molluscs.* http://eprints.cmfri.org.in/8671/1/Marine_Ornamental_Fish_Culture.pdf.

Verhecken, A. The Indole Pigments of Mollusca. *Annls Soc. r. zool. Belg. - T.* **1989**, *119*, 181–197.

Vibrio parahaemolyticus (Wikipedia).

Viera, M. P.; de Viçose, G. C.; Fernández-Palacios, H.; Izquierdo, M. Grow-Out Culture of Abalone *Haliotis tuberculata coccinea* Reeve, Fed Land-Based IMTA Produced Macroalgae, in a Combined Fish/Abalone Offshore Mariculture System: Effect of Stocking Density. *Aquacult. Res.* **2016**, *47*, 71–81.

Vimala, P.; Thilaga, R. D. Antibacterial Activity of the Crude Extract of a Gastropod *Lambis lambis* from the Tuticorin Coast, India. *World Appl. Sci. J.* **2012**, *16*, 1334–1337.

Wang, N.; Whang, I.; Lee, J. A Novel C-Type Lectin from Abalone, *Haliotis discus discus,* Agglutinates *Vibrio alginolyticus. Dev. Compar. Immunol.* **2008**, *32*, 1034–1040.

Woodcock, S. H.; Benkendorff, K. The Impact of Diet on the Growth and Proximate Composition of Juvenile Whelks, *Dicathais orbita* (Gastropoda: Mollusca). *Aquaculture* **2008**, *276*, 162–170.

Wu, F.; Zhang, G. Suitability of Cage Culture for Pacific Abalone *Haliotis discus hannai* Ino Production in China. *Aquacult. Res.* **2013**, *44*, 485–494.

Wu, F.; Zhang, G. Pacific Abalone Farming in China: Recent Innovations and Challenges. *J. Shellfish Res.* **2016**, *35*, 703–710.

Yap, C. K. Shells of *Telescopium telescopium* as Biomonitoring Materials of Ni Pollution in the Tropical Intertidal Area. *Int. J. Adv. Appl. Sci.* **2014**, *3*, 11–18.

Yap, C. K.; Noorhaidah, A.; Azlanb, A.; Azwady, A. A. N.; Ismail, A.; Ismail, A. R.; Siraj, S. S.; Tan, S. G.. *Telescopium telescopium* as Potential Biomonitors of Cu, Zn and Pb for Tropical Intertidal Areas. *Ecotoxicol. Environ. Saf.* **2009**, *72*, 496–506.

Yap, C. K.; Noorhaidah, A.; Tan, S. G. *Different Soft Tissues of Telescopium Telescopium as Potential Biomonitoring Tissues of Zn Bioavailability in Malaysian Intertidal Mudflats.* Nova Science Publishers, 2011.

Zarai, Z.; Frikha, F.; Balti, R.; Miled, N.; Gargouria, Y.; Mejdouba, H. Nutrient Composition of the Marine Snail (*Hexaplex trunculus*) from the Tunisian Mediterranean coasts. *J. Sci. Food Agric.* **2011**, *91*, 1265–1270.

Zhang, D.; Wang, C.; Wu, H.; Xie, J.; Du, L.; Xia, Z.; Cai, J.; Huang, Z.; Wei, D. Three Sulphated Polysaccharides Isolated from the Mucilage of Mud Snail, *Bullacta exarata philippi*: Characterization and Antitumour Activity. *Food Chem.* **2013**, *138*, 306–314.

Zhang, J.; Zhu, A.; Wu, C. Nutritional Composition Analysis and Evaluation of *Monodonta labio* Muscle. *Food Sci.* **2011**, *17*, http://en.cnki.com.cn/Article_en/CJFDTotal-SPKX201117075.htm.

Zhang, D.; Wu, H.; Xia, Z.; Wang, C.; Cai, J.; Huang, Z.; Du, L.; Sun, P.; Xie, J. Partial Characterization, Antioxidant and Antitumor Activities of Three Sulfated Polysaccharides Purified from *Bullacta exarata*. *J. Funct. Foods* **2012**, *4*, 784–792.

Zhou, D.; Zhu, B.; Qiao, L.; Wu, H.; Li, D.; Yang, J.; Murata, Y. *In vitro* Antioxidant Activity of Enzymatic Hydrolysates Prepared from Abalone (*Haliotis discus hann*ai Ino) Viscera. *Food Bioprod. Proc.* **2012**, *90*, 148–154.

Zhou, J.; Wang, R.; Cai, J. Virus-Like Particles Detected in Postlarvae of Abalone (*Haliotis diversicolor supertexta* Lischke) Potentially Associated with Mass Mortalities. *Afr. J. Microbiol. Res.* **2013**, *7*, 1569–1573.

Zhu, B. W.; Wang, L. S.; Zhou, D. Y.; Li, D.; Sun, L.; Yang, J.; Wu, H.; Zhou, X.; Tada, M. Antioxidant Activity of Sulphated Polysaccharide Conjugates from Abalone (*Haliotis discus hannai*). *Eur. Food Res. Technol.* **2008**, *227*, 1663.

INDEX